Towards the Governance of Open Distributed Systems: A Case Study in Wireless Mobile Grids

D i s s e r t a t i o n
zur Erlangung des Grades eines Doktors der Wirtschaftswissenschaft
der Rechts- und Wirtschaftswissenschaftlichen Fakultät
der Universität Bayreuth

vorgelegt
von

Dipl.-Kffr. Univ. Tina Balke

aus Potsdam

Dekan: Prof. Dr. Markus Möstl
Erstberichterstatter: Prof. Dr. Torsten Eymann
Zweitberichterstatter: Dr. Julian Padget (Senior Lecturer)
Zusätzlicher Gutachter: Dr. Marina De Vos (Senior Lecturer)
Tag der mündlichen Prüfung: 23. September 2011

Abstract

New networking technologies such as wireless mobile grids and peer-to-peer middleware are examples of a growing class of open distributed systems whose strength is the absence of a central controlling instance and which function through the cooperation of autonomous entities that voluntarily commit resources to a common pool. The social dilemma in such systems is that it is advantageous for rational users to access the common pool resources without making any commitment of their own. This is commonly known as "free-riding". However, if a substantial number of users followed this selfish strategy, the system itself would fail, depriving all users of its benefits. In this dissertation, we demonstrate how governance decisions can induce cooperation in such systems and how normative frameworks in combination with multi-agent system simulations can be successfully employed to analyse their effects, even at an early development stage.

We show that our approach is not only practical and powerful, but also easily accessible. We demonstrate its functionality by implementing a prototype to explore the impact of enforcement mechanisms on wireless mobile grids, a concept which has been proposed to address the energy issues arising in the next generation of mobile phones and the networks that connect them. We also infer lessons from this example for open distributed systems in general. Simulation experiments quantify the benefits of enforcement mechanisms for wireless mobile grids. We analyse these results with respect to the costs of enforcement as well as further criteria that reflect the interests of the multiple stakeholders in the system. We conclude with some observations on how the lessons learned from both process and outcomes may be applicable to the broader context of open distributed systems. In particular, we highlight (i) the use of simulation using intelligent agents and a normative framework as a means for in silico exploration of complex systems for both business and technological objectives, and (ii) the insight offered into a range of enforcement mechanisms and a better understanding of the conditions and constraints under which they are applicable.

Ausführliche Zusammenfassung

Meine in englischer Sprache verfasste Dissertation beschäftigt sich mit normativen Eingriffen und deren Auswirkungen auf offene verteilte Systeme ohne zentrale Kontrollinstanz, in denen Akteure mit möglicherweise konfliktären und eigennutzenmaximierenden Interessen miteinander interagieren.

Ausgehend von sogenannten "Wireless Mobile Grids"(WMGs) – einer spezifischen Ausprägung eines offenen verteilten Systems – wird in der Dissertation sowohl für das spezifische WMG-Szenario als auch allgemein gültig für offene verteilte Systeme untersucht, inwieweit nicht-technische Durchsetzungmechanismen kooperatives Verhalten von Akteuren fördern und welche Auswirkungen unterschiedliche Mechanismen auf die verschiedenen Stakeholder des Systems haben.

Bei WMGs handelt es sich um ein Kooperationskonzept für die vierte Mobilfunkgeneration, welche sich mit einem Energieverbrauchsproblem konfrontiert sieht. So wandeln sich bedingt durch einen wachsenden Funktionsumfang und Anwendungen wie bspw. große Touchscreens, Digitalkameras, GPS-Empfänger und Videostreaming Mobiltelefone derzeit zu multimedialen „Alleskönnern" und entwickeln hierdurch einen immensen „Energiehunger". Der Energieverbrauch bei 3G-Geräten hat sich im Vergleich zu den Geräten der ersten und zweiten Handygeneration verdoppelt und dieser Zuwachs wird sich in Zukunft nicht verlangsamen (Katz and Fitzek, 2006). Aufgrund entgegengesetzter Kundenwünsche kommt eine Vergrößerung des Akkus als Lösung nicht in Frage (Perrucci et al., 2009).

Aus diesem Grund wurden von Prof. Frank Fitzek und Dr. Marcos Katz WMGs als ein möglicher Lösungsansatz für die erwähnten Energieprobleme vorgeschlagen, prototypisch implementiert und technisch getestet. Die Idee der WMGs stellt sich dabei wie gefolgt dar: ausgehend von traditionellen 3G Mobilfunknetzen, welche durch eine Konzentration der Kommunikation auf die zellulare Kommunikation zwischen den einzelnen Mobilfunkteilnehmern mit einer Basisstation charakterisiert ist, wird die traditionelle 3G Struktur um eine Kurz-Distanz-Verbindung erweitert.

In WMGs verbinden sich einzelne Mobilfunknutzer mittels ihrer unterschiedlichen Endgeräten mit Hilfe einer Kurz-Distanz-Verbindung (bspw. WLAN oder

Bluetooth) in einer Peer-to-Peer-Manier untereinander und nutzen diese Verbindung zur Kooperation, d.h. zum gemeinsamen Verwenden von Ressourcen der Endgeräte wie CPU und Speicher, oder dem gemeinsamen Download / Streamen von Dateien. Der Vorteil der Kurz-Distanz-Verbindung liegt in schnelleren und energieeffizienteren Kommunikationsprotokollen. Die Mobilfunknutzer können entsprechend beispielsweise von schnelleren Datenraten und längeren Akkulaufzeiten im Vergleich zu traditionellen 3G Verbindungen profitieren. Zudem wird aus Sicht der Mobilfunknetzbetreiber die Netzwerkinfrastruktur entlastet.

Trotz des theoretisch hohen Energieeinsparungspotentials gibt es derzeit von Seiten der Mobilfunkbranche Zweifel am Konzept der WMG, welche primär auf befürchtete Kooperationsprobleme zurückzuführen sind. Für die Akteure im System ist es günstiger zu betrügen (d.h. Ressourcen zu beziehen ohne selbst Ressourcen zur Verfügung zu stellen), da jeder Beitrag zum System in Form von Kosten (wie beispielsweise Energiekosten) zu verstehen ist. Deshalb werden Normen benötigt, um Kooperation zu fördern.

Durch die offene dezentrale Struktur des Systems ist es nicht möglich, implizit von einer Durchsetzbarkeit der Normen auszugehen. Folglich stellt sich die Frage, wie mögliche Durchsetzungsmechanismen zur Reduktion des Kooperationsproblems beitragen.

Basierend auf realen Handydaten[1] habe ich in meiner Dissertation eine normative Multiagenten-Simulation entwickelt, die es ermöglicht, verschiedene Durchsetzungsmechanismen zu analysieren.

In diesem Zusammenhang leistet meine Dissertation einen wissenschaftlichen Beitrag durch die Untersuchung der folgenden zwei Hauptthesen:

1. Der Einsatz von Durchsetzungsmechanismen kann die Kooperation in einem WMG steigern.

2. So genannte "Normative Frameworks" in Kombination mit Multiagenten-Simultionen eignen sich sehr gut, um die dynamische Interaktion von Akteuren eines Systems mit den im System vorhandenen Normen darstellen.

Zur Verifizierung dieser beiden Thesen präsentiert diese Dissertation zunächst einen Überblick über die existierende Literatur zu Normen und deren Durchsetzung in Kapitel 2. Kapitel 3 stellt im Anschluss die WMG-Fallstudie detailliert vor und zeigt auf, in wie weit in ihr Kooperationsdilemmata bestehen. Kapitel 4 führt in normative Systeme ein, diskutiert deren Bedeutung für die Modellierung von Normen und präsentiert das in dieser Dissertation verwendete formale Model eines Normative Frameworks.

Dieses formale Model wird in Kapitel 5 mit Hilfe von *AnsProlog* und Inst*AL* in ein Computermodell übersetzt und die Fallstudie mit Hilfe dieses Modells

---

[1]Frank Fitzek und seine Forschergruppe haben bereits Prototypen von WMG-Mobiltelefonen entwickelt und haben mir diese technischen Daten wie beispielsweise den Energieverbrauch bei verschiedenen Empfangs- und Sendezuständen für die Dissertation zur Verfügung gestellt.

implementiert. Die Analyse der Ergebnisse dieses Modells zeigt, dass Durchsetzungsmechanismen helfen können, der beschriebenen Dilemmasituation entgegenzuwirken und es sich lohnt, detailliertere Untersuchungen vorzunehmen.

Als Resultat dessen, widmen sich Kapitel 6 und 7 der Erweiterung des zuvor entwickelten Modells um eine Multiagenten-Simulation. Diese Dissertation beschreibt dabei zum einen die formalen Grundlagen dieser Erweiterung. Des Weiteren wird ein lauffähiger Prototyp des WMG-Beispiels implementiert.

Dieser Prototyp wird in den Kapiteln 7 bis 9 eingesetzt, um drei verschiedene Durchsetzungsmechanismen auf Ihre Wirkungsamkeit bezüglich der Kooperation und der damit verbundenen Energiekosten in WMGs zu untersuchen. Dabei zeigt sich, dass sogenannte Polizeiagenten einen vielversprechenden Ansatz zur Reduktion des Kooperationsdilemmas in WMGs bieten.

Kapitel 10 widmet sich im Anschluss der erweiterten Analyse der Simualtionsergebnisse. Spezieller Fokus wird dabei auf die Interessen der verschiedenen Stakeholder des Systems, sowie der Interpretation der Ergebnisse für offene verteilte Systeme im allgemeinen gelegt.

Zusammenfassend bestätigt diese Dissertation die beiden Eingangs erwähnten Hypothesen. Sie leistet dadurch einen wichtigen Beitrag sowohl aus wissenschaftlicher als auch praktischer Sicht. So gibt diese Dissertation erstmals detaillierte Einblicke in die Kooperation in WMGs und zeigt auf, wie Normative Frameworks und Multiagenten-Simulationen effektiv eingesetzt werden können um Governance-Entscheidungen einfach und schnell zu analysieren.

Acknowledgements

The world ain't all sunshine and rainbows. It is a very mean and nasty place and it will beat you to your knees and keep you there permanently if you let it. You, me, or nobody is gonna hit as hard as life. But it ain't how hard you hit; it's about how hard you can get hit, and keep moving forward. How much you can take, and keep moving forward. That's how winning is done. [Rocky Balboa]

People who have written one, always say that a dissertation is a journey, and indeed my dissertation was an amazing journey for me. Nevertheless it was also a tough fight: fighting the topic, fighting my own doubts, fighting the white pages... In this fight, I would have never reached the stage of submission without a large number of people that stood in my corner and to whom I owe many thanks.

First of all my three supervisors Torsten Eymann, Julian Padget and Marina De Vos, who introduced me to the art of research and to the joys of fighting an idea as well as pinning it to paper. Torsten gave me the chance to explore completely new aspects of my research by relieving me from my teaching duties at the University of Bayreuth, to spend several research visits at the University of Bath. Julian and Marina made my time in Bath very special and in the end I stayed a lot longer than expected. Marina in particular has become more than a supervisor, but I would like to thank her for being a great flatmate and friend who helped me to not go insane in the last couple of months.

My colleagues in Bayreuth and Bath played a big part in making my PhD time very special and I would like to thank everyone in these teams, with special thanks going to three persons in particular: James Mitchell, Marios Richards and Dimitris Traskas. James introduced me to the ideas of unit testing and code elegance and assisted with setting up my experiments in two "clouds": Amazon EC2 and my personal Bath "cloud" consisting of spare CPU cycles graciously donated by James himself, Jim Grimmett, Simon Powers, Kewei Duan, Beate Grawemeyer and Thomas Saunders. Without these my experiments would not have been finished in time, so a big thanks goes to all the nodes. Marios Richards gave me a number of invaluable tutorials in MATLAB and Dimitris Traskas fought against the man with the golden fleece with me. Thanks also go to the people responsible for the man with the golden fleece – Rafael H. Bordini and Jomi F. Hübner – for answering questions really fast and trying to help with the simulation environment and its "special features"[2] whenever possible.

Other people that I want to express my gratitude to are Rainer Hegselmann, Sascha Kurz and Jörg Rambau for valuable discussions of my research ideas in our postgrad research seminar, Erich Schikuta for pointing me to Frank Fitzek (and by doing this giving my dissertation a new direction) and Frank Fitzek for giving me valuable insights into the wireless mobile grid concept that eventually became the case study of this dissertation.

I also would like to thank my family for their everlasting support and for telling me to pursue my dreams, as well as for understanding that a dissertation takes time.

Most of all however I would like to thank "meinem Liebsten" for his never ending support and encouragement. He not only helped me with numerous hours of discussions and gave me a great deal of Corel Draw introduction and inspiration, but most importantly helped me to trust myself that I can go the full distance and "win this fight and kill the beast" and supported me in an endless number of ways through the whole process.

Acknowledgements to Funding Bodies

In addition to the people mentioned above, I also would like to express my gratitude for the funding I received by various sponsors throughout my dissertation time.

First of all I would like to thank the German Academic Exchange Service who supported the first 8 month of my research visit to the University of Bath with a "DAAD-Doktorandenstipendien". The final two month of my PhD time were sponsored by a scholarship according to the "Bayerisches Begabtenförderungsgesetz" for which I am very grateful.

I was able to present my PhD work in a number of conferences due to a large number of travel grants. I would therefore also like to thank the following bodies for the respective scholarships: the EU (COST Action IC0801 – Agreement Technologies), the International Foundation for Autonomous Agents and Multiagent Systems, the Equality Office of the University of Bayreuth as well as the Alumni Organization of the Faculty for Law, Business Administration and Economics of the University of Bayreuth.

CPU power for the simulations in this effort were sponsored in part by the US Air Force Office of Scientific Research, Air Force Material Command, USAF, under grant number FA8655-10-1-3050.

---

[2] "These are not bugs, these are features."

Contents

Abstract   iii

Ausführliche Zusammenfassung   iv

Acknowledgements   viii

List of Figures   xvii

List of Tables   xxi

List of Definitions   xxiii

Glossary   xxiv

1 Introduction   1
   1.1 The Failure of A Million Penguins 1
   1.2 Theses 3
   1.3 Roadmap 5
   1.4 Supporting Publications 7

I Foundations and Related Work   11

2 Norms and Enforcing Cooperation in Open Distributed Systems   13
   2.1 What is a Norm? 13
   2.2 Open Distributed Systems 19
   2.3 Ensuring Compliance with Norms 24
     2.3.1 Regimentation 25
     2.3.2 Incentive-based Enforcement 27
     2.3.3 The Enforcement Process 29
       2.3.3.1 Detecting Non-Compliance 30
       2.3.3.2 Enforcement Determination 32
       2.3.3.3 Enforcement Application 32
       2.3.3.4 Assimilation 33
   2.4 Enforcement and Cooperation – Related Work 34

     2.4.1   Enforcement and Cooperation in Economic Theory 34

          2.4.1.1   Enforcement and Cooperation in Games-Theoretic Models 35

          2.4.1.2   Enforcement and Cooperation in Experimental Economics 37

     2.4.2   Enforcement and Cooperation in Formal Logics Research 39

     2.4.3   Enforcement and Cooperation in Reputation Research . . 39

     2.4.4   Enforcement and Cooperation in Technical Domains . . . 41

          2.4.4.1   Relay Routing 41

          2.4.4.2   P2P Networks 42

          2.4.4.3   MMORPGs 43

          2.4.4.4   Recapitulation 43

2.5   The Institutional Analysis and Development Framework 44

     2.5.1   The Action Arena 46

     2.5.2   Outcomes 48

     2.5.3   Underlying Factors 49

          2.5.3.1   Biophysical-Technical Characteristics 49

          2.5.3.2   Attributes of the Community 49

          2.5.3.3   Rules-In-Use 49

2.6   Summary 50

3   Case-Study: The Cooperation Dilemma in WMG   53

3.1   The Wireless Mobile Grid Scenario 53

     3.1.1   The Mobile Phone Market: Challenges Arising 53

     3.1.2   The Wireless Mobile Grid Scenario 58

          3.1.2.1   A Short-Link Architecture for Wireless Mobile Grids 58

          3.1.2.2   Use Cases 61

          3.1.2.3   The Energy-Advantage of IEEE802.11 WLAN . 62

     3.1.3   Stakeholder-Advantages in the Wireless Mobile Grid Scenario 64

3.2   The Cooperation Dilemma in Wireless Mobile Grids 68

3.3   Summary 71

II   Normative Modelling   73

4   Normative Frameworks   75

4.1   Constitutive Aspects of Normative Frameworks 76

     4.1.1   Brute Facts and Institutional Facts 76

     4.1.2   Normative Power and Permission 80

     4.1.3   Events 81

4.2   Regulative Aspects of Normative Frameworks 81

     4.2.1   Prohibition and Permission 81

     4.2.2   Obligations 82

     4.2.3   Non-Compliance 83

4.3   Further Aspects of Normative Frameworks 84

	4.3.1	The Initiation and Termination of a Normative Framework	84
	4.3.2	Time	84
4.4	A Formal Definition of Normative Frameworks	84
4.5	Summary	87

5 Designing and Testing the Rules-in-Use   **89**
5.1	Approaches for Specifying a Computational Model of Normative Frameworks	89
	5.1.1	The e-Institutions Framework	89
	5.1.2	OperA/OperettA	90
	5.1.3	The *MOISE*-Family	91
	5.1.4	Artikis et al.	92
5.2	Using Answer Set Programming for Design-Time Reasoning in the Action Arena	92
	5.2.1	*AnsProlog*	93
	5.2.2	Representing Normative Frameworks in *AnsProlog*	96
5.3	The Action Language Inst*AL*	101
5.4	Formalizing the WMG Scenario	106
	5.4.1	Negotiation Phase	108
	5.4.2	Downloading Phase	108
	5.4.3	Sharing Phase	113
	5.4.4	Trace and Query	114
5.5	Discussion of the Results of the Design-Time Model	118
5.6	Summary	120

III Analysing Enforcement in Wireless Mobile Grids using Normative Multi-Agent System Simulations   **123**

6 Simulation Design   **125**
6.1	MAS Systems Simulations for Analysing the Action Arena	125
6.2	The Simulation Research Process	131
6.3	Result Evaluation in the Context of Multiple Stakeholders	135
6.4	Experimental Design	141
	6.4.1	Simulation Parametrisation and Statistical Analysis . . .	142
	6.4.2	Input Factors	143
	6.4.3	Measurements	145
	6.4.4	Experimental Design and Results	147
6.5	Summary	148

7 The Basic WMG Scenario   **149**
7.1	Representing the Basic WMG Scenario as MAS Simulation . . .	149
	7.1.1	General Formalizations	149
	7.1.2	The Biophysical Characteristics	151
	7.1.3	Representing the Community	154
		7.1.3.1 User Types	155

	7.1.3.2	The Basic Agent Decision Process	155
	7.1.3.3	The Basic Movement Pattern	159
7.1.4	Rules-in-Use		160
7.1.5	Implementing the Basic MAS Simulation of the WMG Scenario		161
7.2	Connecting Normative Frameworks and the MAS Simulation in the Action Arena		162
7.2.1	The Run-Time Model		162
7.2.2	Monitoring Dynamic State		164
7.3	Simulation Experiments		166
7.3.1	Simulation Hypotheses and Setup		167
7.3.2	Simulation Results		171
7.4	Summary		177

8 Enforcement Mechanisms for the WMG Simulation — 179

8.1	Retrospective: A Taxonomy for Fostering Norm Compliance	179
8.1.1	Utilization of Normative Empowered Entities	182
8.1.2	Normative Framework Assisted Enforcement (Second-Party / Third-Party)	182
8.1.3	Social Control (Second-Party / Third-Party) and Promisee-Enforced Norms	182
8.1.4	Self-Control	183
8.1.5	Selecting Enforcement Mechanisms	183
8.2	Implementing the Enforcement Mechanisms	184
8.2.1	Normative Empowered Agents	184
8.2.2	Image Information	185
8.2.3	Reputation	186
8.3	Simulation Experiments	189
8.3.1	Simulation Hypotheses and Setup	189
8.3.2	Simulation Results	192
8.4	Summary	200

9 On the Effects of Movement in the WMG Simulation — 203

9.1	Extending the Basic Movement Pattern	204
9.1.1	Movement Patterns for WMG	204
9.1.2	The Extended Random Waypoint Pattern	207
9.1.3	The Nomadic Pursue Group Pattern	207
9.1.4	The Sedentary Movement Pattern	208
9.2	Simulation Experiments	208
9.2.1	Simulation Hypothesis and Setup	208
9.2.2	Simulation Results	209
9.3	Summary	212

IV Evaluation 215

10 Further Simulation Analysis 217
10.1 Implications for the Business Case Study 217
10.2 Generalisation of the Results for Open Distributed Systems . . . 225
10.3 Summary 227

11 Summary and Conclusions 229
11.1 Contributions 229
11.2 Future Work 231
    11.2.1 Relaxing Model Assumption 231
    11.2.2 Extending the Simulation Experiments 232
    11.2.3 Adaptive Learning of Norms 233
11.3 Concluding Remarks 234

V Appendix 257

A The Jason Simulation Code 259

B Implementing the Basic WMG Scenario in Java 281

C The Run-Time Inst*AL* Specifications 289

D Comparing Single- and Multi-Threaded Jason Simulations 297

List of Figures

2.1 The Enforcement Process 30
2.2 The Four Basic Types of Goods distinguished by Ostrom 38
2.3 The Institutional Analysis and Development Framework 46
2.4 Dissertation Research Questions in the Light of the IAD Framework 51

3.1 Number of Mobile Phone Users Worldwide 54
3.2 Power Consumption of Mobile Phones 58
3.3 The Cellular Architecture 59
3.4 The Wireless Mobile Grid Scenario 60
3.5 Comparison of Short Range Capabilities 60
3.6 Comparison of Short Range Architectures 61
3.7 Stakeholders in the Wireless Mobile Grid Scenario 65
3.8 Comparison of the Effects of a 10% Transmission Error Rate . . 66
3.9 The Free Riding Problem in Wireless Grids 68
3.10 The Free Riding Problem as cardinal PD 69
3.11 The Free Riding Problem as ordinal PD 70

4.1 The Transition of the Real World and the Normative Framework 77
4.2 Conventional Generation turns Observations into Normative Facts 79

5.1 The Governor Function in the e-Institutions 90
5.2 The Time Translation Rules 97
5.3 The Π^{base} Translation 98
5.4 The $AnsProlog$ Rules for ensuring Observable Traces 99
5.5 The Condition Statement Translation 100
5.6 The $\Pi^{*}_{\mathcal{N}}$ Translation 100
5.7 Normative Specification Components in InstAL 102
5.8 Overview of the InstAL Translation Process 105
5.9 The Domain File of the Design Time WMG Model 108
5.10 Declaration of Types and Events in the Design Time WMG Model 109
5.11 Specification of the Noninertial Rules in the Design Time WMG
Model 109
5.12 Generation and Consequence Relations for Deadline-Countdown
in the Design Time WMG Model 110
5.13 Generation and Consequence Relations for Downloading in the
Design Time WMG Model 110

5.14 Generation and Consequence Relations for Sharing in the Design
      Time WMG Model 111
5.15 Initial State of the Design Time WMG Model, Post-Negotiation
      Phase 112
5.16 The Success Criteria for the Design Time WMG Model 115
5.17 Defining the Trace Output of the Design Time WMG Model . . 115
5.18 Ouput of a Successful Trace 116
5.19 One Trace of the Interaction between Alice, Bob and the Channels
      of the Base-station 117
5.20 The Success Criteria for our Design Time WMG Model 118
5.21 The General Run-Time Model Design 120

6.1   Agent and Environment 128
6.2   Micro-Macro-Level Interaction 129
6.3   The Simulation Research Process 132
6.4   Outcomes in the IAD context 135
6.5   The Gini Coefficient 139

7.1   Activity Diagram of Agent Download Considerations 157
7.2   Distribution of the Locations chosen by Agents in the Basic WMG
      Experiments 160
7.3   Declaration of Types and Events in the Run Time WMG Model –
      Example 163
7.4   The Components of the Run-Time Normative MAS simulation . 165
7.5   Tukey's Test Results of Population Composition Marginal Means 175
7.6   Tukey's Test Results of $|\mathcal{A}|$ Marginal Means 176
7.7   Tukey's Test Results of $\rho_{neighbourhood}$ Marginal Means 177

8.1   The Enforcement Process (Recalled) 180
8.2   A Taxonomy for Ensuring Normative Compliance 181
8.3   Tukey's Test Results Comparing Simulation Experiments with
      and without Enforcement 193
8.4   Analysis of Variance Comparing Simulation Experiments with
      and without Enforcement 197
8.5   Tukey's Test Results of Population Composition Marginal Means
      for Reputation Settings 198

9.1   Tukey's Test Results of Movement Model Marginal Means for
      Experiments with 1% Normative Framwork Empowered Agents . 209
9.2   Tukey's Test Results of Movement Model Marginal Means for
      Experiments with 5% Normative Framwork Empowered Agents . 211
9.3   Tukey's Test Results of Movement Model Marginal Means for
      Experiments with Image Information 211
9.4   Tukey's Test Results of Movement Model Marginal Means for
      Experiments with Reputation Information 211

10.1 Base Station Transmission Savings – No Enforcement 219
10.2 Base Station Transmission Savings – Normative empowered Agents220
10.3 Lorenz Curves of the Simulation Experiments with Enforcement   223

B.1 Class Diagram of Implemented Jason Agent Classes 282
B.2 Code Fragments of the interaction_agent.asl class   283
B.3 Cooperation Decision Code Fragments of the malicious_agent.asl
     class 285
B.4 Cooperation Decision Code Fragments of the util_agent.asl class   285
B.5 Cooperation Decision Code Fragments of the honest_agent.asl
     class 286

C.1 The Domain File of the Run Time WMG Model 289
C.2 Declaration of Types and Events in the Run Time WMG Model   290
C.3 Specification of the Noninertial Rules in the Run Time WMG
     Model 290
C.4 Generation and Consequence Relations for Deadline-Countdown
     in the Run Time WMG Model 291
C.5 Generation and Consequence Relations for Downloading in the
     Run Time WMG Model 291
C.6 Generation and Consequence Relations for Sharing in the Run
     Time WMG Model 292
C.7 Initial State of the Run Time WMG Model, Post-Negotiation
     Phase 293
C.8 The Run Time WMG Model – Part 1 294
C.9 The Run Time WMG Model – Part 2 295

List of Tables

2.1 Classification of Open Distributed Systems 22

3.1 Power Level and Data Rate for Cellular - 100 byte 64
3.2 Power Level and Data Rate for WLAN Broadcast - 1000 byte . . 64

6.1 The Main Simulation Input Factors 144
6.2 The Main Simulation Measurements 146
6.3 Example of an Experimental Setup Description Table 147
6.4 Example of an Experimental Results Description Table 148

7.1 Experimental Setup for Basic MAS Simulation of the WMG Scenario 170
7.2 Analysis of Variance for Hypotheses 2–4 172
7.3 Analysis of Variance in Comparison of the Population Composition
    – Post Hoc Test 173
7.4 Analysis of Variance in Comparison of $|\mathcal{A}|$ – Post Hoc Test . . . 173
7.5 Analysis of Variance in $\rho_{neighbourhood}$ – Post Hoc Test 173
7.6 Basic WMG Simulation Results Description Table 178

8.1 Discrete trust value (Abdul-Rahman and Hailes, 1997, p. 53) . . 187
8.2 Experimental Setup for Enforcement Experiments 191
8.3 Analysis of Variance Comparing Simulation Experiments with
    and without Enforcement 192
8.4 Analysis of Variance in Comparison of the Enforcement Mechan-
    isms – Post Hoc Test 194
8.5 Analysis of Variance for Hypothesis 5.1, 1% Normative Em-
    powered Agents 195
8.6 Analysis of Variance for Hypothesis 5.1, 5% Normative Em-
    powered Agents 195
8.7 Analysis of Variance for Hypothesis 5.2, Image Information . . . 196
8.8 Analysis of Variance for Hypothesis 5.3, Reputation Information 196
8.9 Simulation Results Description Table for Enforcement Mechanisms 200

9.1 Experimental Setup for Movement Pattern Experiments 210

10.1 Comparison of Base Station Transmission Savings 220
10.2 False Positive and False Negative Values 224

List of Definitions

Definition 1: Governance   3
Definition 2: Norms 18
Definition 3: Open System 19
Definition 4: Distributed System 20
Definition 5: Open Distributed System 20
Definition 6: Reputation 28
Definition 7: Image 29
Definition 8: Bounded Rationality 31
Definition 9: Normative Framework 75
Definition 10: Normative Framework (formal) 86
Definition 11: *AnsProlog* program 93
Definition 12: Ground 94
Definition 13: Deductive Closure for Positive *AnsProlog* Programs . . . 95
Definition 14: Gelfond-Lifschitz Transformation 96
Definition 15: Simulation 125
Definition 16: Multi-Agent System (MAS) 127
Definition 17: EBIT 137
Definition 18: Multi-Agent System (formal) 150
Definition 19: Agent (formal) 150
Definition 20: Flat Earth Model 151

Glossary

Symbols

Symbol	Description	Magnitude
δ	Shadow of the Future	$[0; 1]$
$E_{\{3G,WLAN\},\{rx,tx,i\}}$	Energy Consumed for 3G, WLAN in Receiving, Transmitting and Idle States	$[0; \infty]$
$t_{\{3G,WLAN\},\{rx,tx,i\}}$	Time Consumed for 3G, WLAN in Receiving, Transmitting and Idle States	$[0; \infty]$
$P_{\{3G,WLAN\},\{rx,tx,i\}}$	Power Consumed for 3G, WLAN in Receiving, Transmitting and Idle States	$[0; \infty]$
e	Event	$e \in \mathcal{E}$
$\mathcal{E}$	Set of Events	
$\mathcal{E}_{inst}$	Normative Event	
$\mathcal{E}_{ex}$	Exogenous Event	
$\mathcal{E}_{viol}$	Violation Event	
$\mathcal{E}_{act}$	Normative Actions	
$\mathcal{E}_+$	Creation Event	
$\mathcal{E}_-$	Termination Event	
f	Fluent	$f \in \mathcal{F}$
$\mathcal{F}$	Set of Fluents	
$\mathcal{P}$	Set of Permissions Fluents	
$\mathcal{W}$	Set of Powers Fluents	
$\mathcal{O}$	Set of Obligations Fluents	
v	Violation	
$\mathcal{X}$	Set of Possible Conditions	
$\mathcal{G}$	Generation Relation	
Δ	Initial State of the Normative Framework	

Symbol	Description	Magnitude
σ	Signature of an *AnsProlog* program	
Π	Finite set of rules about σ in an *Ans-Prolog* program	
$lm(\Pi)$	Answer Set of Π	
$\mathcal{B}_{\Pi}$	Herbrand Base	
$\Pi_{\mathcal{N}}$	*AnsProlog* Program of the Normative Framework	
Π^{base}	Base Component of the *AnsProlog* Program of the Normative Frameworks	
Π^{n}	Time Component of the *AnsProlog* Program of the Normative Frameworks	
$\Pi_{\mathcal{N}}^{*}$	Normative Framework Specific Component of the *AnsProlog* Program of the Normative Frameworks	
$\mathcal{A}$	Set of Agents	
$|\mathcal{A}|$	Size of the Agent Population	
a	Agent	$a \in \mathcal{A}$
$\mathcal{S}$	State Space of the MAS	
$\mathcal{A}c$	Set of Agents' Actions	
ac	Agent Action	$ac \in \mathcal{A}c$
$\mathcal{S}$	State Space of the MAS	
s_0	Initial State of the MAS	
$\mathcal{I}$	Set of Internal States of an Agent	
$\mathcal{O}b$	Observation Space of an Agent	
γ	MAS transition probability distribution	
κ	MAS Event Function	
ϑ	The Agents' Capability Function	
ξ	An Agent's Decision Function	
ν	An Agent's State Transition Function	
τ	An Agent's Perception Function	
η	An Agent's Utility Consideration Function	
r_d	Movement Distance	$[-1, 1]$
r_α	Nomadic Group Movement Probability Value	$[0, 100]$
$|r_d|$	Step Length of a Move	$[0, 1]$
r_m	Mobility Index of the Agents	$[0, 1]$
r_{sed}	Sedentary Probability	$[0, 100]$

Symbol	Description	Magnitude
r_v	Visibility Radius of the Agents	$(0, 0.5]$
$\rho_{Neighbourhood}$	Average Neighbourhood Density	$[0, \mid \mathcal{A} \mid]$

Acronyms

Acronym	Description
1G	First Generation of Mobile Phones
2G	Second Generation of Mobile Phones
3G	Third Generation of Mobile Phones
4G	Fourth Generation of Mobile Phones
ABM	Agent-Based Model
ABMS	Agent-Based Model Simulation
AHP	Analytic Hierarchy Process
ANOVA	Analysis of Variance
AODV	Ad Hoc On-Demand Distance Vector Routing
ASP	Answer Set Programming
BDI	Belief-Desire-Intention
CBE	Control-Based Enforcement
CDMA	Code Division Multiple Access
CP	Composite Programming
CPU	Central Processing Unit
DES	Discrete Event Simulation
DSR	Dynamic Source Routing
EBIT	Earnings before Interest and Taxes
FMA	Frequency Divison Multiple Access
Gbit/s	Gigabit per Second
HDTV	High-Definition Television
IAD	Institutional Analysis and Development
IBE	Incentive-Based Enforcement
IEEE	Institute of Electrical and Electronics Engineers
iPr	Infrastructure Provider
ITU-R	International Telecommunication Union Radiocommunication Sector
LTE	Long Term Evolution
MANET	Mobile Ad Hoc Network
MAS	Multi-Agent System
MAUT	Multiattribute Utility Theory
Mbit/s	Megabit per Second
MCDM	Multiple Criteria Decision Making

Acronym	Description
MMS	Multimedia Messaging Service
MMORPGs	Massively Multiplayer Online Role-Playing Games
OFDM	Orthogonal Frequency
OFDMA	Orthogonal Frequeny Division Multiple Access
PD	Prisoner's Dilemma
P2P	Peer-to-Peer
RLNC	Random Linear Network Coding
SDY	System Dynamics
SSM	Soft Systems Methodology
QoS	Quality of Service
WMG	Wireless Mobile Grid
WWRF	Wireless World Research Forum

Chapter 1

Introduction

1.1 The Failure of A Million Penguins

On February 1st 2006, Penguin Books in collaboration with the De Montfort University, Leicester, UK, launched the "A Million Penguins" project. The project was based on an idea first expressed by Prahalad and Ramaswamy (2004), that in times of intensified competition and decreasing profit margins, involving customers as a community in co-creating a product is a key to business success. The idea of "A Million Penguins" was to allow customers to create collaborative content in form of novels by using a "wiki" (the technology that supports Wikipedia[1]) which could later on be sold, for example. Despite intensive spending on marketing by Penguin Books (The Institute of the Future of the Book, 2007) as well as high initial participation numbers, within a month after the launch of the project, Penguin Books had to announce its failure. The wiki was closed on March 7th, 2006 and Penguin Books had to amortise the financial losses. In a survey by Mason and Thomas (2008) on the reasons for the failure of the project, two main problems were identified: negatively-perceived *governance decisions* as well as a *contribution inequality*:

1. The "A Million Penguins" project generated a wiki, which is an open distributed system. The term open implies that anyone wanting to join the wiki could do so, without any control on the people joining (with respect to their intentions for joining). Distributed means that in a wiki a network of autonomous users interacts through the wiki platform, which coordinates their actions to generate a product that appears as a single object. After the initial launch of the project, several hundred users registered with the wiki and tried to upload their own completed works to it, violating the concept that the novels on the wiki should be written collaboratively. Other users spammed the website with non-novel related content. That is why – while the wiki was already deployed – Penguin Books started creating a complex set of rules such as the introduction of "reading windows", time

---

[1]www.wikipedia.org

slots at which the wiki would be frozen for several hours each afternoon during which editors could check for violations and readers could catch up with reading through the changes in the novel. Despite the good intentions, these decisions to steer the community into a certain direction were not well-received, and a large proportion of users stopped contributing (Mason and Thomas, 2008).

2. Furthermore, the wiki faced the problem of participation inequality or the so-called 90-9-1 principle (Nielsen, 2008; Hill et al., 1992). This principle suggests that in virtual communities a great majority people (estimates made by Nielsen suggest roughly 90%) will try to free-ride (i.e. benefit from the community without contributing themselves), 9% will occasionally contribute and only 1% of the participants accounts for most contributions, resulting in a participation inequality between the participants. Looking at the statistics of "A Million Penguins" after the wiki closed down, 75,000 different people had viewed the wiki of which 1,476 had registered and in total had produced approximately 11,000 edits. Of these edits over 25% were produced by two users only, with the majority of the remainder being edited by a relatively small percentage of other users. As a result, the initial community idea that one million users could collaboratively work and co-create a product as envisioned by Prahalad and Ramaswamy (2004) failed.

Mason and Thomas (2008) point out that before the start of "A Million Penguins" doubts had already been raised in blogs and newspaper critiques as to what extent the idea behind the project could be successful, and whether in the long run the project could generate the predicted business success. Despite these doubts and critiques being known to the development team of the project, they faced the problem that no similar system existed. They therefore had no experience of how users might react to such a system. Furthermore little testing was undertaken because of problems to analyse how humans might interact with the rules in the system. Thus, for example in the development phase, designers had contemplated limiting edits to 250 words a day per user, but could not perform sufficient tests to predict the possible impact such a rule on users in the system before the system was launched. Testing a system with a large number of human participants can be expensive, a problem which is aggravated by the fact that any increased spending in the development stage automatically results in an increase in the cost commitment to the new project (Riggs, 1982), which increases business risks. As a result of this dilemma, Mason and Thomas (2008) states that "The best they could do was expect the unexpected." From a business perspective investing in a significant project in which, due to a lack of testing, one can only "hope for the best", is likely to endanger to the long-term future of the company and therefore should be avoided.

In this dissertation we look into the question of testing governance decisions for systems which – such as the "A Million Penguins" wiki – have the following characteristics:

- an open distributed nature,

- depend on the interaction of the users in the system and the contribution of the users to the system,

- are in an early stage of conception, and

- relate to new business ideas, where little or no data on the performance of similar the systems exists.

The next section will explain this theme in more detail and formulate the two theses that will be discussed in this dissertation. In order to be able to proof the theses, we first will explain what we understand by the term governance.

1.2 Theses

> **Definition 1: Governance**
>
> The act of a system designer or governing entity in making decisions that specify rules, define expectations, grant power, or verify performance in order to steer the system in a desired direction is called governance.

The word governance itself is derived from the greek verb $\kappa \upsilon \beta \epsilon \rho \nu \alpha o$ (*kubernáo*) which means "to steer". Governance may be done in any organization of any size ranging from individuals to enterprises to complete nations.

As the "A Million Penguins" project demonstrates, trying to estimate the impact of governance decisions in open distributed systems is difficult, especially if humans are meant to collaborate and interact in the system and no previous experience about their behaviour in the system is available. One main reason for this is that humans typically exhibit heterogeneous behaviour and can potentially act selfishly. Furthermore they may only perceive their own environment and not be aware, or care about the governing body's desired system. Instead, they act locally on different stimuli. Nevertheless their decisions culminate in a global result. Testing a system under these conditions, especially before it is deployed, is therefore a difficult but important endeavour.

The aim of this dissertation is to address this state and demonstrate how normative frameworks, in combination with multi-agent system simulations, can be employed to analyse the effect of governance decisions in open distributed systems. We present a methodology which is designed to be employed in the critical time of testing a system before it is fully deployed, or when expensive human user tests are infeasible or inappropriate. We discuss in detail how normative frameworks can be used in these systems to formulate governance decisions in order to steer the system in the desired direction.

In detail, this dissertation focuses on the following two main theses:

Thesis 1:

Governance decisions realized through the introduction of enforcement mechanisms can help to reduce the cooperation and collaboration problems in open distributed systems.

As pointed out above, one major problem in collaborative systems such as the "A Million Penguins" wiki, is participation inequality. Governance in the form of rules that encourage contribution or discourage non-contribution – which is typically referred to as enforcement – is one way of addressing this issue. In this dissertation, we will show how a particular governance decision, namely the introduction of enforcement mechanisms, helps to address the contribution problem in open distributed systems. This is even the case if the entities in the system are heterogeneous with private utility functions and act potentially selfish, i.e. try to enjoy the advantages of the system without contributing to it. One particular focus is on systems that are in an early prototyping stage of development, where testing enforcement mechanisms with real humans does not seem feasible.

For answering this question we have chosen to conduct a specific case study which concerns so-called wireless mobile grids: a concept for user interaction in future generation mobile networks. The reason for choosing this case study is its match with the characteristics of the systems we want to focus on, as well as the fact that whilst wireless mobile grids are still in an early stage of development, the first prototypes of the corresponding mobile phones exist. This allows us to underpin our analysis with empirical transmission data for these phones, resulting in a more realistic analysis. In order to keep the results as widely applicable as possible for any system matching the above mentioned characteristics, despite focusing on the specific wireless mobile grid case study, we try to keep our design decisions as generic as possible and explain each of them in detail. Furthermore we assign one section of this dissertation to the generalization of the results of the case study to open distributed systems in general.

In order to be able to analyse governance decisions for wireless mobile grids at an early prototyping stage, we formulate our second thesis:

Thesis 2:

Normative frameworks and multi-agent system simulations can be used to model and reason about the interaction of users with one another as well as with the norms in an open distributed system.

Whenever testing with real systems is not feasible, models are used to represent them. To the best of our knowledge, no system currently exists that models the interaction of humans with one another as well as their behaviour in respect to the norms (i.e. the rules) in a running system. In this dissertation we present a methodology for developing such a model and test the enforcement mechanisms from Thesis 1 with the help of this model. In doing so, we gain knowledge about the effects of enforcement mechanisms for addressing the problem of free-riding in open distributed systems.

Besides these two theses, this dissertation has one more focus which is not a thesis as such: In a project like "A Million Penguins" several stakeholders (i.e. interest groups) might be involved which could each be affected differently by a governance decision. In this dissertation, we therefore will highlight the impact of the introduction of enforcement mechanisms to an open distributed system in the light of multiple stakeholders. However we view this question as a management question, rather than an actual research question, as there can be no comprehensive answer without a detailed analysis of the different stakeholder interests and such an analysis is not part of this dissertation.

In detail this dissertation will make the following contributions:

- We give insight into wireless mobile grids and the cooperation problem in them.

- We demonstrate how to use normative frameworks to develop a design-time model of a normative system, in order to be able to reason about governance decisions at design-time.

- We provide a methodology on how to generate a run-time model from the design-time model and how to combine it with a multi-agent systems simulation. This allows us to have a run-time focus on the analysis of governance decisions.

- We give insight into how different enforcement mechanisms affect cooperation in open distributed systems and how different stakeholders perspectives need to be incorporated in governance decisions.

1.3   Roadmap

In order to study the two theses formulated, following this introduction, this dissertation is divided into four parts consisting of ten chapters, plus four appendices. Of these four parts, part one serves to lay out the foundations of this dissertation by defining the main concepts as well as referencing related work. Part two focuses on normative modelling and presents the formal model, its general implementation as well as a first model of the wireless mobile grid case study. Part three then illustrates how so-called multi-agent system simulations in combination with normative frameworks can be used to model user interaction in open distributed systems with norms. It furthermore presents the simulation experiments of this dissertation and analyses the results of these experiments with respect to Thesis 1. Part four concludes this dissertation with the summary of the main results as well as a description of their implications. To explain the outline of this dissertation in more detail, we now briefly present the contents of each chapter.

In Chapter 2 we outline the state of the art of ideas and concepts relevant for the particular governance decision we are focusing on in this dissertation, i.e. the introduction of enforcement mechanisms to open distributed systems. Therefore, we first of all define the terminology relevant in this context. We furthermore explain how the contribution inequality experienced in the "A Million Penguins" wiki manifests itself in form of cooperation dilemmas in these systems. After presenting related work that attempts to address this problem as well as highlighting the weaknesses therein, we introduce the institutional analysis and development framework. This framework will be used as a guide to structure the research that proves our theses stated above.

Chapter 3 then presents the wireless mobile grid case study. Wireless mobile grids are one example of the open distributed systems studied in this dissertation. They are still in an early prototyping stage of development and are expected to suffer from free-riding problems. This case study serves as the example for showing how to model and analyse governance decisions by means of enforcement mechanisms throughout the remainder of this dissertation.

In Chapter 4 introduces the concept of normative frameworks and presents requirements for them to be applicable to portray normative aspects of governance decisions. The chapter also provides a formal (i.e. mathematical) model of normative frameworks.

Chapter 5 builds upon the idea of normative frameworks discussed in Chapter 4. As pointed out before, when presenting our research questions, we intend to use multi-agent system simulations in combination with normative frameworks. However, in order to be applicable in a simulation, the formal model of a normative framework presented in Chapter 4 needs to be realised as a computational model. Chapter 5 explains how such a computational model can be realised using *AnsProlog* and presents a first implementation of the case study introduced in Chapter 3. Here we present what we call a "design-time" model in which we can examine all the possible states the system can reach.

In Chapter 6 we present multi-agent system simulations as one means to model autonomous heterogeneous decision makers that interact with the normative framework. In addition we outline the general structure of the chapters that present the simulation experiments and explain the statistical foundations for the analysis of the simulation results.

In Chapter 7 integrates the normative framework model used in Chapter 5 and the multi-agent system simulation. Using this new simulation model, we simulate the case study presented earlier without any enforcement mechanisms, i.e. before any governance decisions have been made. This simulation model and its results then serve as reference points for the analysis, once we start testing the impact that different enforcement mechanisms have on the system.

In Chapter 8 these enforcement mechanisms are then added to the multi-agent system simulation and tests are conducted to explore how they alter the earlier results of the simulation experiments without any enforcement.

In Chapter 9 we extend the multi-agent systems simulations with different movement models and analyse their effect on the simulation results.

Chapter 10 summarizes the simulation results from the previous chapters and focuses on their implications for the specific wireless mobile grid case study as well as for business concepts for open distributed systems with human interaction which might face free-riding problems.

Chapter 11 summarizes the dissertation and discusses future work.

1.4 Supporting Publications

We have published the following papers relating to topics presented in this dissertation. All these publications have been fully peer-reviewed by international experts if not indicated differently.

Balke and Villatoro (2011): Balke, T. and Villatoro, D. (2011). "Operationalization of the Sanctioning Process in Hedonic Artificial Societies." In Workshop on Coordination, Organization, Institutions and Norms in Multiagent Systems @ AAMAS 2011.

In this paper we present our definition of enforcement as well as a generic enforcement process. Particular emphasis thereby is put on the roles associated with the different stages of the enforcement process. Several paragraphs in Section 2.3.3 originate from this paper.

Balke (2009): Balke, T. (2009). "A Taxonomy for Ensuring Institutional Compliance in Utility Computing." In G. Boella, P. Noriega, G. Pigozzi, and H. Verhagen, editors, Normative Multi-Agent Systems, Number 09121 in Dagstuhl Seminar Proceedings. Schloss Dagstuhl - Leibniz-Zentrum füur Informatik, Germany. http://drops.dagstuhl.de/opus/volltexte/2009/1901.

This paper presents the first draft of the taxonomy of enforcement mechanisms presented in Section 8.1.

Balke et al. (2009): Balke, T., König, S., and Eymann, T. (2009). "A Survey on Reputation Systems for Artificial Societies." Bayreuther Arbeitspapiere

zur Wirtschaftsinformatik 46, University of Bayreuth. `http://opus.ub.`
`uni-bayreuth.de/volltexte/2009/616/pdf/techreport_final.pdf.`

This paper is a survey paper about the different reputation mechanisms
employed in artificial societies. It has not been peer-reviewed. Text
passages in Sections 2.3.2, 2.4 and 8.1 originate from this paper.

Balke and Eymann (2010): Balke, T. and Eymann, T. (2010). "Challenges
for Social Control in Wireless Mobile Grids." In T. Doulamis, editor, Grid-
Nets2009, volume 25 of Lecture Notes of ICST, pages 147–154. Springer.

In this paper we present the cooperation dilemma inherent in our wireless
mobile grid case study for the first time and discuss challenges for social
control mechanisms to address these issues.

Balke et al. (2011): Balke, T., De Vos, M., Padget, J. A., and Fitzek, F.
(2011). "Using A Normative Framework to Explore the Prototyping of
Wireless Grids." In M. De Vos, F. Nicoletta, J. V. Pitt, and G. Vouros,
editors, Coordination, Organizations, Institutions, and Norms in Agent
Systems VI, volume 6541 of Lecture Notes on Computer Science, pages
95–113. Springer.

This paper presents the first computational model of the case study used
in this dissertation with the help of normative frameworks and *AnsProlog*.
The model presented in this paper is a design-time model. In Chapter 5
we present a modified version of the model presented in this paper.

Balke et al. (2011): Balke, T., De Vos, M., and Padget, J. A. (2011). "Analysing Energy-Incentivized Cooperation in Next Generation Mobile Networks
using Normative Frameworks and an Agent-Based Simulation." Future
Generation Computer Systems Journal, 27(8): 1092–1102 `http://www.`
`sciencedirect.com/science/article/pii/S0167739X11000574.`

A first attempt to model the wireless mobile grid case study with both a
normative framework model as well as a multi-agent system simulation is
presented in this article. In the article the two components are developed
as separate models. It is shown which benefits each of the models has and
to what extent these two models could possibly contribute to a combined
modelling approach. Parts of this article are incorporated into Section 7.1.

Balke et al. (2011b): Balke, T., De Vos, M., Padget, J. A., and Traskas, D.
(2011). "On-line reasoning for institutionally-situated BDI agents." In K.
Tumer, P. Yolum, L. Sonenberg, and P. Stone, editors, Proceedings of 10th
International Conference on Autonomous Agents and Multiagent Systems
(AAMAS 2011), pages 1109–1110.

In this short paper we present the run-time model of the wireless mobile
grid case study consisting of both a multi-agent system model as well as
a normative framework component. Parts of this paper are presented in
Section 7.2.

Balke et al. (2011a): Balke, T., De Vos, M., Padget, J. A., and Traskas, D. (2011). "Normative Run-Time Reasoning for Institutionally-Situated BDI Agents." In Proceedings of the 13th International Workshop on Coordination, Organization, Institutions and Norms in Agent Systems (COIN) @ WI-IAT2011.

This paper is an extended version of Balke et al. (2011b). In the paper we explain the link between the design and the run-time model in more detail as well as present code fragments of both models. Several paragraphs of this paper can be found in Section 7.2 as well. However, it has to be noted that the model presented in this paper differs from the one described in this dissertation. The difference between the two models is that we chose to model the agreements between actors using obligations in this dissertation, which is not the case in Balke et al. (2011a).

As these papers are our own original work, when directly citing them in this dissertation we forgo the utilization of quotations marks, but reference the papers in the respective sections.

Part I

Foundations and Related Work

Chapter 2

Norms and Enforcing Cooperation in Open Distributed Systems

Having established the theses that this dissertation will answer, this chapter lays the foundations for doing so by defining the most important terms and explaining the most important concepts associated with them. In detail, we will start by examining *norms* which are one of the major instruments of governance. This will be followed by an explanation of the notion of open distributed systems used in this dissertation as well as an explanation of the cooperation problems inherent in them. Afterwards the focus will shift to enforcement of norms as one means to address the cooperation problem. In detail we will present a general view on enforcement first before reviewing the existing literature on the cooperation problem and norm enforcement. One of the major aims of this dissertation is to compare governance decisions, with a special focus on enforcement concepts. For this reason the last part of this chapter is devoted to the Institutional Analysis and Development (IAD) framework, which provides a methodology for analysing and comparing the implications of governance decisions.

2.1 What is a Norm?

Norms have been studied in a variety of research domains including natural sciences, social and formal sciences as well as applied sciences. As a consequence a universal definition of the term does not exist in literature (Horne, 2001). Reviewing the existing literature on norms numerous topologies (Morris, 1956), categorizations (López y López, 2003) and specialized definitions of the term (Gibbs, 1965) can be found. Each of these has its own merits, however they were all written from the perspective of a specific research discipline, thus stressing points that are particular to that discipline. This makes it hard to find a universal definition of the term and attempts usually end in a very broad definition which is difficult to apply to specific problems at hand (Hollander and Wu, 2011). When looking up the term "norm" in Websters Electronic Dictionary (2011) for example, several explanations of the term can be found reflecting the different views on the term in different scientific disciplines such as social science (e.g. sociology, psychology, philosophy, economics and legal theory), natural science, formal science (e.g. deontic logic), applied sciences (e.g. computer science and engineering), etc.:

1. an authoritative standard

2. a principle of right action binding upon the members of a group and serving to guide, control, or regulate proper and acceptable behaviour

3. average as:

4. a: a set standard of development or achievement usually derived from the average or median achievement of a large group

   b: a pattern or trait taken to be typical in the behaviour of a social group

   c: a widespread or usual practice, procedure, or custom <standing ovations became the norm>

5. etc.

Although its broad focus makes this definition impractical for the purpose of this dissertation it is a useful starting point to review the existing literature on norms in order to define the term afterwards. Keeping the above definition in mind, in the next paragraphs this dissertation will look at the different research domains and the typical notion of norms that is being used in each of them. The goal of this undertaking is to find similarities and differences in the usage of the term across the various domains, as well as to point out concepts relevant to the theses addressed in this dissertation. As this dissertation has its background in the business and social science domain, special emphasis will be put on the ideas discussed in the respective literature. The intention of this review is not to provide an exhaustive overview on all the work that has been done with regard to norms, but rather to highlight the different perspectives on norm research by briefly presenting the best known or most representative research on the topic.

Norms in Social Science: Social science is an umbrella term for a number of research disciplines that study different aspects of human society and the relationship of the individual members within the society to one another and to the society. Against this background, in the social sciences norms typically are viewed as means to define these relationships, and portraying as well as regulating the behaviour of the members of a group or society as well as the society as a whole (Neumann, 2008).

Research on norms in the social sciences tends to focus on the *social function* of norms (see the ideas by Merton (1968) for example), the *social impact* of norms (e.g. North (1993)), or mechanisms leading to the *emergence and creation* of norms (e.g. Gilbert (1995)).

With regard to the aspect of the social function, norms are often concerned with the behaviour members of a social group should (or should not) perform regardless of the possible consequences (also referred to as obligations). Furthermore they deal with the expectations resulting from the anticipation of the other actors in the system with regard to this behaviour.

This idea is formalized by Tuomela and Bonnevier-Tuomela (1995) for example, who specify a social norm as a norm having the following form: "An [individual] of the kind F in group G ought to perform task T in situation C."[1] and thereby highlight the four major aspects of norms in social science: (i) that individuals can be different in their behaviour and might perform different roles in a society (i.e. F), (ii) the importance of the relationship of the individual to a group or society (i.e. G), (iii) the obligation to perform (and indirectly not to perform) certain tasks (i.e. T) as a result of a norm, as well as (iv) the notion of context-dependency (i.e. C) of a norms, i.e. that norms might only be valid in particular situations.

Another important aspect Tuomela and Bonnevier-Tuomela stress is the distinction between so-called "r-norms" and "s-norms". The former are norms that are "created by an authority or body of agents authorized to represent the group (this body can also be the entire group)" (Tuomela and Bonnevier-Tuomela, 1995). This authority might for example be a governing or legislative body, the operator of a platform, a chosen leader, etc. S-norms are norms emerging as a feature of a (social) normative context, i.e. the result of mutual beliefs about the way a particular situation should be handled, general codes of conduct or conventions. Thus, in contrast to r-norms, s-norms are not based on rules defined by any authority, but tend to highlight the social aspect of norms. The inclusion of s-norms in this distinction is a rather important feature of the social science. Based on the idea of s-norm, in addition to the notion of norms being guidance on what actors are ought or expected to do, in the social sciences norms are also viewed as an information source of what is perceived to be "normal" in a group or population (Therborn, 2002). In the social sciences, this "normal" behaviour is explained as emerging as a general pattern of behaviour by the actors of a society making choices without any centralized planning (Andrighetto et al., 2007).

With regard to research on the social impact of norms, the focus is placed on the utility provided to or taken away from the actors involved in an interaction. Utility is defined here as the relative (both positive and negative) satisfaction achieved by the actors. This utility can be either internal, such as emotion levels, energy, etc., or external, in the case of money, etc. Social impact research analyses the effect of utilities of the different stakeholders in a society resulting from specific norms as well as on the society as a whole (North, 1990).

Besides these works on the social function, the social impact as well as

---

[1]Despite the strong emphasis of this statement on obligations i.e. what one ought to do, Tuomela and Bonnevier-Tuomela in the course of their definition broaden it to include permission (i.e what one may do) as well.

the emergence and creation of norms, in the social sciences (in particular in philosophy) further important scientific contributions have been made dealing with normative positions (Sergot, 2001). The two normative positions we have talked about in this chapter so far are permissions and obligations. Although these describe norms to a large extent, within the world of legal theory several more positions can be found, including power, duty, right, liability, disability, claim and immunity for example (Hohfeld, 1913). With regard to this dissertation, the position of *power* is of particular interest as a restraint on the (physical) power of autonomous entities. In social science, power can have two different forms, which are both described by Makinson (1986): (i) legal power, and (ii) physical power.

Whereas the former specifies whether an actor is "empowered" to perform a certain action in a legal sense, the latter establishes whether he is physically able to carry out the actions necessary to exercise his legal power (Jones and Sergot, 1996). Whereas this distinction between physical and normative power is relatively easy to make, it is important not to confuse normative power and the term "permissions" explained earlier. To explain the difference between normative power and permission (Makinson, 1986, p. 409) gives an illustrative example:

> ...consider the case of a priest of a certain religion who does not have the permission, according to instructions issued by the ecclesiastical authorities, to marry two people, only one of whom is of that religion, unless they both promise to bring up the children in that religion. He may [in his function as priest] nevertheless have the *power* to marry the couple even in the absence of such a promise, in the sense that if he goes ahead and performs the ceremony, it still counts as a valid act of marriage under the rules of the same church even though the priest may be subject to reprimand or more severe penalty for having performed it.

What is important to note besides this difference between permission and legal power, is the difference in legal and physical power. Thus, despite the priest having the legal power, physical power does not follow automatically. Thus, the priest might be incapacitated by being sick for example, and as a consequence is physically not able to perform the normative action of marrying two people, despite having the legal power to do so. In the opposite direction the two notions of power also do not have a coercive relation: having the practical possibility and power to act does not necessarily imply that a legal power is also existing. To give an example, someone might not have the legal power to conduct a marriage, despite being physically able to do so (e.g. by not being physically incapacitated and knowing the procedure etc.).

Norms in Natural Science: In science, the term natural science refers to the sciences such as biology, chemistry, or physics that use a naturalistic approach in order to study objects and phenomena of the physical world, which is understood as obeying rules or laws of natural origin. As a result of this focus on laws of nature or natural constants, the term norm in the natural science is typically used when speaking about proven phenomena and generally accepted standards (i.e. "is-statements" (Kelsen, 1960)), rather then being viewed in the "ought to" or "is permitted to" context such as in the social sciences. That does not mean that no research is being done on what is understood by the term "norms" in the social sciences (in natural science research is being done on topics such as group behaviour and the interaction of individuals in a group or kin selection for example). However, despite this research often including normative considerations similar to the ones on in the social sciences, researchers in this area do not commonly refer to the term "norm".

Norms in Formal Science: Formal sciences such as mathematics, logics, theoretical computer science or statistics, are research domains that use formal systems to generate knowledge. Thus, unlike other sciences, the formal sciences are not concerned with the validity of theories based on observations in the real world, but instead with the properties of formal systems based on definitions and rules. Research topics on norms in the formal science therefore include the formal representation of the deontics of norms (such as the modality $\mathcal{O}(X)$ for obligations) (von Wright, 1951), the reasoning about the consistency of normative specifications and the formal validity of norms in specific situations (Vasconcelos et al., 2004), as well as the determination and verification on whether certain normative states can be reached or not (Viganò and Colombetti, 2007). Of particular note in the field is dynamic deontic logic (Meyer, 1988). Dynamic deontic logic is a re-formulation of standard deontic logic in the dynamic logic syntax. It differs from classical logics in that it (i) includes early attempts to reason about temporal aspects of norms, (ii) allows for the explicit treatment of actions and events as well as their consequences, and (iii) allows for concurrent, sequential and non-deterministic actions (Cliffe, 2007). Based on these features, dynamic deontic logic includes two types of formulae: (i) action formulae that focus on the execution of actions and the order of actions, and (ii) propositional formulae that (combined with each other) feature complex properties relating to the sequence of actions and their consequences. This make it possible to reason about the sequence of actions and their consequences as well as about temporal aspects of norms, and analyse how particular system states can be achieved. This inclusion of temporal aspects is important, as it allowed for the formal modelling of deadlines, i.e. normative statements that something needs to be done by a certain point of time or before a certain other action can be performed (Broersen et al., 2004; Dignum et al., 2004).

Norms in Applied Science: Applied sciences apply and draw inspiration from the scientific knowledge of other scientific fields and transfer it into a physical environment to solve practical problems. Examples of applied science are applied computer science, artificial intelligence or engineering for example. One major focus of norms in applied sciences is – based on the knowledge from other domains – their application and applicability to real world problems. In computer science and engineering for example norms are mainly dealt with in terms of their application to computing and engineering processes in which they tend to be used as tools to complete the task of regulating and controlling a systems or system components (in very much the same manner contracts, protocols, etc. do) (Minsky, 1991b).

Norms (as defined in this dissertation): Looking at how norms are considered and treated in the different research domains, despite the varying perspectives, some common features can be found, such as the idea that norms are rules that define what is considered right or wrong by the majority of a population. The definition of the majority of the population as well as of the particular features of the population (i.e. its size or composition) thereby are domain dependent (Hollander and Wu, 2011). Norms furthermore spread; that means they are acquired and communicated through direct (e.g. communication between the actors in a system) and indirect means (such as adaption and learning processes by the actors).

Taking these common features into account and extending them with the main idea of the social science literature on norms which has influenced this dissertation the most, in this dissertation the following definition of norms will be used:

Definition 2:   Norms

Norms are restrictions on patterns of behaviour of actors that are actively or passively transmitted. These patterns are sometimes represented as actions to be performed (Axelrod, 1986). In principle they dictate what actions (or outcomes) are permitted, empowered, prohibited or obligatory under a given set of conditions as well as specify effects of complying or not complying with the norms (Balke and Villatoro, 2011). Norms should be contextual, prescriptive and followable (Schimanoff, 1980).

The notion of contextuality refers to norms only being applicable in a specific context and not in general. To give an example: despite being generally valid, rules for driving a car are only applicable in traffic settings and these traffic settings provide the context for the application of the respective norms. This notion of contextuality also sets a scope to the time a norm is in force and consequently requires the definition of the activation, as well as existence and deactivation of a norm. That norms should be prescriptive refers to the fact that "those who are knowledgeable of a rule also know that they can be held accountable if they break it" (Schimanoff, 1980, p. 41)[2]. Thus, norms specify what actions an actor "must not" perform (prohibition), "is empowered to perform" (legal power), "must perform" (obligation) or may perform ("permission") if the actors

---

[2]This does not necessarily imply the opposite to be true, i.e. that one can only be held accountable if one knew about a norm.

want to avoid sanctions for non-compliance with the norms being imposed on them. Finally, rules should be followable in the sense that it should be physically possible for the actors in a system to both perform and not to perform prohibited, obligatory or permitted actions, as well as to obtain the legal power to do so.

This dissertation focuses on explicit[3] "r-norms" as defined by Tuomela and Bonnevier-Tuomela (1995); Tuomela (1995) only, hence norms that are "created by an authority or body of agents authorized to represent the group (this body can also be the entire group)". Despite the broader definition of the term above, when speaking of "norms" this dissertation refers to "r-norms" and if not specified otherwise does not consider s-norms.

2.2 Open Distributed Systems

As pointed out in the introduction, the focus of this dissertation is open distributed systems. In order to create an understanding of what is meant by this term, in this section the term will be defined and be explained in more detail. Looking at the term, it consists of two objectives specifying the system term: "open" and "distributed". For this reason, in order to explain the term "open distributed" system, this dissertation starts from the definitions of its objectives – i.e. open systems and distributed systems – and then combine the two terms to arrive at an integrated definition as well as explain the relevance of norms to these systems.

> **Definition 3: Open System**
>
> An open system is a system in which autonomous heterogeneous entities with incomplete knowledge pursue their own goals and possibly interact with other entities (López y López, 2003, p. 2). The system is "open" in the respect that entities can freely join and leave the system (Davidsson, 2001).

As a result of the openness and the autonomy of the entities, several problems arise. One is that entities do not only pursue their own goals in a system, but perform actions because of what they expect other entities or the system to do. Thus, it needs to be established how expected behaviour is defined. This is normally done with the help of norms. However, the specification of expected behaviour with the help of norms does not necessarily imply that the entities in the system will exhibit the expected behaviour. Thus in order to participate effectively in the system the entities need to be able to take into account possible malicious and unexpected behaviour of other entities and adjust accordingly.

---

[3]In general, r-norms are *explicit*, i.e. they have been brought into existence by a proper authority (e.g., group members with authority) and have been formally articulated, specified and written down. Examples of formally articulated norms are laws and regulations for example. If norms have not been written down, but only passed on orally or assumed to be understood by all actors in a system without any formal specification, one talks about "implicit norms". In this dissertation the focus will be on explicit norms only.

In contrast to the term "open system", that mainly focuses on the system composition and entry/access components, the term "distributed system" is mainly used for describing a system's appearance from the outside. The term itself tends to be used in computer science in particular. It is defined as follows:

Definition 4: Distributed System

A distributed system is an application such as a piece of software that executes a collection of protocols to enable a collection of autonomous entities in the system that are connected through a network and some middleware to coordinate their actions, as well as share resources to appear as a single coherent facility from outside the system (Tanenbaum and van Steen, 2007).

Two particular features are important to note about this definition. The first feature is that similarly to open systems, distributed systems perceive the entities acting in the system as autonomous, i.e. independently choosing their actions. These entities possess different resources. These resources might initially only be accessible to specific entities and can be made available to other entities via communication, coordination and/or cooperation mechanisms. This coordination results in the second particular feature of distributed systems: the appearance as a single coherent system from the outside. Hence, despite the underlying distribution of resources and interaction of the entities of the system, from outside the system, this distribution is not visible. An abstraction level is added that covers the distribution and lets the system appear as one component.

Combining the notions of open systems and distributed systems, open distributed systems are defined as follows for the purpose of this dissertation:

Definition 5: Open Distributed System

Open distributed systems are systems in which autonomous entities with some form of social relationship with one another are free to join and leave the system as well as perform actions in the system such as interacting with other entities. They base their decisions and actions on their own goals as well as their expectations about the system and the behaviour of the other entities. The result of the combined individual decisions and actions is a global system emergent behaviour that – in contrast to the individual decision making processes – can be perceived from the outside of the system.

From an economic perspective, this definition is rooted in the ideas of the Austrian School, especially the concept of "methodological individualism" (Schumpeter, 1908). Methodological individualism perceives system-wide developments as the aggregation of decision and actions by individual entities (Hayek, 1996) and tries to explain emergent system behaviour in a bottom-up approach (Lukes, 1968).

From an analytical perspective, methodological individualism can take two forms. It can explain emergent system properties: (i) in terms of individuals alone, and (ii) in terms of individuals plus relations between individuals.

As Arrow (1994) argues, the first version is not achievable in practice, as the social structures and the environment individuals are embedded in will always have an impact on the individuals and therefore should not be neglected. That is why he argues that the broader view should be taken. Following his line of argument, we focus on the latter form of methodological individualism when researching the effects of governance decisions in open distributed systems. Thus, despite analysing system-wide changes as a result of norm changes, in this dissertation the system-wide changes are viewed as result of the individual entities' decisions and their interaction between them. As a consequence the effects of norm changes on individuals, their actions as well as their social structures will be the starting point of the normative analysis conducted in this dissertation.

To elaborate in more detail on what kind of systems this dissertation refers to when speaking of open distributed systems, let us look at some examples of these systems and their applications. Table 2.1 provides an overview of major open distributed systems, grouping them by their intended application.

The first group of open distributed systems we consider are those – such as mobile ad hoc networks (MANETs) – that have been designed with relay routing as their primary application. Typically in a MANET bandwidth is shared in order to allow communication between individuals (referred to as mobile nodes). If two mobile nodes are out of reach of one another's mobile antennas, for example, and cannot directly communicate with each other, they can use other nodes as a communication channel, routing their communication through them. Currently, two big challenges have been identified for this kind of systems: the first is the mobility of participants, which the routing protocols need to account for. The second challenge is the participation of the mobile nodes as relays. Relaying information costs energy. The problem with this is that battery-powered mobile nodes have a limited power life cycle, which is drained every time they relay messages. As a consequence, MANET users might be tempted to free-ride, i.e. to enjoy the benefits of the MANET, but not to contribute to the MANET by routing other nodes' communications (Buchegger and Chuang, 2007). This, however, puts the whole network at risk. The reason for this is simple: as pointed out, network users can exhibit strategic behaviour and are not necessarily cooperating by making their resources available without the prospect of rewards for their good behaviour. However, unreciprocated, there is no inherent value in cooperation for a user. A lone cooperating user draws no benefit from their cooperation, even if the rest of the network does. Guaranteed cost paired with uncertainty or even lack of any resulting benefit does not induce cooperation in a (bounded) rational[4], utility-maximizing user. Without any further incentives, rational users therefore would not cooperate in such an environment and all be worse off than if they cooperated. This

---

[4]The term "bounded rationality is defined on page 31 in detail.

Table 2.1: Classification of Open Distributed Systems (Oram, 2001; Wrona, 2005)

application	system	shared resources	example
relay routing	mobile ad hoc networks	bandwidth, buffer memory	ad hoc on-demand distance vector routing (AODV), dynamic source routing (DSR)
	wireless mesh networks	bandwidth, CPU	RoofNet, Rooftop
P2P networking	file-sharing	bandwidth, disk space	Gnutella, Napster, BitTorrent
	anonymous publishing	bandwidth, disk space	Publius Publishing System, FreeHaven, Freenet
	grid computing	CPU	SETI@home
	communication & conversation	bandwidth, CPU	Skype, Jabber
	anti-censorship news accessing	bandwidth, disk space	Red Rover
cloud computing	service provision	software, platforms, infrastructural components (CPU, bandwidth, memory, ...)	Amazon S3, Amazon EC2, Eucalyptus Cloud Computing Service
virtual teams	MMORPGs	character abilities, resources, manpower, ...	guilds, clans or teams in World of Warcraft, Guild Wars, EverQuest

phenomenon is referred to as "Tragedy of the Commons" in literature (Hardin, 1968; Ostrom, 1990)[5]. As a result, enforcement mechanisms are required in open distributed relay routing applications to reduce the prevalence of free-riding.

Another example of open distributed systems are Peer-to-Peer (P2P) networks. They are a way to organize a group of participants/nodes with equal privileges to allow for the partition of tasks or workloads between these participants/nodes. These participants/nodes are referred to as peers (Milojicic et al., 2002). The best known example of task partition is probably file-sharing (e.g. BitTorrent or Gnutella). File-sharing comprises the idea that the individual users (called peers) offer files to share across a P2P network. On request by another peer these files can be sent to the requesting peer. The problem in this kind of systems is that sending and providing files does not have any immediate benefit for the providing peer, resulting in the already described tragedy of the commons and free-riding issues. Furthermore, P2P networks typically are large scale networks and as a consequence, individual participants or nodes are unlikely to have global knowledge of the whole network, but only local information. That is why one of the main issues in P2P networks is to provide mechanisms to help to search for information. This is normally done by routing information requests through the P2P network. With regard to this routing one major issue is the robustness of the system, which is required in several forms. Firstly, due to the open nature of the system (i.e. participants/nodes can join and leave at any point), its general functionality needs to be ensured. Thus, the system's functionality and the protocols (i) should not rely on single nodes, (ii) need to able to handle the entrance and leaving of any number of nodes, and (iii) need to be able to cope with malicious nodes (i.e. nodes that on purpose intend to harm the system's functionality).

Secondly, robustness against free-riding is required, i.e. that users do not only use resources, but share their own resources in return. Thus, if too many free-riders exist in a P2P system, the performance of the whole network could be at risk, depriving all nodes from the network's intended advantages. Just like in MANETs and other relay routing applications, taking actions against free-riding is therefore essential for the success or even the survival of the network.

Cloud computing – the penultimate application presented in Table 2.1 – is a newer paradigm that is often associated with the concept of service provision. The general idea behind this is to envision a system in the form of different scalable modules or resources that are coupled together. These modules/resources are defined as services which can be commodities and are provided/acquired over wide area networks (such as the Internet) via standardized interfaces or technologies. The services can range from simple resources such as CPU or disk space to software applications or whole infrastructures and can theoretically be offered and developed by different, possibly competing organizations[6]. Currently

---

[5]The problem described by Hardin as well as Ostrom et al. originally – as the name indicated – referred to common pool resources (which will be explained later in this chapter) only. However, in recent publications the term has also been used for other types of goods such as public goods. In this dissertation we will use this more recent broader view of the term.

[6]An overview over current cloud computing providers can be found in Leavitt (2009) for

only few cloud computing providers exist, hence the described scenario is still a vision. If one follows the ideas of this vision however, predicts a situation where several service providers might compete with one another for the attention of possible customers, especially if the services provided are being paid for. From a service consumer's point of view it is important to choose a provider that satisfies his requirements the best. Enforcement mechanisms are one tool for ensuring a better average user satisfaction under these circumstances and therefore, also play an important role in current cloud computing research.

The last application presented in Table 2.1 is often not considered a classical open distributed system, despite having all the features of it: Massively multiplayer online role-playing games (MMORPGs). MMORPGs are video/computer games in which a very large number of players interact with one another within a virtual game world. The human players therefore often assume the role of a virtual character and guide it through the virtual world. The virtual worlds normally are persistent, i.e. constantly existing and evolving, however the player can log in and out at anytime. One basic feature of MMORPGs is that they require social interaction between the players. In order to develop their character a player may have to cooperate with other players, as certain tasks in a MMORPG require a certain amount of teamwork[7]. These tasks usually require players to combine the specialized skills of their virtual characters and take on roles/tasks (based on these skills) in the team. Examples of such tasks are protecting other players from damage (called "tanking"), "healing" damage done to other players or damaging enemies in a battle for example. The players in these settings rely on each other to have a better chance of succeeding. Very often the reward for succeeding then needs to be shared by all the members of the group who have survive the task. This situation results in a dilemma for the player. Despite needing a strong group for better chances of success, the more players survive at the end, the more have to share the reward. Furthermore, a player might steal the reward and run off with it hoping that the players that have used their strength in the course of fulfilling the task do not have enough strength left to pursue them. As a result – although wanting to benefit from the task completion (and resulting reward), players are discouraged from contributing to save their own resources, which in turn however weakens their group. Consequently, in MMORPGs the tragedy of the commons can be found resulting in the need for mechanisms to encourage cooperation.

2.3   Ensuring Compliance with Norms

In the last section, we examined open distributed systems and their particular features. One dominant theme was that they face the tragedy of the commons, i.e. free-riding problems due to their open, distributed and large scale nature.

---

example.

[7]The most common types of these teams are called "Pugs" (Pick up groups). Pugs are small groups with a short lifespan and often have the focus of completing a specific quest. Larger groups with a longer lifespan are called Guilds (Johansson and Verhagen, 2010).

If not solved, these free-riding problems can endanger a complete network and deprive all users of the possible benefits. Enforcement mechanisms are one way of addressing the free-riding problem and we now examine them more closely.

So far, although mentioning enforcement in the definition of norms, little was said about enforcement itself, but indirectly the assumption was made that norms are generally enforceable, i.e. it can be made sure that everyone follows the norms. This assumption is rather unrealistic in open distributed systems and one aim of this dissertation is to address this problem by looking more closely into the effects that different enforcement mechanisms have on these systems. Therefore in this section the theoretical foundations are laid for the concept of enforcement as one means to ensure compliance with norms.

In general two mechanisms for ensuring compliance with norms can be found in the literature: (i) In the first it is assumed that the actors (and/or their actions) in a system can be controlled (this includes the control of the physical power of actors) and any non-normative actions can be stopped before they can take place and affect the system. This notion is typically referred to as regimentation (Grossi et al., 2007) or control-based enforcement (CBE) (Perreau de Pinninck Bas, 2010, p. 14). (ii) The second point of view portrays the complete opposite: it assumes that only the actors in a system themselves have control over their physical powers. Hence, they can choose and control the actions they perform and therefore enforcement must be executed after the actions have taken place with the help of incentives such as sanctions or rewards. This form of enforcement is referred to as incentive-based enforcement (IBE).

To make the distinction between the two forms of enforcement easier, for the purpose of this dissertation, when talking about "enforcement" we refer to the notion of IBE; otherwise the term "regimentation" will be used.

2.3.1 Regimentation

As explained earlier, regimentation tries to bring about a state in which the deviation from a norm is physically not possible, or where the deviation from the norm does not have any effect on the rest of the participants in a system. This idea is especially popular in artificial societies, i.e. societies such as multi-agent systems in computer science (Weiss, 1999; Russell and Norvig, 2002; Wooldridge and Jennings, 1995) – that are inhabited by artificial entities that perform tasks on behalf of their owners. Minsky (1991b) for example distinguished two modes of regimentation which together comprise the regimentation idea: regimentation by *interception* and regimentation by *compilation*.

Regimentation by compilation assumes that all participants' mental states are accessible to the system (closed systems), and can be altered to be in accordance with the normative framework. Thus, participants are treated as a "white box" whose content can be analysed and altered as needed by the regimenting entity of the system (Balke, 2009). Minsky gives an example of a computer program component that is supposed to join a running system. In his example the source code of this new program is accessible to the regimenting entity of the system and this entity can check whether based on the source code the program can conduct any actions that deviate from the norms of the system. If this is the case it can either stop the program from joining the system or urge a change to the program. This concept is applied for example in the KAoS architecture (Bradshaw et al., 1995).

In the case that the mental states are not accessible to the system (i.e. the inner states of the participants are a black box to the system) norm compliance is ensured by indirectly constraining the actions of the individual participants. Thus, regimentation by interception uses a regimentation component (typically a piece of middleware in computer systems) that at run-time intercepts all actions (in computer systems this normally refers to messages being sent to the rest of the components) and dismisses those that are not in accordance with the norms of the system. This idea is used in systems such as ISLANDER, that uses so-called "governors" as regimentation entities (Esteva et al., 2002). Thus, when two participants want to interact with each other they have do so by using the governors as a proxy for their communication. The governor parses their messages and verifies their conformance with the norms of the system. If the messages conform to the norms it routes the corresponding messages to the appropriate recipients and interaction can take place. Otherwise the messages are dismissed and no non-norm-conforming interaction will be initiated.

Looking at the implementation side of interception-based regimentation, in artificial societies, according to (Perreau de Pinninck Bas, 2010, p. 15) this approach is relatively easy to implement and deploy. Nevertheless it exhibits a number of problems which act as drawbacks to the idea.

Interception-based regimentation relies heavily on messages being sent and checked for unwanted behaviour. The problem with this, however, is that the basic mechanism cannot deal with norm-deviations by messages not being sent, such as in the case of obligations. Minsky (1991a) has proposed a way of solving this issues by giving its regimentation components (so-called controllers) the task to keep track of the controlled participant's state and the obligations it has to fulfill. The controller then sends a message on behalf of the participant it observes, if that participant misses its obligation. However, in order to do so, it needs to know all the parameters necessary for creating the respective message, which causes implementation difficulties and reduces the number of unfulfilled obligations a controller can cope with. Furthermore, the additional computing overhead produced by the interception of messages as well as the book-keeping of the participants' states may degrade the system's performance. As a result, interception-based regimentation can be a serious bottlenecks for (especially real-time) norm enforcement.

One problem both kinds of regimentation have with regard to their applicability to open distributed systems is that they limit the autonomy, as well as the physical power of the participants (or at least assume this to be easily possible). In reality however – especially if human actors are involved – this may not be the case. In reality often situations arise, where from a business perspective regimentation is not possible because consumers will not accept being monitored all the time or where from a technical perspective regimentation is simply not feasible. Thus, not only do participants frequently join and leave

open distributed system where their mental states cannot be checked (this is especially the case if humans interact with one another), but situations might arise in which the participants have agreed to conduct the actual transaction outside the monitored environment (i.e. outside the interception area). To give a real world example for such a situation, let us have a look at eBay[8]. The eBay platform is used to match potential buyers with potential sellers with the help of auctions. However, once the matching has taken place and according to eBay regulations the transaction partners agreed to live up to their transaction tasks (e.g. deliver a good after the money has been sent), the actual exchange of good and money takes place outside eBay. Despite having several indirect mechanisms in place that should encourage participants to stick to the agreements, eBay cannot directly force them to do so, as it has no direct control over the actions or mental states of participants outside the space of the website and consequently makes completely effective regimentation impossible.

2.3.2 Incentive-based Enforcement

IBE has been proposed to solve compliance issues where regimentation is not feasible or sensible. In contrast to regimentation approaches that pursue the idea of ensuring complete compliance, IBE aims to reduce norm-deviating behaviour to the smallest amount possible.

With regard to IBE, Ellickson (2005) proposes the usage of both "carrots" and "sticks" – i.e. rewards and sanctions – as incentives and categorizes three types of actions by participants that are of relevant with regard to IBE:

1. good behaviour that is to be rewarded,

2. ordinary behaviour that warrants no response (as giving a response to the most common behaviour only tends to increase the costs of administering sanctions), and

3. negative behaviour that is to be punished.

Ellickson's approach uses the game-theoretic idea of utility (see Section 2.4 for more details). This implies that the participants' goals, motivations and decision-making processes take into account many issues that may be combined into a single value – *utility* – which is computable through a "utility function" (Balke and Villatoro, 2011). Thus, every situation in an interaction is attributed certain utilities. Depending on their choice of actions the participants in an interaction can achieve these utilities. Sanctions are normally conducted in form of giving negative utilities to the norm-deviating participant, whereas rewards come in the form of positive utilities. Thereby it is normally assumed that the participants try to maximize their utilities, i.e. that possible sanctions or rewards can help to induce norm-conformal behaviour. What is important to note is that utility functions may be different for different individuals, thus

---

[8]For more information see www.ebay.com or the respective country-specific websites.

individuals might judge the same situation differently with regard to the utilities involved and consequently might act differently given the same setting.

One of the roles of norms in this context is to give information about which actions are valid and which utilities might result from different actions and outcomes, however they have no regimenting physical power, i.e. it is the participants choice whether to follow a norm or not.

With regard to the actual enforcement actions, two different options are available: *direct* and *indirect* enforcement (Vázquez-Salceda et al., 2005). Direct enforcement affects the individuals immediately and is directly observable (like bans, fines and physical punishment). Indirect enforcement affects only the individuals' future actions and may be directly visible or not (e.g. warnings, reputation).

Besides the classification into direct and indirect mechanisms, enforcement is typically distinguished as to who enforces. IBE can be either *self-enforced*, enforced by a *second-party*, enforced by (a group of) *third-party observers*, or enforced by special *enforcement entities* that are being given special powers to fulfill this task (Balke and Villatoro, 2011).

Violations can be purposeful or accidental. For self-enforcement to be effective the violator needs to realize its own violation – even if it is accidental – and enforce a sanction on itself.

In contrast, second-party enforcement refers to enforcement being executed by the parties directly involved in a transaction. Imagine an actor A interacting with another actor B. Second party enforcement then means A using incentives to improve the likelihood that B's actions will conform with the norms and vice versa (Yarbrough and Yarbrough, 1999).

One problem with second-party enforcement is that the actual enforcement has the form of an iterated game in the game-theoretic sense. Thus, if being cheated on by actor B, actor A can only wait for the subsequent encounter between the two to apply the enforcement (e.g. punish B by not interacting with it again). This concept of responding to an action in kind (i.e. returning benefits for benefits, and responding with non-cooperation or cheating to cheating) on the current or next possible encounter is referred to as *reciprocity* (Axelrod, 1984).

In enforcement situations with third-party observers enforcement is executed by entities that are not directly involved in the transaction where the violation has taken place. In the literature the most common examples of third-party enforcement are *reputation* concepts. In Balke et al. (2009) we presented a survey on reputation and its usage in artificial societies and gave an in-depth definition of the terms relevant to the concepts. The following sections on reputation are taken from this survey.

Definition 6:   Reputation

Reputation is the process and the effect of transmission of a target's (i.e. the actor being evaluated) *image*.

> ### Definition 7:   Image
>
> Image is a global or averaged evaluation of a given target on the part of an individual. It consists of a set of evaluative beliefs (Miceli and Castelfranchi, 2000) about the characteristics of a target. These evaluative beliefs concern the ability or possibility of the target to fulfil one or more of the evaluator's goals, e.g. to behave responsibly in an economic transaction. An image, basically, tells whether the target is "good" or "bad", or "not so bad" etc. with respect to a norm, a standard, a skill etc.

In the third-party enforcement example above, reputation refers to an actor C observing an interaction between A and B, formulating its own image about the two interacting entities and circulating that image to other system participants[9].

One advantage of third-party observer enforcement is that it does not necessarily need a centralized instance that collects all enforcement information, but rather a large number of decentralized mechanisms exist, such as the ones described in Regan and Cohen (2005); Zacharia et al. (1999); Yu and Singh (2002); Sabater-Mir and Sierra (2002); Sabater-Mir et al. (2006). One problem with regard to any reputation model however is that in general they rely heavily on information from the actors, which might on purpose pass on false information. In the earlier example, C might lie to others about A and B because it might be beneficial to him.

Another technique that might be employed for third-party observer enforcement is the utilization of so-called enforcement entities. Enforcement entities are participants in a system that are empowered by the system to monitor normative behaviour and act upon violations. Police officers are an example of enforcement entities: in contrast to "normal" citizens, having witnessed a norm-violation they are (legally) empowered by the state – in their function as police officers – to carry out actions like arrests, etc. and thus indirectly enforce norm-conformance.

2.3.3   The Enforcement Process

The preceding paragraphs, have mainly focused on the source of the enforcing action and distinguished enforcement based on it. However, despite its high relevance for enforcement in general, this is not the only essential part when it come to enforcement – enforcement is a far more comprehensive process. Given the dissertation's focus on enforcement mechanisms as one example for

---

[9]The evaluation circulating as reputation may concern a subset of the target's characteristics, e.g. its willingness to comply with socially accepted norms and customs. More precisely, referring to Conte and Paolucci (2002) we define reputation to consist of three distinct but interrelated objects: (i) a cognitive representation, or more precisely a believed evaluation (any number of actors in the system may have this belief as their own); (ii) a population-level dynamic, i.e., a propagating believed evaluation; and (iii) an objective emergent property at the actor's level, i.e., what the actor is believed to be as a result of the circulation of the evaluation.

governance decisions and the resulting changes in a running system, this section looks more closely into the idea and process of enforcement, using work that has been published in Balke and Villatoro (2011) and Balke and Eymann (2009). In this section we present our conceptualization of the enforcement process and try to give a complete view on the possible facets of enforcement. This process will be explained in more detail to first of all give a detailed overview of the design options available for enforcement and to furthermore complete the foundations for the enforcement concepts to be discussed and implemented later in this dissertation.

As shown in Figure 2.1, this dissertation conceptualizes enforcement as a four-stage process, with its phases – although being interlinked – being designable with a degree of independence. The first three stages correspond roughly to the conventional processes of arrest, trial and conviction of transgressors, while the fourth is the process of learning and adaptation that ensues. Each stage involves distinctive activities whose performers are actors executing particular roles. Although in general terms most activities and roles are present in every system that includes enforceable norms, they need to be adapted to the particularities of the normative context where norm deviations take place. When reading this section particular focus to the roles involved in the enforcement process presented should be given. These roles will be of importance at a later point of this dissertation when different enforcement mechanisms are being discussed and tested in a specific setting of an open distributed system.

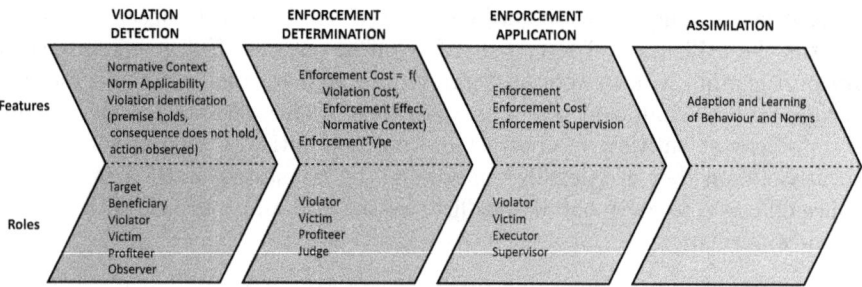

Figure 2.1: The Enforcement Process

2.3.3.1   Detecting Non-Compliance

The first stage of the process is the detection of non-compliance. This stage has two goals: the ascertainment of a violation and the identification of participants involved. There are two obvious types of actors (or roles) involved in the stage: *violator* and *observer*. One should note that, depending on the structure of the norm that is violated, the observer may need to gather enough evidence to ascertain what violation actually took place before any punitive actions can take place. Hence, in order to bring about charges, observers may need to assess *damages*, to assign *blame* and to identify *victims* and *profiteers* that may be affected by the non-compliance.

Each of these roles may be performed by more than one actor in the system and even performed by the same actor. In the particular case of utilitarian actors, the observer role may take different forms: (i) a first-party observer, who controls its own compliance, (ii) a second-party observer, who observes the misbehaviour of his own transaction partner(s) or finally a (iii) third-party observer. Third-party observers can have different forms. They can for example come in the form of "normal" system participants that observe norm deviations from other system participants they are not directly interacting with, or they might come in form the of especially empowered entities (i.e. entities that are employed by the system itself) that supervise the behaviour of other actors in the system (Balke, 2009). In contrast to regimentation concepts, these entities cannot observe the whole system, but only the actions in their vicinity.

One question that is very important with regard to the success of enforcement in the detection stage is to what extent norm deviation may be detected and thus might determine strategic decisions of the actors in the system with respect to norm fulfillment. Whereas complete knowledge and information by the actors is often assumed in many papers, in this dissertation we will make use of the "bounded rationality assumption". In contrast to complete knowledge and information, it is a much weaker assumption about the actors' capabilities. In this dissertation we us Rubinstein's definition of bounded rationality (Rubinstein, 1998):

Definition 8: Bounded Rationality

Bounded rationality of actors implies that – in their decision processes – these actors are limited by the information they have, the cognitive limitations of their minds, and the finite amount of time they have to make decisions. When trying to act rationally they cannot base their decision on complete (system) knowledge, but these "rational" decisions are only rational with regard to their limited scope instead, hence they are acting accordingly to "bounded" rationality.

Furthermore, in this dissertation we assume that actors will only deviate from a norms if they consider that the benefits of a deviation are higher than the possible penalty resulting from the deviation[10]. As a consequence it is assumed that in scenarios where most deviations are likely to be observed and thus penalties are likely to be imposed, participants are more likely to stick to the norms, whereas in the opposite case – giving equal enforcement penalties (or rewards) in both cases – participants might take the reduced probability of detection into account and therefore are more likely to cheat.

---

[10]The benefit or penalty of deviation thereby do not have to be defined in monetary terms, but could correspond to the satisfaction the actors receives from his actions for example.

2.3.3.2 Enforcement Determination

After the initial observation of a norm violation, the enforcement determination takes place. This involves the appraisal of the applicability of the norm within the normative context of the non-compliant action and, if applicable, activation of the normative consequences of infringement and determination of the actual enforcement to be applied. In this stage, *observers* bring charges to a *judge* who should decide if the violator deserves a sanction and then determines the appropriate reward/sanction. The judge may also command the execution of reparatory actions as a consequence of the infringement. This sub-process may involve intense argumentative processes to establish the applicability and severity of enforcement options.

Sanctions or rewards as means of enforcement are usually calculated as a function of the violator, the victims, the effects produced by the violation, and the normative context where the violation was detected. This enforcement calculation might imply a cost which is absorbed by the judge. Some may need to be enacted either automatically or commissioned by the judge.

During enforcement calculation for utility maximizing actors, one should not assume that all actors have the same utility function, and therefore might react differently towards the same enforcement. A problem when wanting to take this factor into account, is that often (and especially in open distributed systems) determining the individual utility functions of the actors is impossible. The reason for this is that normally this information is private (i.e. only known by the respective actor). Hypothetically assuming that all utility functions were accessible, the enforcing entities would be able to calculate a perfect enforcement to decrement the utility function of the violator and/or increment the one of the norm-follower, even if they do not share the same function. However as this is normally not the case, enforcement can be imperfect in that respect.

One final decision has to be made before applying the enforcement action, and this is how it should be applied. The way an enforcement action is applied can be decided by the judge or be imposed by the normative context.

2.3.3.3 Enforcement Application

The enforcement application phase is composed of the actual execution of an enforcement and the assessment of its proper application.

The outcome of the previous stage is a sentence to be carried out by the *executor*. As elaborated earlier these executors may be (Balke, 2009):

The violator ("first-party"): The actor responsible for the violation himself can act as an executor. This might for example be the case if – after violating a norm unintentionally – the violator might want to repay the damage done (maybe to avoid a loss in reputation or to reduce the chances of being punished in another way).

A second-party victim: Depending on the types of norms that are controlling the society, one actor can be directly affected by another actor's norm violation. If this is the situation, the victim can act as the observer, calculator and applier of the enforcement on the violator.

A third-party observer: If an actor is seen violating a norm, the observing actor can have the right to apply enforcement to the violator, even if this actor did not suffer from the violation.

A group of third-party observers: The act of enforcement can be distributed amongst a group of actors. This type of enforcement act is often used in indirect enforcement concepts such as reputation (which has an aggregated effect such that, the more actors use it, the more powerful it becomes).

Normative empowered entities: Normative empowered entities such as authorities are entities that are not capable to observe all actions but only the ones that are within their vicinity. They have the power (designated by the normative framework that they all belong to) to perform enforcement actions, i.e. applying sanctions as well as rewards.

It is important to note that enforcement may not come for free but may have a cost associated to its application that may be borne by *executors* or the system. Sometimes the costs are directly associated with the enforcement. A straightforward example of *costly enforcement* is imprisonment, where the state has to support prisoners. Reputation is normally one of the most effective *relativized-cost* enforcement mechanisms. Transmission of (bad) information regarding an actor has a relative cost (depending on the degree of responsibility of the information transmitted and the retaliation level of the members of the society) to the actor transmitting the information, however, it might also affect the target of the rumour.

After the *executor* has acted, the *supervisor* is in charge of assessing that the *violator* has indeed received the enforcement, and that the enforcement has served the purpose for which it was designed. The *supervisor* is also responsible for ensuring that other actions that are associated with the infringed norm are properly triggered and carried out. In case of compensational enforcement, in particular, it is the *supervisor*'s role to ensure that the *victims* are compensated.

2.3.3.4 Assimilation

Once the actual enforcement act has been performed, the last stage of the enforcement process takes place: the assimilation. Assimilation is the processes through which individuals or the normative system itself take advantage from enforcement to modify subsequent behaviour.

As we have seen, by performing the corresponding roles, violators, observers, victims, judges, supervisors and executors come into contact with information about norm-compliance that they could ideally incorporate in their decision-making process and may hence shape their own future behaviour. The normative system however, may specify ways that information about the enforcement process (besides the ones which participants obtain directly through interactions) may become available to them. By so doing, norms about enforcement may give shape to a space for individual evolution where specific aspects of compliance are given more relevance than others and therefore facilitate evolution of the system along different lines.

While enforcement is intended as a motivation for specific actions, it may furthermore have an ostensible objective (compensation, retaliation, deterrence, exemplarity), but the actual effect of enforcement on an actor's motivations is a private matter of convictions and thus directly unobservable for the system. Nevertheless, the subsequent behaviour of actors is observable, hence the system as a whole, or its *legislators*, may use evolution of the behaviour of individuals as input for *purposeful* evolution of the normative system. In either case, modifications of behaviour need to be observable somehow. The natural means are probes and indicators that must be aligned with performance parameters accessible to legislators or the system's dynamic features. This way, the choice and balance of those conventions that determine the availability and form of information about enforcement and norm-compliance have significant effects on the overall collective behaviour – collective behaviour that results from the aggregate behaviour of informed individuals, as well as collective behaviour determined by new norms that result from legislative or autonomic adaptation.

2.4 Enforcement and Cooperation – Related Work

In Section 2.2 we described the cooperation dilemma in open distributed systems which comes in the form of free-riding problems. The dilemma described was that in order for the total utility of an open distributed system not to deteriorate, it depends on a certain amount of individual altruism from the actors in the system with regard to their cooperation. This conflict between the interest of the individual actors in the short run as well as the systems as a whole in the long run, is not new, but exists in various forms in different systems and due to its social impact is often referred to as social dilemma.

The question of free-riding and reciprocity with regard to the tragedy of the commons dilemma and the cooperation issues related to it have been studied for a long time in various research disciplines. In this dissertation we will draw inspiration from these previous works. We will therefore give an overview of the works relevant for this dissertation and highlight the respective limitations in the applicability for the free-riding/cooperation problem in open distributed systems as outlined earlier.

2.4.1 Enforcement and Cooperation in Economic Theory

In economics-inspired works enforcement is usually studied as an incentive for rational behaviour of utility-based agents. It is seen as an amount taken from an agent's benefits and the effectiveness of enforcement is usually measured against system equilibria, in line with the theory of Becker (1968). The topic has been framed mostly in terms of mechanism design and the issues that economists have studied more thoroughly are the information about infraction and enforcement (Fehr and Gächter, 2000), as well as the amount and pervasiveness of enforcement (Dreber et al., 2008). Methodology has been either game-theoretic (see Coleman (1998) for example) or experimental (e.g. Gurerk et al. (2006)).

2.4.1.1 Enforcement and Cooperation in Games-Theoretic Models

One of the classical domains in which cooperation dilemmas and their solutions are studied is game theory. Game-theoretic models typically make the following assumptions:

- The actors are economically rational. That means they try to maximize their expected utility based on their knowledge about possible payoffs as well as assumptions about the actions of other actors.
- The game is symmetric, i.e. the payoffs for playing a particular strategy depend only on the other strategies employed, not on who is playing them.
- With regard to preferences of solutions, actors prefer defection over unilateral cooperation and bilateral cooperation over defection (Snidal, 1991).

As a result of these assumptions, six strategic cooperation problems have been identified, which are all presented in Wrona (2005)[11]. Of these games, the cooperation problem inherent in open distributed systems is best described by the prisoner's dilemma (PD), which therefore is a useful starting point for future analysis.

The PD was first described and discussed in the 1950s by Merrill Flood and Melvin Dresher as part of the RAND Corporation's investigations into game theory[12] (Stanford Encyclopedia of Philosophy, 2007). As in the cooperation setting of open distributed systems described earlier, the general idea of the PD is that two actors have to make decisions independently and cannot be sure of what the other actor does. Facing that decision, for an individual actor it is always better to defect (no matter what the other actor does) resulting in overall worse utilities for both actors then they could have achieved in a cooperative strategy (Poundstone, 1992). The intrinsic problem of the PD is that neither actor can do any better in the game by switching unilaterally (Allen et al., 2010) as proven by the Nash equilibrium of the game (Nash, 1950). As a result, in the PD's classical form as one-shot game, no interaction is achieved. Since single interactions consequently always pose a problem, in a next step it was analysed whether repeated instances of the PD – also know as iterated prisoners dilemma (IPD) – help to induce positive reciprocity in cooperation situations.

In the IPD, the PD is played repeatedly between two actors, inducing a so called "shadow of the future" (Axelrod, 1984). Thus, if a PD is played repeatedly between two actors and two requirements are fulfilled it was shown

---

[11]The six games are the Harmony Game, two forms of the Assurance (or Stag Hunt) Game, the Coordination Game, the Chicken Game and the Prisoner's Dilemma. For a detailed description of all these games please refer to (Wrona, 2005, p. 44ff.)

[12]Despite Flood and Dresher first expressing the ideas of the PD, it was named and formalized by Albert Trucker in the same year.

that reciprocity can be established in the cooperation dilemma. These two requirements are (i) The actors must be able to recognize previous cooperation partners and remember the outcome of the interactions with them, and (ii) the number of interactions between the actors is not allowed to be fixed in advance.

These requirements need to be fulfilled in order for an actor to be able to distinguish between good (i.e. formerly cooperating) and bad (i.e. formerly cheating) cooperation partners and punish the former by not cooperating with them again. The reason for the second requirement is that the dominant strategy for the final interaction is to cheat. Anticipating this, it is beneficial for the actors to cheat in the next-to-last interaction to not be cheated on in the last one, etc. This results in a spiral that goes back to the very first interaction and therefore stops any reciprocity effects from taking place. The shadow of the future is the probability δ of a repeated interaction between actors in the future. δ is often also interpreted as a discount factor by which future utility (such as payments) are discounted to the present equivalent. In the literature it is often assumed to be influenced by two factors: (i) the average life-span of an actor, which is proportional to δ, as well as (ii) its mobility that is indirectly proportional to δ.

According to the folk theorem (Friedman, 1971) cooperation is a feasible outcome in an IPD if the shadow of the future (i.e. δ) is sufficiently large enough. Looking at the examples of open distributed systems described earlier this initially positive sounding result poses several problems: First of all, the IPD is classically an two-person game, i.e. it looks at a pair of persons interacting only. However open distributed systems often have multiple interacting actors at the same time. This is often helped by portraying the interaction between each pair of actors in a group as a single game. The problem however with this approach is that it can highly distort utility considerations in situations where the contribution of one actor is received by several other actors. This problem is partially solved in n-person games (Axelrod and Dion, 1988), however further problems remain with regard to the applicability of game-theoretic approaches to open distributed systems. Secondly, the IPD (both in the two-player and the n-player version) heavily relies on a high value of δ. As pointed out, most open distributed systems have large numbers of participants and possibly can show a high mobility of the individual actors (e.g. in the relay routing applications). As a consequence the probability of repeated interaction between two actors is rather low and the folk theorem therefore is not applicable. This however means that the IPD and game-theoretic approaches are not sufficient for analysing and addressing the cooperation issues in open distributed systems. Finally, one of the initial PD (and consequently also IPD) assumptions was the rationality of the actors in the system. We have already explained that this assumption is a rather difficult one with regard to human behaviour, causing another applicability problem of game-theoretic approaches to the problems at hand.

2.4.1.2 Enforcement and Cooperation in Experimental Economics

As outlined earlier, one of the problems with regard to game-theoretic approaches is that their are based on rather rigid assumptions, such as the "Homo Economicus" rationale that actors in a system always act in a way which is rational and utility-maximizing. Although this simplification is helpful in developing first ideas and theories about cooperation situations, especially with human actor involvement, it has been criticized for its over-simplification and possible invalidity of results. Experimental economics tries to address this issue by placing humans as decision-makers in the center of their analysis (Smith, 1990, p. 1). The aim of experimental economics is two-fold: the testing of theoretical findings in "real-world settings" and/or the generation of new scientific theories. This originally natural science based approach in social science bears the challenge of including human behaviour and decision making in the experiments. Humans act consciously and unconsciously. The decision calculus, on what they take into account when deciding about something is different from person to person. As a result a multitude of relevant decision criteria exists and the analysis of cause-and-effect relations is difficult, both with regard to initially identifying the relevant (set of) parameters as well as identifying the interdependencies between them.

One of the first experimental economics experiments was conducted by Chamberlin (1948) who analysed strategic behaviour with regard to demand and supply of a homogeneous product with the help of laboratory experiments. Since his article, extensive research has been conducted in experimental economics. With regard to the cooperation dilemma and enforcement some of the most cited articles are on the research of R. Selten[13], G. Bolton and A. Ockenfels (see Bolton (1991) or Bolton et al. (1998) for example) as well as E. Ostrom (who – together with Oliver E. Williamson – was awarded the 2009 Nobel Memorial Prize in Economic Sciences for "her analysis of economic governance, especially the commons" (The Nobel Foundation, 2009)).

In Bolton and Ockenfels (2000), the authors described results from classroom experiments on cooperation in basic game-theoretic settings and were able to show that humans act strategically. However in contrast to game-theoretic ideas not only pecuniary payoff is relevant for the decision making but also relative payoff standing as well (this is referred to as ERC theory by the authors) and in later works they were able to show that in small groups, a propensity to punish non-contributors exists. They conclude that this propensity might be one way of (partially) solving the free-riding problem inherent in systems like the open distributed systems described earlier. Similar results are shown in Güth (1995) and Fehr et al. (1997) for example. One weakness of their approach that is pointed out by Güth and Ockenfels is that the ERC theory focuses on "local behaviour in the sense that it explains stationary patterns for relatively simple games, played over a short time span in a constant frame" (Bolton and Ockenfels, 2000, p. 198). As a consequence it poses some problems in the applicability to

---

[13]R. Selten was awarded the 1994 Nobel Memorial Prize in Economic Sciences (shared with John Harsanyi and John Nash) for his research on this topic.

open distributed systems, where larger time-frames and variable settings (e.g. with regard to the number of actors in a group) are dominant.

Whereas Bolton's and Ockenfels' research was mainly conducted in form of classroom experiments, Ostrom is particularly known for her field studies concerning common pool resources. Examples of her studies include communal tenure in high mountain meadows and forests in Switzerland, *Zanjera* irrigation communities in the Philippines, Sri Lankan and Nova Scotian fishery grounds, Californian groundwater basins, etc. (Ostrom, 1990).

One of the results of her research was a more specific classification of cooperation dilemmas with regard to the goods they concerned as well as a good-specific problem discussion. Her classification is shown in Figure 2.2.

	Substractability of Use	
	low	high
Difficulty of excluding possible Beneficiaries — low	Toll Goods	Private Goods
Difficulty of excluding possible Beneficiaries — high	Public Goods	Common-Pool Ressources

Figure 2.2: The Four Basic Types of Goods distinguished by Ostrom (2005) (originally published in (Ostrom and Ostrom, 1977, p. 12))

Ostrom distinguished the difficulty of excluding possible beneficiaries as well as the substractability of use. The former is particularly interesting with regard to the free-riding issues described. If exclusion of others is difficult or costly to achieve, strong incentives exist to benefit from the goods without contributing. The latter criterion – substractability – refers to the question whether the consumption of a resource limits the quantity available for others (e.g. when withdrawing water for irrigation purposes from a water basin) or not (e.g. watching a DVD that can be watched by others afterwards).

In her works, amongst other findings, Ostrom has shown how to address cooperation dilemmas for both free-riding types of goods (i.e. public and common-pool goods) and stated very general rules about her findings. These rules include the application of reciprocal monitoring techniques, the requirement that norms need to be defined by the community they are effecting, etc. Despite Ostrom's success of solving most of the free-riding problems in her field studies, with regard to the open distributed systems issues at hand, applying Ostrom's results proves difficult. As stated by Ostrom herself, her results are based on field studies with small groups sizes only, whereas open distributed systems are often large scale systems with thousands of participants. Furthermore, a second problem exists that is common to all experimental economic approaches with regard to the early prototyping focus taken in this dissertation. As pointed out in the introduction, in an early prototyping phase of development it is crucial to arrive at initial ideas about how interactions in the open distributed systems might work and how issues arising from these interactions could be resolved. By definition it is not yet possible to test a running system, but from a business perspective one wants to test the system/product before putting it onto the market. Even though this could be done by testing the respective open distributed system ideas with few participants only and extrapolating the results, as discussed by Ostrom for example, sample sizes have a significant impact on the results in cooperation dilemmas, possibly causing problems to the results interpretation. Finally, testing with human subjects is costly. Especially at an early prototyping stage of development, companies might not want to spend these incremental costs without knowing whether their investment might result in positive returns.

2.4.2 Enforcement and Cooperation in Formal Logics Research

Besides economics, cooperation and enforcement have been studied in formal logics research as well. In this domain, some works deal with enforcement, and incentives in general, as a component of the notion of norms and thus study the structural relationships of those components, for instance the relationship between target and victims[14], the syntax of activation and deactivation conditions or links between infractions and reparatory actions (Perreau de Pinninck Bas et al., 2010). Others are interested in the dynamics of norm-compliance and thus deal with enforcement as events triggered when an infraction occurs García-Camino (2009). Finally there are works that take enforcement as a feature that depends on the type of norm (conventions, social, regimented, functional,... (López y López and Luck, 2003)). These authors are interested, for instance, in the operational semantics of compliance and enforceability (Grossi et al., 2007), or in the class of incentives most naturally associated with different norm types (Andrighetto et al., 2010).

2.4.3 Enforcement and Cooperation in Reputation Research

In Section 2.3.2 we presented the definition of reputation (and image) as well as pointed out the role that reputation plays with regard to the establishment of trust, which is an important prerequisite for enforcement and cooperation. In the literature, both trust as well as reputation mechanisms can be found when it comes to the topic of enforcement. The former mainly build upon image information and consequently pose problems when it comes to application in

---

[14]Detailed description of our understanding of the terms is given in Section 2.3.3

open distributed systems as due to their open large scale nature the chances for repeated interactions are rather reduced. In contrast, reputation mechanisms augment the image information by using circulated information from other participants and consequently reduce the problem of limited repeated interaction probabilities.

Looking at the literature on reputation mechanisms, one can find two types of reputation systems: the ones that store information and calculate reputation metrics centrally (such as eBay for example) and the ones that are decentralized. Although easier to implement, centralized systems typically face two problems: firstly – especially in large scale systems – a central entity can become a bottleneck slowing down the entire system; and secondly the system needs to be trustworthy (as well as trusted by the participants) as otherwise it could take advantage of everyone else. Both these issues are addressed by decentralized systems (e.g. (Zacharia et al., 1999), (Yu and Singh, 2002), (Eymann, 2000), (Sabater-Mir and Sierra, 2001) and (Sabater-Mir et al., 2006)), which however pose face problems themselves, e.g. conflicting goals of efficiency to information loss on the one side and robustness on the other side. Furthermore, as information from third persons is involved in reputation systems, the systems must be able to handle false or malicious information, to stop participants from artificially changing their own or others reputation. Typical examples of this problem are: (i) badmouthing, (ii) ballot stuffing, (iii) colluding, and (iv) whitewashing.

Badmouthing refers to situations where an interaction partner is giving negative information about his partner despite being satisfied with the interaction. Reasons for this might be possible competitive situations between the two. Thus, by evaluating a competitor negatively the badmouther's own reputation rises in comparison, which might for example make him more attractive to possible customers.

Ballot stuffing typically refers to reporting feedback for interactions that did not take place. The problem is usually solved using cryptographic mechanisms.

Colluding refers to participants faking interactions and giving themselves positive feedback for these in order to increase their own reputation.

Whitewashing portraits the idea that participants change their identity if the reputation of a new identity is higher then their old one. This is often "solved" by assuming that a change of identities is not possible. If this assumption is not feasible, one potential solution is to add costs to all newcomers (Sun et al., 2005).

All of these issues have been addressed by different researchers and we will rely on their work with regard to the reputation mechanisms presented in this dissertation. Thus, despite the problems mentioned, reputation has been proven to increase enforcement in large scale systems with uncertainty about the cooperation partners' intentions. In the latter part of this dissertation we will compare one reputation mechanism in more detail to other means of enforcement. The choice with regard to the mechanisms will be based on the existing literature (e.g. Balke et al. (2009); Paolucci et al. (2006, 2009)).

2.4.4 Enforcement and Cooperation in Technical Domains

Besides the more general research on enforcement and cooperation in the domains just highlighted, many attempts exist that try to find solutions for specific open distributed systems by focusing on the specific features of the respective systems. This section will look more closely at the solutions presented for the applications mentioned in Table 2.1.

2.4.4.1 Relay Routing

As briefly explained in Section 2.2, relay routing applications in open distributed systems feature nodes (or participants) that are dynamically and arbitrarily located in such a way that the communication between the nodes does not rely on a static underlying network infrastructure (Royer, 1999). As a result of this lack of a static infrastructure the nodes need to rely on the other nodes in the network to route their communication (in form of packets) to the destinations, i.e. the network nodes need to cooperate to ensure general functioning of the communication.

Typically two types of uncooperative behaviour are distinguished in these settings: *faulty/malicious* and *selfish* behaviour. The former refers to the broad class of misbehaviour in which nodes are either faulty and therefore cannot follow the norm of routing, or where nodes on purpose try to attack the network (Michiardi and Molva, 2003). The latter also refers to intentional non-cooperation by the nodes. In contrast to malicious behaviour, reasons for non-cooperation are personal utility considerations rather then the explicit purpose of harming the system. In relay routing applications the main threat from selfish nodes is the non-routing of data ("blackholes") which may result in performance degradation or communication breakdowns (Hu, 2005).

Scanning relay routing literature, with regard to the cooperation problem, very often the problem is assumed to be non-existant (see Johnson et al. (2001); Perkins and Royer (1999) for example). Research that does take the problem into account and tries to address it, typically has one of the two following foci: (i) virtual currency schemes (e.g. Nuglets (Buttyán and Hubaux, 2000) or Sprite (Zhong et al., 2003)), or (ii) reputation mechanisms (e.g. CONFIDANT (Buchegger and Boudec, 2002), CORE (Michiardi and Molva, 2002), OCEAN (Bansal and Baker, 2003), Watchdog (Marti et al., 2000) or LARS (Hu and Burmester, 2006)).

Looking at the currency schemes first, they typically either use currencies (or some form of tokens) that the forwarding nodes get and the sending nodes need to pay; or use receipts which a forwarding nodes gets and can exchange for the virtual currency (or tokens) with a credit clearing service[15]. In both cases tamper-proof hardware is required so nodes cannot falsify receipts or increase their virtual currency. This dependence on tamper-proof hardware

---

[15]This credit clearing service normally is in addition used to determine the value of the payments and the charges.

presents one big problem in the applicability of open distributed system, as the hardware to join and interact in the system cannot be checked centrally, but is located with the individual participants. Introducing a central component does not seem appropriate for a truly open and distributed system, as it would be counterproductive to the initial idea of such systems and would reduce their benefits dramatically. Based on the same reasoning – despite being discussed at large in literature – currency systems using a central server for determining the charges and credits involved in a transmission of communication messages (this is done by Sprite for example) cause problems.

One further general problem of currency-based systems for relay routing applications is that these systems according to Wang et al. (2004) suffer from the so-called "location privilege problem". This problem stresses that nodes in different locations of a network have different chances to earn the virtual currencies. Thus, systems tend to be unfair towards nodes on the periphery of a network, who have less chance of being required for routing and thus have a low chance of earning virtual currency.

Learning from the problems of currency-based systems, reputation mechanisms proposed for relay routing applications in open distributed systems tend to try to avoid the centrality and local privilege problems just mentioned by relying on neighbourhood information. As is common to reputation mechanisms in general, in these settings, systems participants judge the reliability of possible transaction partners based on the experience of other the other individual in the system. As this information is stored with the individual participants it consequently theoretically suits the decentralized idea of the system better. Nevertheless the problems described in the previous section (i.e. badmouthing, ballot stuffing, colluding, and whitewashing) often exist and cause difficulties for current relay routing reputation mechanisms.

2.4.4.2   P2P Networks

P2P networks are another application where the free-riding problem exists due to their large-scale open distributed nature. To give an example of the relevance of the free-riding problem in P2P systems several studies have been conducted. Adar and Huberman (2000) for example examined the Gnutella network, which – at the point of its launch in March 2000 – was the one of the first decentralized peer-to-peer networks, pioneering later networks adopting the model. In their experiments Adar and Huberman showed that already soon after Gnutella's launch, a large proportion of the user population – upwards of 70% – were free-riders (i.e. enjoyed the benefits of the system without contributing to its content), nearly 50% of all responses in Gnutella were returned by the top 1% of sharing hosts and the files shared where not necessarily those desired by most users. Adar and Huberman summarized this participation inequality problem as the tragedy of the digital commons and concluded that for P2P systems to function in the desired way, mechanisms against free-riding have to be found. In literature and later P2P systems various approaches to encourage peers to cooperate and to avoid free-riding issue have been presented. Similar to relay routing applications, the solutions range from game-theoretic analysis (Courcoubetis and Weber, 2006), to direct reciprocity (i.e. image-related approaches) such as employed in the BitTorrent protocol (Cohen, 2003), reputation concepts (e.g. KaZaA (Leibowitz et al., 2003)) as well as currency-based systems (e.g. MojoNation and Karma (Vishnumurthy et al., 2003)). With the system structure and the enforcement concepts being very close to the ideas presented with regard to the relay routing applications, the proposed enforcement mechanisms for P2P systems exhibit similar problems as the ones proposed for relay routing applications.

2.4.4.3 MMORPGs

One central element of MMORPGs is the social interaction of the participants in the virtual world with one another in form of teams. The most common form of these teams have a short term nature and are formed with the purpose of fulfilling a certain task. The general idea thereby is that the individual team-members contribute their character's special skill, etc. to the team and get a share in the reward if the team succeeds. One major problem for players in this context is to find team-members that are willing to contribute to the team and will not just join a team for the possible reward without contributing any of their resources or try to cheat the team otherwise by trying to steal the reward.

In order to solve this problem, three approaches can be found in MMORPGs ranging from central services (so-called matchmakers) over entities with special rights (called game moderators) to simple image-related approaches.

The first approach – matchmakers – works similar to centralized reputation mechanisms. In general, when matchmakers are being used, participants searching for team partner need to register with the matchmaker. This matchmaker then teams up players according to their skills and furthermore keeps track of their social performance. In the matchmaker of Halo 3 for example, players are awarded experience points for winning a match. Should they lose, no experience points will be given. However, if the player quits the game during a match, one experience point will be deducted from their account. The problem with this kind of mechanism is the centralized component and the resulting additional data traffic. For this reason other approaches than the central mechanism have become popular in MMORPGs. The already mentioned game moderators are one of them. Game moderators (or game masters) may be paid employees or unpaid volunteers who attempt to supervise a virtual game world. For this purpose they are often given special rights and information such as access to special features in the game that are not available to "normal" players (i.e. non-game moderator).

2.4.4.4 Recapitulation

Summing up the work conducted with regard to enforcement – both from a theoretical perspective as well as with regard to other technical domains – an unsatisfying picture can be drawn. Whereas the theoretical solutions make assumption that are incompatible with large scale open distributed systems, the proposal made for specific systems large scale open distributed system, either do make system specific assumptions as well, or are based on specific heuristics and are therefore hard to compare with each other and so far no comparison can be found in literature. The goal of this dissertation however is to compare the effects of different governance decisions (especially enforcement concepts) for one system in an early prototyping stage. This is why it is essential to be able to compare different approaches in a given setting, instead of contrasting them against different ones. This requires a framework that gives guidelines on what to take into account when aiming at comparing and analysing the effects of a governance decision.

In the business related literature the soft system's methodology (SSM) (Checkland, 1999) is often used for this purpose. SSM was developed in the 1960's at Lancaster University. It aims at analysing the problem situations within a company – such as the change or introduction of a product – by examining the influencing factors individually and combining the results afterwards in an integrated analysis. One advantage of SSM methodology is that it specifically includes social and human factors.

For their methodology the Lancaster team proposed several criteria that should be specified to ensure that a given analysis is rigorous and comprehensive. These criteria are summarized in the mnemonic CATWOE (Customer, Actors, Transformation process, Weltanschauung (worldview), Owner, and Environment) (Smyth and Checkland, 1976).

Other techniques that are used when analysing business change are PESTLE, HEPTALYSIS or MOST (Abea et al., 2008). Most of the techniques take human actors and their behaviour into account and are in that respect suited for the analysis of the open distributed systems described earlier. However one problem all of these techniques (including CATWOE) have in common, is their limited focus on norms. This makes them unsuitable for this dissertation.

2.5   The Institutional Analysis and Development Framework as a Tool for Analysing Governance Decisions

As concluded in the last section, one of the main requirements when trying to investigate the effects of governance decision in systems, where humans repeatedly interact within norms that guide their choice of strategies and behaviours, is a tool or framework that actually helps to compare the different decisions and guides the analysis of them. One tool that has been popular with scholars from diverse research domains when is comes to answering this question, and has the advantage of both focusing on norms as well as including (human) actors and social aspects, is the Institutional[16] Analysis and Development (IAD) framework.

---

[16]In the current literature norms and institutions are often differentiated. As Ostrom's usage of the term institution is very broad and fits the definition of norms given earlier in this chapter of the dissertation, the framework is considered applicable for norms and consequently

The IAD framework was originally designed and published 30 years ago by the group of Elinor Ostrom (Kiser and Ostrom, 1982) and has since been applied to analyse a multitude of different normative settings (Ostrom, 2005, p. 9). Examples of its application for diverse types of research questions are:

- the impact of norms on creating effective monitoring and evaluations in government development projects (Gordillo and Andersson, 2004);
- the regulations of the telecommunication industry (Schaaf, 1989);
- the effects of norms on the outcome in common-pool dilemmas (e.g. Oakerson (1990); Ostrom et al. (1994); Ostrom (2000); Ostrom and Walker (2003); Ostrom (2005)); or
- the effects of norms on knowledge (Hess and Ostrom, 2007)[17].

The wide acceptance and the sophistication of the IAD framework as well as its focus on the problems addressed in this dissertation, namely the study of effects of governance decisions on open distributed systems, make it an ideal analytical scaffolding for structuring our analysis. In order to be able to use it effectively for the dissertation, in the next paragraphs the framework will be explained in more detail. Special focus thereby will be given to the components of the framework relevant for the normative analysis.

The methodology of the IAD framework itself is not designed as a static model and consists of the three clusters of components shown in Figure 2.3: the underlying factors affecting the normative design (represented by the three boxes on the left of Figure 2.3), the patterns of interaction occurring within action arenas and the outcomes.

The *underlying factors* comprise a checklist of "those independent variables that a researcher should keep in plain sight to explain individual and group behaviour" (Gibson, 2005, p. 229). These variables are structured into a causal theme allowing for a detailed analysis of specific situations and resources in the so-called "action arena". The outcomes of the analysis as well as so-called "patterns of interaction" (that can have the form of any emergent behaviour) are used to evaluate the performance of the system by employing evaluation criteria relevant for the system.

There are three ways to enter the framework when studying a question, namely at each of the horizontal sections: one can start at the left-hand side with the underlying factors (the physical characteristics, the attributes of the respective community, and the rules-in-use at several levels), in the middle with the action arena, or at the right-hand side with the outcomes. Beginning with the outcomes makes sense when existing outcomes are available and questions concerning the specific outcomes should be asked. This is not the case in situations in which ones wants to find out how normative changes could influence a system, as the changes have not taken place at that stage of early prototyping. For predictive first prototyping settings therefore, starting at one of the other two components of the framework seems more appropriate. In the explanation of the IAD framework we will start with the action arena, as it is the central

---

being used in this dissertation.

[17]A more detailed account of applications of the IAD framework can for example be found in (Ostrom, 2005, p. 9) or (Hess and Ostrom, 2007, p. 42–43).

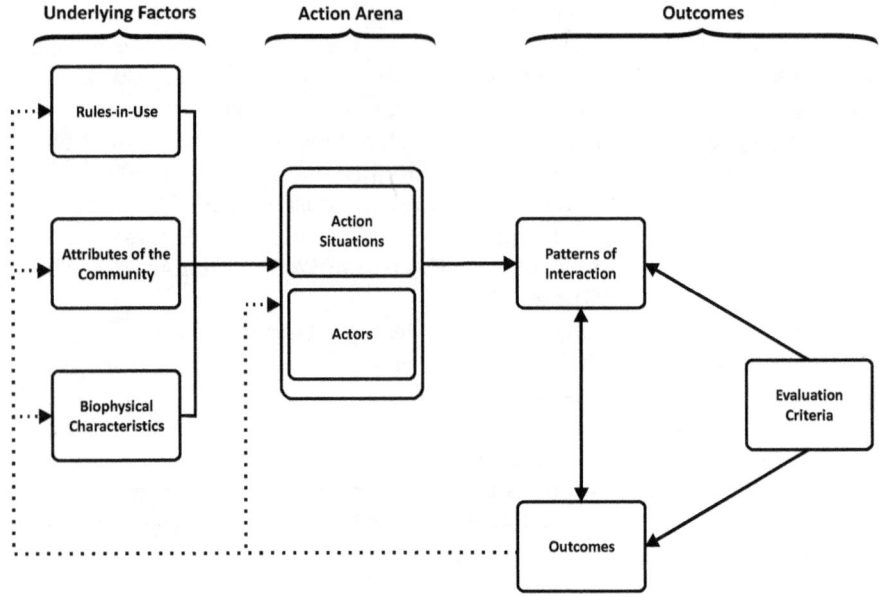

Figure 2.3: The Institutional Analysis and Development Framework (Kiser and Ostrom, 2000, p. 59) (originally published in Kiser and Ostrom (1982))

element of analysis (or "focal unit" in the terminology chosen by Ostrom (2005)) within the IAD framework.

2.5.1   The Action Arena

The action arena consists of two components (Ostrom et al., 1994):

1. an action situation and

2. the participant(s) in the situation.

Action situations refer to a distinct social space where the participants interact in some way. From an analysis point of view, recurring structures of situations are of particular interest, making it important to distinguish the different action situations. For this purpose, Ostrom et al. (1994) defined seven clusters of variables that constitute and characterize action situations and as a consequence allow to identify unique settings:

1. participants,

2. positions,

3. actions,

4. outcomes,

5. transformation functions,

6. costs and benefits, and

7. information.

The first cluster is comprised of the *participants* in the action situation. The participants are the actors in the action arena who are actively involved in an action situation. This cluster therefore links actors (i.e. the second component of an action arena) to the action situation.

The second cluster of the action situation are so-called *positions*. What Ostrom et. al refer to as positions are roles such as "boss", "a first mover", "police officer" or "judge" associated to participants in an action situation setting. These roles are attributed specific rules such as authorized set of actions.

These *actions*, which a participant can take in a stage of an interaction process, are the third element of the action situation. Examples of action are to cheat or cooperate, or even to do nothing at all. For analysis purposes it is important to note that in many situations it is impossible to analyse the complete array of actions available to participants, as this might exceed the capacity of the theoretical models and distract from the initial research question. For this reason, most analyses attempt to identify the actions relevant to the action situation and its underlying research problem as well as the actions that are most likely to cause a significant difference in the outcomes of the action arena.

The fourth cluster of the action situation are the *outcomes* that the participants can potentially achieve with their actions. Ostrom et al. focus mainly on the outcomes being directly under the control of the participants, their definition however also leaves room for accounting for the impact of the individual participants' actions in the system as a whole (i.e. the macro level).

The fifth element of the action situation is a set of *transformation functions* that link the participants and their actions with outcomes and thus define the combination of input factors that can lead to certain outcomes. The nature of the transformation functions can either be stochastic or deterministic. It is important to note however, that the degree of certainty and completeness of the transformations functions can vary significantly. Thus, very often only partial information about the factors affecting the relationship between participants' actions and certain outcomes are available, making predictions about possible outcomes hard if not impossible.

The sixth cluster comprises the *costs and benefits* assigned to (the combinations of) actions and outcomes. Thus, despite the outcome of an action being the same at two different points of time, the costs and benefits attributed to the outcomes at each specific time can give different positive or negative weights to them as well as the actions leading to the outcomes[18]. As a consequence they can act as sources of influence on the participants' decision making, assuming

---

[18]These weights can be different for each participant.

that participants try to optimize the weights in their favour (e.g. by maximizing the benefit/costs ratio).

The seventh cluster is closely linked to the type of transformation functions. It is the *information* available to the participants when deciding about their actions. This information can include one or more of the following: the information about the actions available to the participants, information about the transformation functions, information about the possible payoffs of certain outcomes as well as information about the possible outcomes themselves. Whereas it is often assumed that participants have complete information, in reality this is seldom the case.

In themselves these seven clusters are relatively complex leading to a large number of unique combinations to analyse. Furthermore, each of the parts is further comprised by a combination of input parameters from the underlying factors that will be discussed in section 2.5.3.

The action situations are not the only components of the action arena, but are needed for the actors to become effective. To predict how actors behave and thus whether and how their interaction with the system will take place, Ostrom et al. distinguish four clusters that researchers need to keep in mind when analysing an action arena: (i) the preference evaluations (and resulting preference order) actors attribute to potential actions and outcomes (including their possible payoffs) in an action situation, (ii) the individual information-processing capabilities of the actors, i.e. how the they acquire, access, process, use and retain information, (iii) the individual resources (such as money and time) available to the actors, as well as (iv) the selection criteria distinctive to each actor.

2.5.2   Outcomes

After the interaction has taken place in the action arena, the results of this interaction need to be analysed. Ostrom et al. (1994) divides this part of the IAD framework into three sub-components. Thus, she views the actual results of the interaction in terms of emergent behaviour in accordance with the notion of Hayek (1996), as any pattern of interaction (including chaos). Only once these patterns of interaction have been evaluated with the help of criteria relevant to the systems, does she talk about outcomes. Defining relevant evaluation criteria is one of the most problematic parts when it comes to the outcomes analysis of the IAD framework. Very often concepts of fairness, the efficiency of the system, etc... are mentioned in this context, however as soon as several stakeholders with potentially competing interests are linked to the system that is being analysed the task of balancing between these competing interests and defining appropriate criteria becomes a lot more difficult. Sections 6.3 and 10.1 will deal with this issue of criteria for competing stakeholder interests in more detail.

2.5.3   Underlying Factors

The underlying factors, as pointed out before, are the input parameters for implicit and explicit assumptions about the action arena and serve as a checklist for variables that researchers should keep in mind when analysing norms and institutions. The underlying factors group the way of thinking about the action arena into three components, i.e. the *rules* that are used to define right and wrong and thereby also define relationships between the actors, the characteristics of the *physical world*, as well as the *attributes of the community* within which the action arena occurs. The implicit or explicit assumptions made with regard to these three categories all influence the way in which the seven elements of the action situation and the actors in the action arena are conceptualised. For this reason they typically form the starting point for any normative analysis and the factors are used to identify some typical action arena outcomes, resulting from a particular combination of the factors.

2.5.3.1   Biophysical-Technical Characteristics

The first set of components, affected by the variables of an action situation are the biophysical-technical characteristics of the relevant physical world. These components most importantly refer to the physical possibility/physical power of performing actions (either biophysical or technical) and thus the basic underlying fixed physical attributes of a system. Imagine, for example, that in order to point out that someone has violated a rules, an actor is obliged to communicate his observation to the relevant authorities. However, if no medium of communication is available to him (e.g. he is not granted any access to a police station), despite having the intention to fulfil his obligation, he might no be able to do so. Hence, the specific attributes of the physical world (no communication available) physically limit the action space of the actors.

2.5.3.2   Attributes of the Community

The second set of components that affects the structure of an action arena relates to the attributes of the community (often referred to as "culture") in which the action situation is located. Thus, communities can vary significantly with regard to the level of the common understanding of norms or the action arena, with the extent to which the preferences of the actors are heterogeneous (Taylor, 1976), or concerning the distribution of resources among the actors for example, or even the general acceptance of norms.

2.5.3.3   Rules-In-Use

The final component of the underlying factors, the rules-in-use are the focus of this dissertation. Similar to norms, rules have been studied in a variety of diverse research domains ranging from philosophy, to social and legal science and economics for example. The way in which Ostrom and her group use the term with regard to the IAD framework is similar to the r-norms by Tuomela and Bonnevier-Tuomela (1995). As a consequence it corresponds very well to the definition of norms used in this dissertation.

Recalling the earlier given definition of norms and applying it to the rules-of-use component in the IAD framework, two important observations can be made. Firstly, as pointed out in the definition of norms (and as a consequence the rules-in-use) they specify what actions an actors "must not" perform (prohibition), "must perform" (obligation) or may perform ("permission") if the actors want to avoid sanctions for non-compliance with the norms being imposed on them. This distinguishes behaviour explained with reference to a rule from behaviour based on the biophysical-technical characteristics of a situation.

Secondly, attention should be drawn to the contextuality criterion. Thus, the rules-in-use are contextually dependant on the action arena as well as the other underlying factors.

The combination of rules with the physical world as well as the community thereby generates particular types of situations and influences the way the seven components of the action situation described earlier are conceptualized. Thus, Ostrom et al. (1994) point out that a typical normative analysis for example might work in the following way: starting off with the underlying factors, typical action situations resulting from particular combination of the factors are identified and analysed with regard to the outcomes if they are combined with the actors in the action arena. The purpose of this dissertation is to study the impact of a change in norms in an open distributed system (with particular focus on the enforcement of norms). That is why in the course of this dissertation we will analyse open distributed systems in general and a specific open distributed system scenario in particular. Hence, starting off by identifying the underlying factors, their impact on the action arena as well as its results will be one central element of the research in this dissertation. Thus, by altering the rules (and in particular the enforcement rules) of the system as well as observing and analysing the changes in the action arena and the outcomes of the interaction in the action arena as a result of the rule-changes, the suitability of different enforcement mechanisms for achieving certain system behaviour will be studied in depth.

2.6 Summary

In this chapter we laid out the foundations on norms and open distributed systems by presenting the related work for the dissertation's theses. In detail we explained the particular features of open distributed systems and their specific requirements with regard to the questions of cooperation and enforcement as well as analysing governance decisions in them. One particular finding when analysing the related work was that neither game-theoretic or empirically-based economic work are suitable for analysing the theses in this dissertation. In the final section we presented the IAD framework. This framework serves as a guide for the research to be conducted in this dissertation, by giving guidelines on which specific aspects for analysing norm are important for which question. In detail, starting off with the underlying factors of the IAD framework, the theses formulated in Chapter 2 can be mapped to the framework as shown in Figure 2.4.

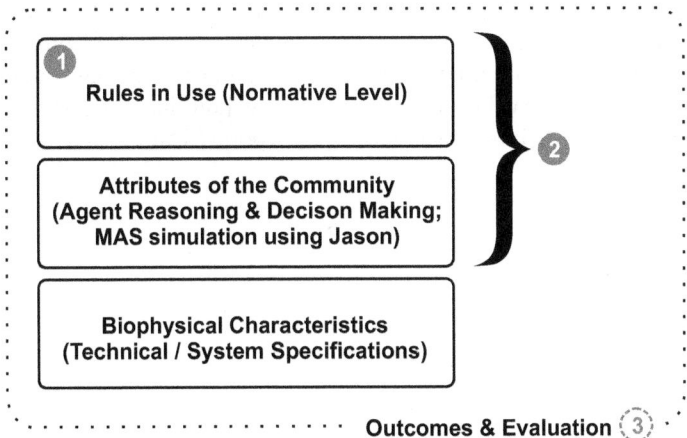

Figure 2.4: The Theses of this Dissertation in the Light of the IAD Framework

The first thesis (highlighted by the red circle with the number one) focuses on the rules-in-use in the IAD terminology. This dissertation investigates which effects a change in these rules of use has on the action arena and its outcomes. The particular rule-changes or governance decisions we are interested in are changes in the enforcement mechanisms in open distributed systems. With respect to these enforcement mechanisms, particular focus will be given to the question of whether and to what extent dynamic movement of the actors in the system has an impact on the enforcement mechanism. Thus, one particular feature of open distributed systems is that as a result of the movement of the actors in the system, the social (and possibly physical) network structure dynamically changes. As this structure to a large extent determines the patterns of interaction (e.g. who has a chance to interact with whom), we hypothesize that it will also effect the results of the enforcement mechanisms and analyse this hypothesis accordingly. The dissertation deals with the analysis of the interaction of the community (i.e. the actors in a system) and norms and with each other in the action arena. As we will outline in Chapter 5, we use a computer simulation when analysing Thesis 1. So far, no framework that allows for including reasoning of the actors about the norms of the system in respect to their actions at run-time of a simulation exists. This motivates the second thesis (highlighted by the red circle with the number two): how to incorporate such a run-time reasoning of actors interacting with a normative system? Finally, there is one additional aspect that this dissertation will address, but not treat as a separate thesis, but rather as side aspect of the analysis of the enforcement mechanisms. This aspect is concerned with the outcomes of the interaction in the action arena. The aspect is indicated by the red encircled number three in Figure 2.4. The complex open distributed systems this dissertation is concerned with have multiple stakeholders with potentially heterogeneous interests with regard to the system results. In terms of the IAD framework, this dissertation will have a look at how evaluation criteria for these multiple stakeholder problems can be composed and used to evaluate a system. In detail, we do not aim to provide an answer on what optimal configurations of the norms in the systems would be – this would be unfeasible. Instead, we aim at pointing out how the multiple stakeholders and their interests could be incorporated in the result evaluation.

Chapter 3

Case-Study: The Cooperation Dilemma in Wireless Mobile Grids

3.1 The Wireless Mobile Grid Scenario

The last chapter presented related work focusing on the problem of governing open distributed systems with respect to enforcing desired system behaviour. This chapter focuses on presenting a case-study of a specific open distributed system. It will serve as an example for analysing the general research questions throughout the rest of this dissertation.

3.1.1 The Mobile Phone Market: Challenges Arising

The case study has its roots in the mobile phone market – a market that is currently experiencing a fundamental structural upheaval. Whilst mobile phones already have been available since the 1960's, the current ubiquity mobile telecommunication services is a relatively recent development. During the 1990's several changes took place resulting in a tremendous boom in the mobile phone industry (Gruber and Verboven, 2001). Amongst these changes were the liberalisation of telecommunication regulations, the introduction of mobile telecommunication systems with larger frequency band spectra[1], and the standardization of technologies such as transmission protocols. This boom along with falling costs of mobile handsets resulted in mobile phones being commoditized. In 2010 more

---

[1]One example of a network with the new frequencies is the "D-Netz" in Germany. The term D-Netz refers to a cellular digital mobile telecommunication system that utilizes the GSM 900 MHz frequency band. It was introduced in 1991 and allowed for multi-service provision, i.e. voice, text and data.

than five billion subscribers[2] worldwide were registered, as shown in Figure 3.1.

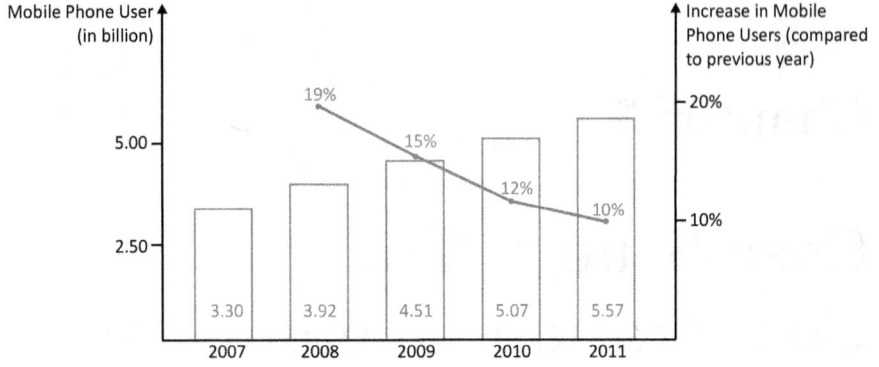

Figure 3.1: Number of Mobile Phone Users Worldwide (in billion) (EITO, 2010).

Following the recent growth of the number of mobile phone subscribers, the mobile phone industry is now facing a problem of market saturation. This becomes apparent when looking at the percentage numbers illustrated by the line in Figure 3.1. In the last five years the number of mobile phone users has almost doubled, however the rate of growth of the number of users has declined. Looking at the situation in more detail the following conclusion can be drawn: In Germany – the country with the highest number of mobile phone contracts in the EU – market penetration has reached 132%, resulting in increased competition (Bundesnetzagentur, 2010, p. 89-90) and a consequent decline in the profit margins of telecommunication providers. This trend is unlikely to be reversed in the near future. A study by management consulting firm A.T. Kearney predicts an average profit margin of -7% for all European telecommunication providers in 2012 (Hastenteufel et al., 2009).

Reasons given for this negative profit margin are a competitive pressure on prices for mobile phone contracts, paired with high fixed costs for investments in mobile phone infrastructure. The latter has become a necessity due to increasing data volume being transmitted in mobile phone systems. Cisco Systems (2010) for example, predicts an annual doubling in the data volume transmitted for the years 2009–2014. This corresponds to an increase from 91 petabytes per month in 2009 to 3.5 exabytes in 2014. Similar growth rates are predicted by the German Engineering Trade Association VDI (2010). AT&T, a network provider, reported that the 3% of their customers having an Apple iPhone, accounted for 40% of their network traffic (Reardon, 2009). Extrapolating this number with current predictions for sales of mobile phones the resource problem becomes apparent. Cellular mobile telecommunication systems have a limited bandwidth. The more

---

[2]The number of subscribers refers to all fixed-term contracts and prepaid cards by the end of the respective years. The numbers for 2011 are predicted numbers.

data that is transmitted at the same point of time, the more this bandwidth has to be divided between the individual transmission requests. As a result the Quality of Service (QoS)[3] for each individual transmission decreases and can, at worst, cause complete network breakdowns. For network providers, the resulting necessary investments in the improvement of the network infrastructure lead to high costs that are not necessarily counterbalanced by additional profits due to the reasons described earlier.

Currently two options are being discussed by the mobile phone industry to address the problems just described: (i) increasing the profit by attracting new users with the help of improved service offerings, and (ii) reducing costs by safeguarding the resources in the system.

Lately, of these two options the first one has been looked at more closely (Andersson et al., 2006, chapter 2). The reason for this focus on the first option is a result of the commercial underperformance of the current (i.e. third) generation of mobile phones.

Looking back, when in the mid 1980's the first generation (1G) of commercial mobile phones was introduced, the first main service was voice and users were separated with regard to frequency by using Frequency Division Multiple Access (FDMA). Since this time roughly 20 years have passed and in the mainstream mobile phone market the transition from the second (2G) to the third generation[4] (3G) is about to finish. This third generation offers a lot more than voice, ranging from much improved data services to higher network security. Besides these new features and being successfully introduced to users across the world, 3G was not the commercial success that was expected by many (Capgemini Consulting, 2005). This is partly attributed to the extent that, during the development and deployment of the current 3G mobile communication generation, the industry particularly focused on capabilities for enabling voice and basic data communications (Katz and Fitzek, 2006, p. 467–468) and thereby moved the users as well as the investors out of focus[5].

That is why, as the next generation networks (also referred to as Fourth Generation or 4G in short[6]) are being conceived and developed, a paradigm shift

---

[3]In telecommunication networks QoS typically refers to a set of metrics and techniques to measure and manage network resources. This includes the ability to provide different priority levels to different applications, users, or data flows, or to guarantee a certain level of performance to a data flow. QoS is especially important for applications with high data transmission rate such as real-time streaming Voice over IP as these applications often require fixed bit rates and are delay sensitive.

[4]The third generation of mobile network systems is also known as International Mobile Telecommunications-2000 (IMT-2000) networks.

[5]Another reason was the high prices paid for 3G spectrum licences in some markets.

[6]Despite several efforts by industry and academia to define 4G (see (Kim and Prasad, 2006; Khan et al., 2009, chapter 1) for example) so far no widely accepted definition of the term exists. For the purpose of this dissertation we define one based on its important characteristics and capabilities – namely *heterogeneity* and *convergence*. Thus, we perceive 4G both from a vertical perspective as linear extension to current 3G systems aiming at high data rates (100 Mbit/s wide coverage and 1 Gbit/s local area coverage) and at the same time pursue a horizontal perspective focusing on the integrative role of 4G as a convergence platform or wireless ecosystems for multiple different wireless networks. This convergence most importantly enables users to independently or simultaneously access different networks with a

in wireless communication is taking place (Katz and Fitzek, 2006). As a result
of the commercial problems of 3G, the telecommunication industry is pursuing
a different development strategy for 4G phones and networks. Instead of the
technology-centric-view of 3G, in the fourth generation the user was placed in the
center of interest (i.e. a user-centric view) (Wrona and Mähönen, 2004). Mobile
phone providers and network operators are trying to determine the consumers'
problems with 3G and learn to rectify these for 4G. So what are problems of 3G
from the consumers' point of view?

Some work has already been done to determine user preferences, and identify
future usage scenarios as well as developing new services. Several discussions
concerning the question of consumer requirements have taken place at the
Wireless World Research Forum (WWRF, 2011) for example. However so far
these have produced little results. One of the few studies to uncover which
aspect of 3G is seen as problematic and would stimulate consumer demand for
mobile devices (including mobile phones, PDAs and laptops) in the future if
being solved, was conducted by TNS among 15 countries in mid-2005 (TNS,
2004). This study revealed that the lack of coverage in some areas as well as high
prices for 3G in some countries have an negative impact on consumer take-up of
3G. Furthermore rather than being attracted by figures like high throughput
numbers, consumers are more interested in useful, convenient and enjoyable
services[7]. In addition the study revealed that two days of battery life during
active use topped the wish list of key features in 14 of the 15 countries surveyed,
indicating that insufficient battery life is a major problem for consumers around
the globe.

Thus, although new mobile phones are designed to allow their users access
to ubiquitous wireless connection and communication, as well as offering them
new services and features, the battery capacity has not risen at an adequate
rate, making the top wish a major challenge.

As batteries can only store a fixed amount of energy, they set a limit to the
operational time a user is able to use a phone within one charging cycle, i.e. the
battery life. At the same time power consumption in 4G telephones is expected
to increase due to several reasons (Katz and Fitzek, 2006, p. 479–480):

- One of the main ideas of 4G networks is that users shall have the ability
  to be connected all the time. Consequently, this means that the energy

---

single terminal and thus allows not only for higher QoS but also the enhancement of existing
3G services like mobile broadband access, Multimedia Messaging Service (MMS), video chat,
mobile TV, but also new services like high-definition television (HDTV). 4G may allow roaming
with wireless local area networks, and may interact with digital video broadcasting systems.

As far as the modulation and multiple access schemes for this kind of convergence are
concerned, multi-carrier techniques using Orthogonal Frequency Division Multiplexing (OFDM)
(combined with access techniques such as CDMA) and its multi-user extension Orthogonal
Frequeny Division Multiple Access (OFDMA) are the main component techniques for 4G.
Current pre-4G technologies such as Long Term Evolution (LTE) are aiming into the 4G
direction, however so far they do not fulfill the original 4G requirements of data rates
of approximately up to 1 Gbit/s specified by the International Telecommunication Union
Radiocommunication Sector (ITU-R) and are therefore not regarded as 4G technologies.

[7]Although these services could certainly exploit high data rate capabilities, it is the services
not the data rates that consumers are drawn to.

stored in the batteries is constantly drawn upon.

- One of the mayor technologies behind 4G are multi-antenna techniques. Having multiple antennas in one device however also requires higher amounts of power for their operation.
- 4G networks allow for lower energy per bit transmission. However, as energy per bit decreases, the transmitted power needs to be increased to maintain the same acceptable signal-to-noise ratios.
- The spectrum allocation in 4G networks results in higher frequencies being used. In these higher frequencies, attenuation is significantly higher. As a result more power for transmissions using these frequencies are required.
- New services and capabilities of 4G mobile phones such as video and audio processing, large amounts of mass memory or the utilization of localization services using triangulation requires higher data rates as well as greater on-board processing power. This increases power consumption significantly.

In summary, with the increase of complexity of mobile phones and their transition to smart phones offering Internet, digital photo cameras, mp3 players, access to new services and more, the energy consumption of mobile phones has increased significantly. This results in higher power consumption and consequently lower stand-by times in which a user can use his mobile phones without recharging it. Furthermore the problem arises that batteries get too hot without active cooling (Perrucci et al., 2009).

The problem is further increased by the relatively low improvement of battery performance, which has only increased by 80% within the last 10 years, whereas processor performance doubles every 18 month according to Moore's law (Andersson et al., 2006).

To give an example of the battery consumption problems in mobile phones, Figure 3.2 (Fitzek, 2007, p. 442) shows the power (and cooling) requirements of different mobile phone generations, as well as the approximate power consumption - temperature threshold of 6 Watts, above which active cooling is required for mobile phones.

The values shown in Figure 3.2 are maximum values and none of the current generation of mobile phones uses all the shown components to a maximum at the same time (the current energy value for an average smartphone is approximately 5 Watts which is below the 6 Watts threshold). Nevertheless it can clearly be seen that in the course of mobile phone development, the power consumption of 1G (i.e. around 2 Watts (W)) phones has tripled within the last years and that the basic communications and signal processing capabilities of the mobile phones, account for roughly 50% of the power budget. As explained before, this is a problem with regard to the battery life cycle. Another indirect problem of the increased power consumption is the battery temperature. The more power that is consumed in a given amount of time, the more physical energy is produced. This energy is partially transformed into heat, such that the more power is consumed the more heat is produced. As a result, a general increase in mobile phone power consumption might lead to active cooling being required to stop them from overheating. Active cooling however produces noise and furthermore

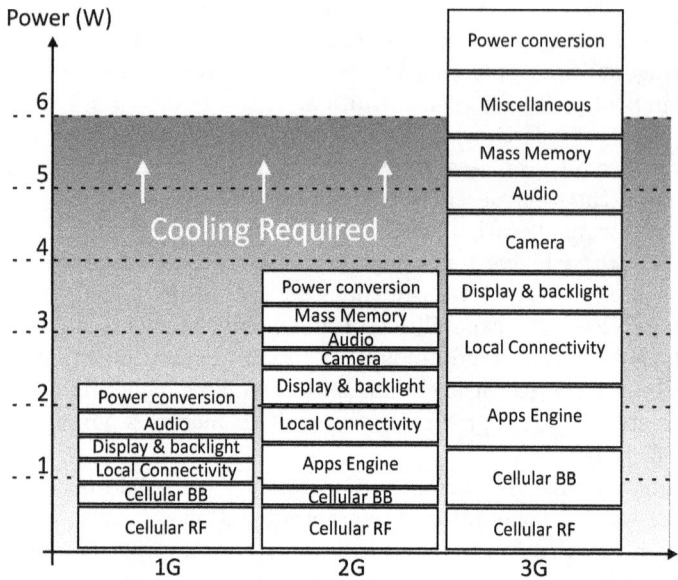

Figure 3.2: Power Consumption of Mobile Phones (Fitzek, 2007, p. 442)

it consumes additional space. This would result in size increases, which are also undesirable.

3.1.2 The Wireless Mobile Grid Scenario

3.1.2.1 A Short-Link Architecture for Wireless Mobile Grids

How might we solve the conflicting demands of more complex devices and services on the one hand and limited battery capacity that is strained by the energy consumption of the devices needed for future mobile services on the other hand?

Fitzek and Katz (2007b) propose one way of tackling these issues, by developing and testing a new distributed communication infrastructure for 4G networks. They start from the conventional cellular architecture that can be seen in Figure 3.3.

This communication architecture is characterised by the focus on the communication between the mobile devices and a base station which is the *only* access point for the mobile devices to the network services. In this architecture the mobile devices work as terminals, as they only communicate with the base station through a battery-intensive cellular link and all services terminate in the mobile devices. The hard- and software solutions for this architecture are more or less static, i.e. they do not change in response to different situations or requests. Whereas this was sufficient in the past where the focus was mainly on voice-centric services, in 4G networks, with their increasing number of different services, this becomes a problem as the components of the mobile devices need

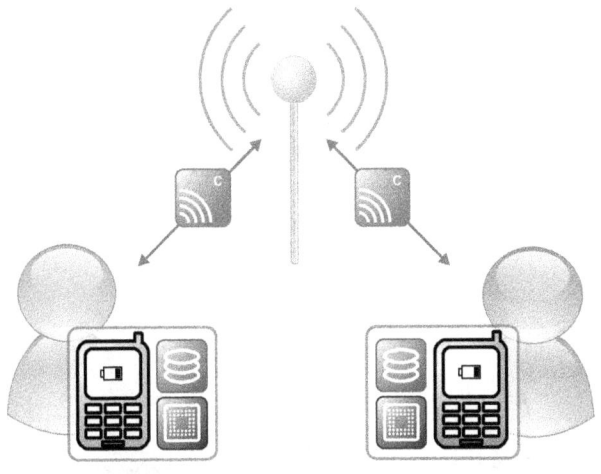

Figure 3.3: The Cellular Architecture (Fitzek and Katz, 2007a)

to be designed for worst-case scenarios. In Figure 3.3 these devices are battery, storage capacity and processing power for example.

To overcome these energy problems, Fitzek and Katz (2007a) proposed the concept of *wireless mobile grids* (WMG) as shown in Figure 3.4.

In these grids, mobile devices with potentially different capabilities are envisioned to cooperate, in jointly trying to achieve their individual goals. The idea is that the participants in the WMG share their limited resources (such as computational power or bandwidth) for their own (e.g. reduced battery consumption) as well global system-wide benefits. These benefits are especially on the side of the resource consumption (e.g. battery) (Perrucci, 2009), but further benefits exist. Section 3.1.3 will highlight the WMG benefits in more detail.

The cooperation between the mobile devices is enabled with the help of a short range communication link. So far, this dissertation has only talked about short range transmission very generally, however when considering the number of existing technologies that could be used with regard to the WMG idea. Figure 3.5 gives an overview over the current short range options and compares them with regard to transmission speed and transmission range.

Aiming for the optimization of both speed and range, IEEE802.11 WLAN and Bluetooth v2.0 seem especially interesting. Furthermore, both technologies have the advantage of already being included in modern phones. Given that Bluetooth v2.0 can be employed in two different ways, in total three different technologies for the short range transmission need to be looked at:

- IEEE802.11 WLAN,

- Bluetooth v2.0 without broadcast, and

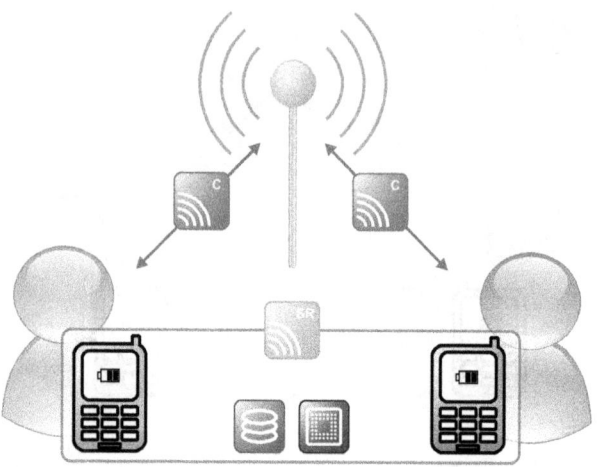

Figure 3.4: The Wireless Mobile Grid Scenario (Fitzek and Katz, 2007a)

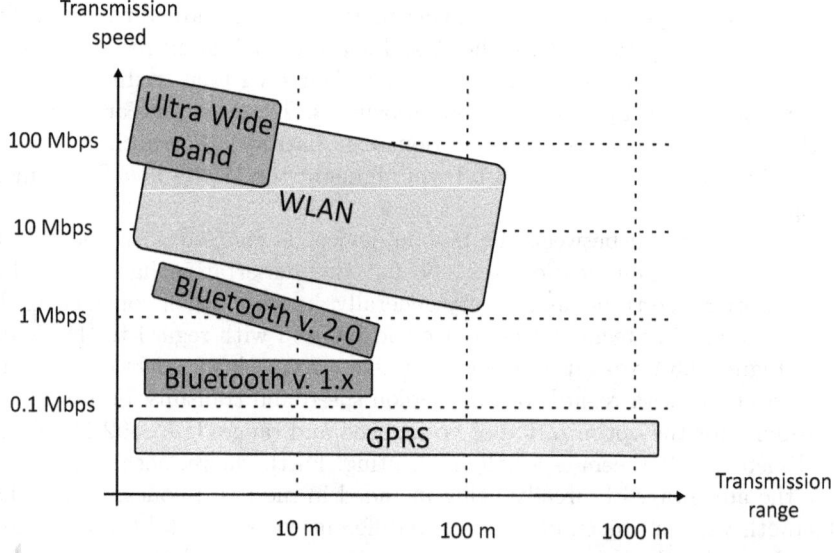

Figure 3.5: Comparison of Short Range Capabilities (Fitzek and Reichert, 2007)

- Bluetooth v2.0 with broadcast.

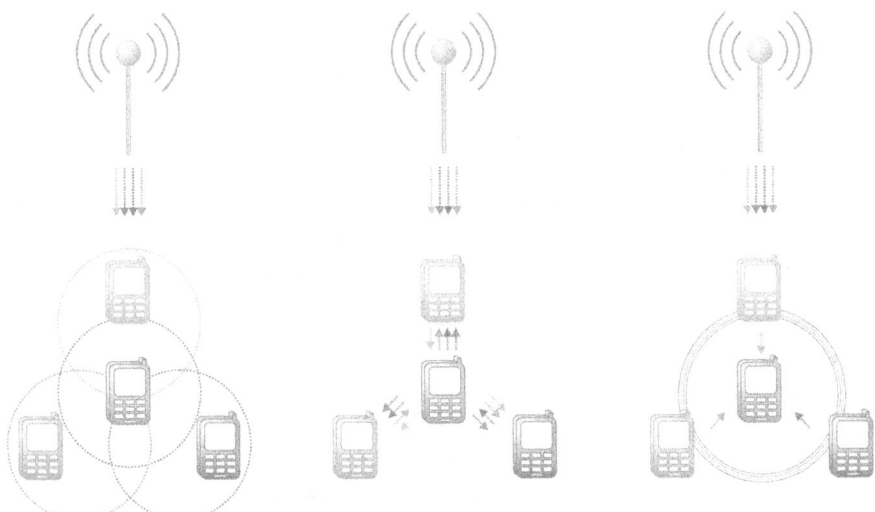

Figure 3.6: Comparison of Short Range Architectures

Figure 3.6 shows these three different technologies and highlights their important differences with regard to the WMG. What is shown in the figure is that using IEEE802.11 WLAN (left part of the figure) mobile devices can communicate directly with each other. This is not the case for Bluetooth, which adopts a master-slave architecture with the master functioning as a relay for the slaves. Thus, the slaves can only communicate via their master with each other. As a result more communication and consequently more data exchange is needed. How much data needs to be exchanged for the WMG cooperation thereby depends on the broadcast capability of the master. In the case that the master cannot broadcast packets to its slaves (middle), obviously the number of transmissions is larger than for the case when the master is able to broadcast (right). Based on these considerations as well as measurement of the (battery) costs associated with the different technologies Perrucci et al. (2009) showed that IEEE802.11 WLAN is the most promising one for WMGs. That is why this dissertation we focus on the IEEE802.11 WLAN technology.

3.1.2.2 Use Cases

For the better understanding of the WMG idea we will now briefly present a scenario that indicates how WMGs could be employed in the future.

We consider a multicast scenario in a busy financial district such as the City of London or the Frankfurt (Main) financial district. During their work day many individuals are likely to be interested in stock market data or financial and general news. When a major news event takes place, if not sitting in front of a computer with Internet access (e.g. on the way to or from work or during a lunch break), these people might use their mobile phones in order to get the latest data or information they want.

Currently this scenario causes severe problems for infrastructure providers. Thus, at peak times in the City of London for example network providers experience severe capacity problems resulting in long waiting times for requested data or even network failures (Bingham, 2010). Hence, due to the large density of people in metropolitan areas such as the above deployed network infrastructures are not sufficient to meet user demands because base stations cannot cope with the amount of requests. The reason for this is straightforward: assuming that the base station is using the normal multiplexing technique in which the bandwidth is divided into several sub-slots ("channels") and each mobile phone is allocated one slot. As total bandwidth is fixed, the more slots are allocated, the smaller the bandwidth that can be assigned to each channel. As a result download times increase leading to both more battery consumption and lower quality of the downloading service.

Using a cooperative approach, if mobile phone users are interested in the same news items, users could share the task by receiving a subset of the multicast channels over the cellular link from the base station and exchange the missing pieces over the short range link.

3.1.2.3   The Energy-Advantage of IEEE802.11 WLAN

To understand the IEEE802.11 WLAN WMG idea and its energy implications better, this section focuses on the technical aspects of the scenario (especially the WLAN transmission) in more detail. First we state the basic definition of energy $[E]$:

$$Energy = Power * Time \ [Joules] \tag{3.1}$$

Battery consumption depends on two aspects: the power $[P]$ consumed per connection type and the time $[t]$ needed for the actual transmission. The power required for sending data differs greatly for different connection types, as does the transmission rate (i.e data units conveyed per unit of time), which determines how long a transmission of a sub-slot of a specific size takes and consequently indirectly influences the *time*-value in the energy-definition. More specifically (assuming equal power values), if the transmission rate is low and a transmission takes longer, more energy is consumed for the transmission.

The total energy consumption is composed of two components: the energy consumed over the traditional cellular 3G connection (E_{3G}) plus the energy consumed over the short link (i.e. WLAN) connection (E_{WLAN}). In case of no cooperation the latter is 0, i.e. it is assumed that the WLAN connection is turned off and the mobile phone user has to download the complete news information he wants alone using the 3G connection. In case of WMG cooperation it is assumed that both connections (WLAN and 3G) are turned on and the devices help one another in a peer-to-peer-like fashion. Assuming x_{Coop} cooperating agents for example, each agent only needs to download only a part of the total news information from the base station (i.e. $\frac{1}{x_{Coop}}$ in an ideal scenario) and exchange the missing parts with the other $x_{Coop} - 1$ cooperation partners, using the short link connection (i.e. WLAN in our scenario). Therefore the energy consumption per device in the cooperation case (E_{Coop}) is comprising four parts:

1. the energy consumed for downloading from the base station using the 3G link ($E_{3G,rx}$) (plus the energy consumed while the 3G connection is idle ($E_{3G,i}$)),

2. the energy consumed for receiving the remaining chunks of the news information on the WLAN connection ($E_{WLAN,rx}$),

3. the energy consumed sending the device's own chunks to the other participants via the WLAN connection ($E_{WLAN,tx}$), and

4. the energy consumed by WLAN in idle phases (i.e. when not transmitting or receiving anything but waiting for the next interaction) ($E_{WLAN,i}$)[8]:

$$E_{Coop} = \underbrace{E_{3G,rx} + E_{3G,i}}_{\text{3G consumption}} + \underbrace{E_{WLAN,rx} + E_{WLAN,tx} + E_{WLAN,i}}_{\text{WLAN consumption}} \quad (3.2)$$

The individual energy costs depend on the power level of that connection type as well as on the corresponding time it is in use. Hence the formula can be expanded to:

$$
\begin{aligned}
E_{Coop} \quad = \quad & \underbrace{t_{3G,rx} * P_{3G,rx} + t_{3G,i} * P_{3G,i}}_{\text{3G consumption}} \quad\quad (3.3) \\
+ \quad & \underbrace{t_{WLAN,rx} * P_{WLAN,rx} + t_{WLAN,tx} * P_{WLAN,tx} + t_{WLAN,i} * P_{WLAN,i}}_{\text{WLAN consumption}}
\end{aligned}
$$

Representative power and time values for the transmission measurements in different states using 3G and WLAN connection have been determined by Perrucci et al. (2009, p. D10) for example. Their results are reproduced in Tables 3.1 and 3.2.

The numbers by Perruci et. al are based on measurements taken from a Nokia N95. These numbers indicate that although the power needed for the WLAN and the 3G state are about the same, the data rate for the 3G link (0.193 Mbit/s for the receiving state) is significantly lower than that of WLAN (5.115

---

[8]Given new network coding protocols such as random linear network coding (RLNC) (Ho et al., 2006), this kind of interaction even does not necessarily require the coordination of the different mobile phone users on who is to download and share which chunk, resulting in possibly no energy being used for coordination purposes Fitzek et al. (2010). Therefore any form of negotiation costs are neglected in this dissertation.

Table 3.1: Power Level and Data Rate for Cellular - 100 byte

state	power value [W]	data rate [Mbps]
receiving	1,314	0,193
idle	0,661	-

Table 3.2: Power Level and Data Rate for WLAN Broadcast - 1000 byte

state	power value [W]	data rate [Mbps]
sending	1,629	5,623
receiving @ 3m	1,375	5,379
receiving @ 30 m	1,213	5,115
idle @ 3m	0,979	-
idle @ 30m	0,952	-

Mbit/s, receiving state, 30m distance) leading to a significantly worse energy per bit ratio for the 3G link.

As a consequence, the cooperation scenario has an energy advantage compared to the conventional cellular communication architecture, especially if the number of cooperating mobile phones is high and a large proportion of the data traffic can be carried via the short-link (i.e. WLAN) connection.

3.1.3   Stakeholder-Advantages in the Wireless Mobile Grid Scenario

After having described the WMG scenario in the previous section, this section focuses on the question of the system's stakeholders' interests. This section aims to identify the main stakeholders of a WMG as well as their relations to one another and furthermore outline the benefits the stakeholders can draw from a WMG.

Figure 3.7 shows the major stakeholders of a WMG. These are mobile phone users, mobile phone manufacturers, the manufacturers of infrastructure components (such as base stations, etc.) as well as network and service providers that are subsumed as infrastructure providers (IPr) in this dissertation[9]

The infrastructure manufacturers are firms such as Alcatel-Lucent that provide and maintain the infrastructural components of a mobile phone network. They include the base stations and mobile services switching centers (the network element that performs the telephony switching functions of the mobile phone network). The mobile phone manufacturers are firms such as Nokia or Sony Ericsson that develop mobile phones. Normally users have contracts with the infrastructure providers (e.g. Orange in the UK or T-mobile in Germany), who sell the phones together with network usage contracts to the users. In WMGs,

---

[9]For reasons of simplicity and because several examples exist in which the two roles are congruent we refrain from viewing network providers and service providers separately. Examples of these congruent cases are T-mobile or Telephónica for example.

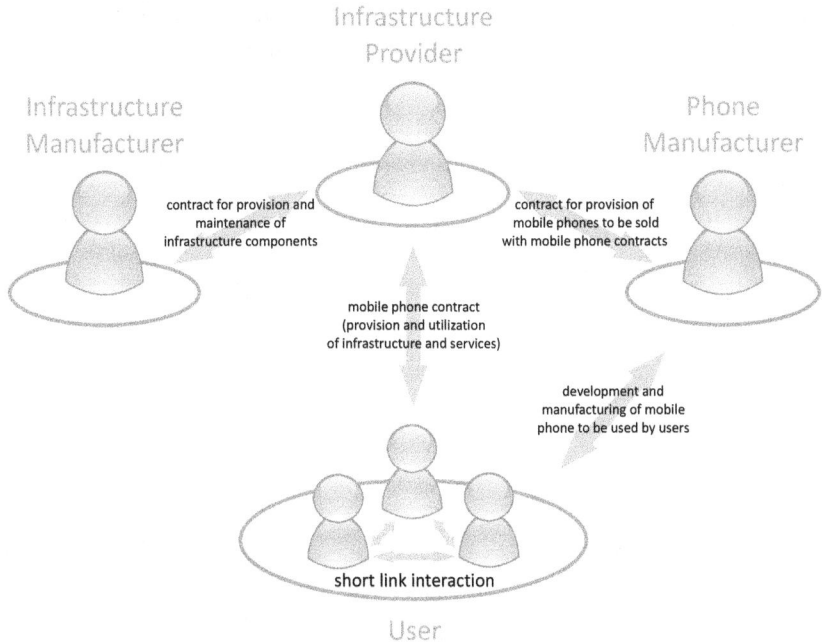

Figure 3.7: Stakeholders in the Wireless Mobile Grid Scenario

users depend both on the transmission costs resulting from the contracts with the IPr as well as on the cooperation of other users to benefit of the WMG.

With regard to the business perspective taken in this dissertation, the obvious question with regard to the stakeholders is what they can gain from a WMG, i.e. why they should be interested in its success (Fitzek and Katz, 2007a, p. 56–57)?

For the user, besides energy saving, two further advantages can be identified: potentially lower service costs and robustness.

Energy Saving: By using the short link interface instead of the cellular one, the overall energy consumption is reduced, because as indicated in Tables 3.1 and 3.2 the energy per bit ratio is significantly better on the short link.

Low Service Costs: As only a part of the total file or service is downloaded via the cellular link by each user, for the individual user the download-times and consequently the costs associated with download can be reduced. This is especially relevant for users with limited bandwidth allowances.

Robustness: As the service is provided over different paths one can furthermore expect diversity gains. The user is not dependent on a single connection with the base station but possibly can rely on a number of different user (and their mobile phones) within his vicinity.

From the IPr point of view several advantages can be found, with the most important one being the ability of offering better quality transmissions without further investments in infrastructure components such as base stations. This quality increase stems from increased transmission rates and reduced effects from transmission errors.

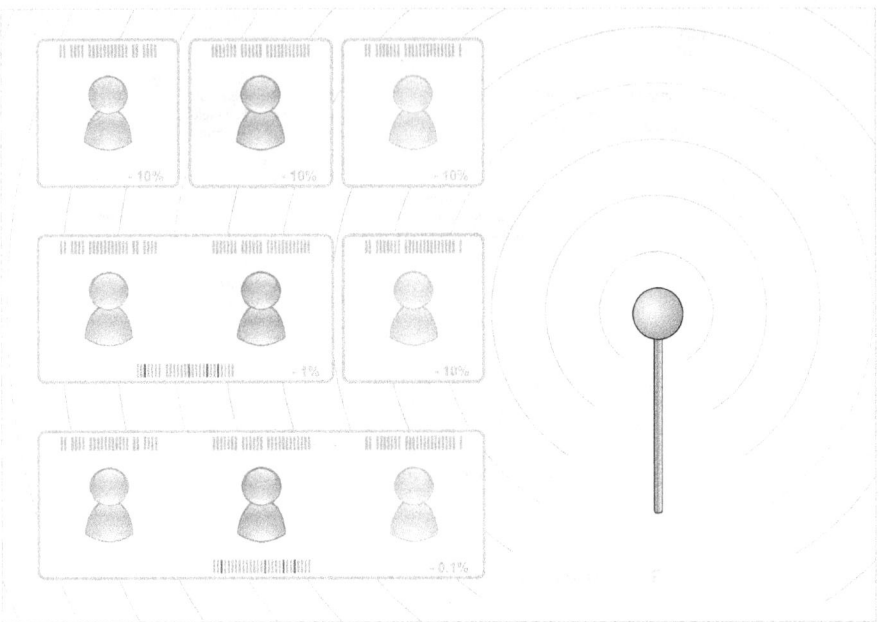

Figure 3.8: Comparison of the Effects of a 10% Transmission Error for Non-Cooperation as well as Cooperation with two or three Cooperating Mobile Phone Users

Lower Transmission Error Rates Figure 3.8 illustrates the error rate effects in more detail, by showing the consequences of a 10% error rate for the transmission from a base station to mobile phones (Vingelmann et al., 2010).

If no cooperation takes place and mobile phone users want to download the same file, the base station transmits using multicast (top row). Each mobile phone user thereby receives approximately 10% faulty chunks, or putting it differently, a 10% faulty file (represented by the respective missing chunks in the figure). What is important to note is that these 10% errors are not necessarily the same for all mobile phone users, but are statistically independent. As a result of the faulty file, the mobile phone users requests the file or at least the faulty chunks again. Resulting from the respective requests the base station will resend the faulty chunks/file. This process is repeated until all mobile phone users have all correct chunks of the file.

Looking at what happens if cooperation takes place, the effects of the error rates drop significantly. Thus, if several mobile phone users cooperate for example, due to the statistical independence, their individual error rates can be multiplied to arrive at a group error rate. Hence, for two cooperating mobile phone users the combined group error rate equals $10\% * 10\% = 1\%$ and in case of three mobile phone users this rate drops down to 0.1%. As a result, from the base-station's (i.e. the IPr's) point of view, less resend requests occur and its resources are drawn upon less. This effect intensifies if higher error rates are assumed[10].

New Services and Higher Service Scalability As explained earlier, co-operative networks using IEEE802.11 WLAN have higher data rates at lower infrastructure resource costs. As a result, IPr can offer existing services at a better quality or even introduce new services that require high bandwidth and data rates and are currently infeasible.

Larger Market Penetration and Increased Revenue This in turn can attract a higher number of potential customers which can be transformed into additional revenue. As WMGs do not require investments into the infrastructural resources, the additional revenues are not accompanied by fixed investment costs.

From the mobile phone manufacturers' point of view the impact of WMGs is especially interesting with regard to battery manufacture. As outlined briefly earlier, with regard to sales and consequently financial success currently two main problems for manufacturers exist: the device price and the energy consumption, with the latter being split into three sub-problems, namely: the battery life cycle and connected operating time of mobile phones, as well as the cooling required for the batteries.

Low Costs for Energy Saving Devices To be more precise, as outlined earlier, the higher energy consumption expected from the next generation of mobile phones may result in the need for larger batteries if the stand-by times of the mobile phones are to be kept at the same level. Larger batteries however do not only pose a size issue, but may require active cooling, which would negatively impact consumer interests. WMGs offer a way to support high data rates through cooperation without the additional needs on the side of the energy consumption and consequently the cooling of the device. Given that WMGs rely on existing technologies, the concept offers a further advantage as development costs needed to reap these benefits, can be expected to be low.

The only stakeholder group for which no advantages can be found with regard to the WMG idea are the infrastructure manufacturers. In contrast, they even have to fear profit setbacks if IPrs demand fewer infrastructural components.

---

[10] According to Zhang et al. (2010) this effects still hold even if new transmission technologies like LTE are applied. Thus LTE currently multicasts 125% of the packages demanded by the mobile phone users to reduce transmission error resend requests. As the authors show in their paper, using cooperation the LTE overhead of 25% can be reduced by 80-94%, making the concept advantages even if new technologies are considered.

Since the WMG idea is especially triggered by the IPrs, on which they are dependant as sellers of their mobile phones in combination with mobile phone contracts, their leverage and consequently the importance of their interest is low.

3.2 The Cooperation Dilemma in Wireless Mobile Grids

In the last section the advantages of the WMG scenario with regard to the stakeholders were presented. Although the WMG shows advantages to its stakeholders with regard to the resource consumption or transmission rates, it features a problem that is very common to distributed cooperative architectures in general: it depends on the cooperation of the users.

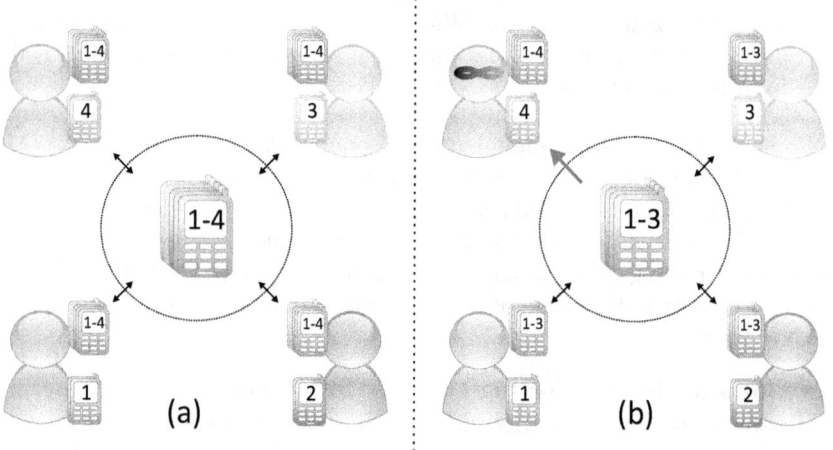

Figure 3.9: The Free Riding Problem in Wireless Grids

The cooperation idea in the WMG is that, as shown in Figure 3.9(a) the users voluntarily commit their resources, forming a common pool which can be used either by all of them in order to achieve a common goal or individual ones. The utility which users can obtain from the pooled resources is much higher than they can obtain on their own. For example, in the financial district WMG scenario they can download files faster and in addition reduce battery consumption.

However, the problem in this combination is that commitment comes at a cost in the form of battery consumption for sending file chunks, i.e. $E_{WLAN,tx}$. As a consequence, (bounded) rational users would prefer to access the resources without making any commitment to cooperate, as shown in Figure 3.9(b). Thus the user in the top left corner(with the mask) can enjoy the full benefits from the common pool without committing anything himself, i.e. he cheats on the three other agents.

However, if a substantial number of users followed this selfish strategy, the network itself would be at risk, depriving all users from the benefits (Wrona and Mähönen, 2006). The reason for this is straightforward: network users can have strategic behaviour and are not necessarily obediently cooperating by making their resources available without the prospect of rewards for their good behaviour. Unreciprocated, there is no inherent value to cooperation to a user. A lone cooperating user draws no benefit from its cooperation, even if the rest of the network does. Guaranteed cost, paired with uncertainty or even lack of any resulting benefit does not induce cooperation in a (bounded) rational, utility-maximizing user. Without any further incentives, rational users therefore would not cooperate in such an environment and would all be worse off than if they cooperated (Ostrom, 1999; Axelrod, 1981).

Looking at the problem from a game-theoretic point of view and considering battery consumption only (i.e. neglecting other aspects of WMGs), the individuals are caught in a prisoners dilemma (PD) that is shown in Figure 3.10[11].

		Cooperation partner n	
		cooperation	no cooperation
User	cooperation	$E_{no\ coop} - E_{coop}$ $E_{no\ coop} - E_{coop}$	$E_{no\ coop} - (E_{coop} - E_{WLAN,tx})$ $-E_{WLAN,tx}$
User	no cooperation	$-E_{WLAN,tx}$ $E_{no\ coop} - (E_{coop} - E_{WLAN,tx})$	0 0

Figure 3.10: The Free Riding Problem as cardinal PD

Figure 3.10 shows the relative payoffs for a WMG user and one of his potential cooperation partners. The payoffs are shown in terms of energy savings for each transaction partner in comparison to the option of a partner downloading everything himself. Three cases can be distinguished:

1. In case of case of cooperation by all sides the energy saving potential is equivalent to $E_{no\ coop} - E_{coop}$ (i.e. simply the energy gained by cooperating in contrast to down everything self) for each partner[12].

2. In case one partner defects, he gains an additional payoff $E_{WLAN,tx}$,

---

[11] The described PD also exists when non-battery related aspects of WMGs are considered. We chose to neglect these non-battery related aspects issues in order not to distract from the PD analysis unnecessarily.

[12] The absolute size of this energy difference depends both on the size of the file to obtain as well as on the number of cooperation partners, which is why Figure 3.10 only uses relative payoffs.

because he saves energy for not sending his file chunks to his partner. The other partner at this point as a consequence does not receive the file chunks and has to download the remaining file chunks himself. As a result he is still downloading everything himself (i.e. has no energy saving with regard to the download) and in addition has additional energy costs for sending his file chunks.

3. The last case shows the payoffs for both partners defecting. In this case they both have to download everything themselves leading to a payoff of 0 for both partners.

Using these values to develop an ordinal order of preferences for both players the PD looks as shown in Figure 3.11. For both players the ordinal order of preferences is $1 \succ 2 \succ 3 \succ 4$.

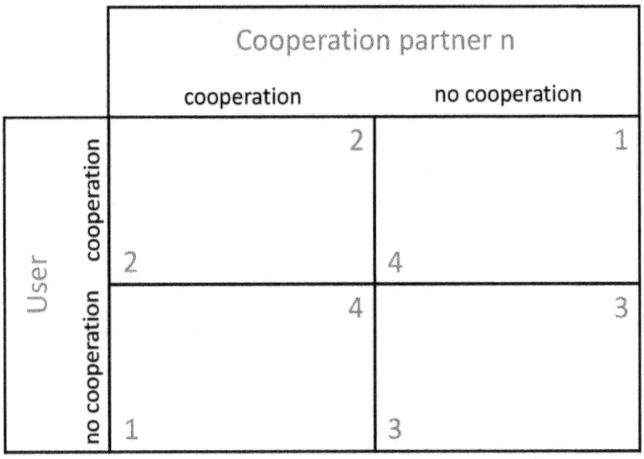

Figure 3.11: The Free Riding Problem as ordinal PD

Hence the dominant strategy for both players is strictly not to cooperate. If the other one cooperates, no cooperation is better because he can gain $E_{WLAN,tx}$ and if the other player does not cooperate, no cooperation is still the better option as one would not have the advantages deriving from cooperation (the chunks the other one is sending) and still has to carry the transmission costs $E_{WLAN,tx}$.

Looking at the total welfare of the interaction (i.e. the combined payoffs of both partners) it becomes obvious that the result from the individual preferences is the worst case, as the total welfare is 0, compared with potentially $2*(E_{nocoop} - E_{coop})$ in the case of cooperation. Hence, although an overall beneficial result is possible, individual preferences lead to a non-beneficial state.

As pointed out before, one consequence of this non-beneficial state is that all stakeholders are deprived of the potential benefits that can be gained of an WMG. As a result of this risk as well as the business interests associated with the concept, it is essential to analyse how to deal with the problem at hand well in advance, in an early prototyping stage.

3.3   Summary

In this chapter we presented a case study of one particular open distributed system that is currently in an early prototyping stage of development, so-called WMGs. We outlined its high relevance for the mobile phone industry by presenting current industry data and analysed the WMG idea (and its advantages) with regard to the stakeholders in a WMG. One particular aspect of the WMG we focused on in our presentation was the cooperation issues inherent in WMGs, namely that it is advantageous for mobile phone users to take advantage of a WMG, but not to contribute to it. These issues were discussed analytically as well as mathematically in the form of a PD. As a result of the cooperation issues, we concluded that if these issues are not resolved the whole WMG idea is at risk. For this reason – in the early prototyping stage of WMGs – it is important to analyse how possible governance decisions such as the integration of enforcement mechanisms into a WMG could affect the system. In the remainder of this dissertation, this WMG case study will serve as an example for the analysis of how normative framework modelling can be applied in the early prototyping stage to analyse the impact of possible governance decisions on a system.

Part II

Normative Modelling

Chapter 4

Normative Frameworks

At the end of Section 2.4 we briefly outlined the components of the IAD framework. One important conclusion from the description given is that although the IAD framework does consist of several components that are relevant for any analysis, the central "focal unit" where all aspects are being combined or rely on is the action arena. As a consequence, the next question that arises when wanting to analyse the effects of governance changes in systems at an early prototyping stage is how one can analyse this particular part of the framework. In this chapter we will present the method this dissertation uses for modelling normative information in the action arena, namely *normative frameworks*. The approach to normative frameworks in this dissertation is strongly based on the works of Cliffe (2007) (as well as the works leading to and resulting from his dissertation) which present a formalization of normative frameworks in order to be able to specify and analyse them. In the later chapters of this dissertation Cliffe's formalization is used for representing the action arena and analysing it with the help of computational simulations. That is why, in the later part of this chapter (e.g. when presenting regulative aspects of normative systems) we will comment on Cliffe's formalization and outline some design choices he made with regard to the representation of normative frameworks.

Before going into detail, let us start by defining what is understood by a normative framework:

Definition 9: Normative Framework

In this dissertation we use the term *Normative Framework* to describe a *set of norms* and the operational semantics which specify and regulate specific interaction aspects of an (open distributed) system. A normative framework may consist of different types of norms (Rawls, 1955; Alchourrón and Bulygin, 1971): (i) *constitutive norms* that regulate the creation of normative facts and the modification of the normative system itself as well as define how the system (and its participants) should interpret events (Grossi,

> **Definition 9:   Normative Framework (cont.)**
>
> 2007), and (ii) *regulative norms* that define what is perceived as normatively correct and incorrect behaviour by means of obligations, prohibitions and permissions (Searle, 1997). Normative frameworks also define the *scope* of the norms they contain, in terms of when (in time), under which conditions, and to which participants and actions they apply.

In the Chapter 2, we outlined the benefits resulting from the utilization of norms for regulating systems in such a way that undesirable system states can possibly be avoided. They may also help to reduce uncertainty in a system and thereby lower transaction costs. Normative frameworks – being a collection of norms for a system – have very much the same benefits. As pointed out above, this chapter focuses on the modelling of the normative aspects of the action arena. For these reasons, in this chapter we turn away from the utility of norms and normative frameworks and focus on their specification and formalization instead.

4.1   Constitutive Aspects of Normative Frameworks

4.1.1   Brute Facts and Institutional Facts

To start with the specification and formalization, this dissertation first examines how normative frameworks relate to the real world (i.e. the abstraction level on which the actors in a system interact) and how both evolve over time.

Figure 4.1 shows our view of the normative framework and the real world in a system. Both views can represented in terms of facts (i.e. system states) at a particular point of time. It is important to note in Figure 4.1 that both the real world and the normative framework present different views on the action arena (that represents the interaction in the system). Whereas the real world view is concerned with the events physically taking place in the action arena, the normative framework view is concerned with the normatively relevant aspects only. The difference between the two views on a system is perhaps best illustrated by an example: the example of a murder. In simple terms, a murder starts with an event in the real world that results in a dead body. Whether the deceased was killed legally or was murdered is a fact that may only be established by a judicial process, which is a normative framework. If someone is convicted and punished for the murder, it is the normative framework that establishes their guilt and determines the sanctions of the convicted person. This then serves as information for the real world, where the actual enactment of the sanction is carried out (e.g. by the imprisonment of the murderer). What is important to note is that despite offering different perspectives on the same events and being

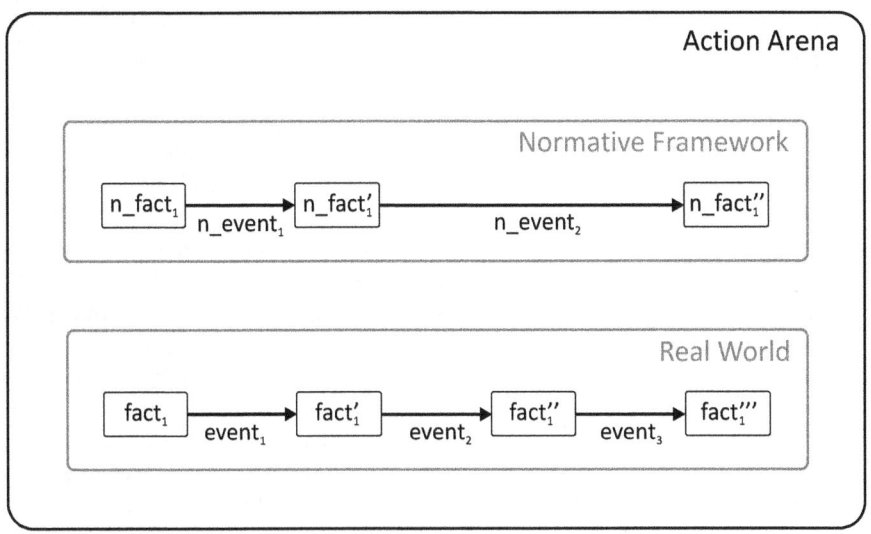

Figure 4.1: The Transition of the Real World and the Normative Framework

displayed separately in Figure 4.1 the two views are not interdependent, but are logically interlinked.

As a consequence of this two questions arise that need to be dealt with when modelling normative frameworks:

Question 1: Assuming that events in the real world trigger the normative framework events, how does this process work, i.e. how is the logical link between the views specified and how do the normative framework events come into being?

Question 2: The second question is linked to the first. It is concerned with the real world events that should be considered from the perspective of a normative framework. How is it determined which real world events are relevant for the normative framework? The case of the dead body is a simple example; the killing event resulted in a state of a dead body which is obviously the important event for the normative framework. Nevertheless questions arise on what other events might be important, which events can be neglected and how to distinguish between the two?

The second thing to notice in Figure 4.1 is the evolution of the two perspectives on the action arena over time. Both perspectives start from an initial state (or facts) when the system is initialised and subsequently change state as a result of events. The facts of the normative framework view are referred to as normative facts (n_fact_i) whereas the real world facts are referred to as $fact_i$ in Figure 4.1. In Figure 4.1 these states (facts) change to new states when respective events take place, either normative events (n_event_i) in the normative framework view or real world events ($event_i$) in the real world view.

Question 3: With regard to the specification and formalization of the normative framework, the question resulting from this evolution from one state (fact) to another is, how can this transition best be described?

Keeping the three questions resulting from Figure 4.1 in mind, this dissertation now turns to related work on these questions in order to answer them.

The work that is most relevant for answering the questions stated above comes from philosophy, where in 1997 Searle published a book entitled "The Construction of Social Reality" (Searle, 1997). In this book Searle discusses constitutive norms[1] as a mutually understood norm basis for interpreting regulatory norms and distinguishes *brute facts* and *normative facts* with regard to them. To Searle, brute facts are facts that follow from a common-sense understanding of the world (e.g. "the cup is empty" or "on average the sun has a distance of approximately 93 million miles to the earth"), whereas institutional facts are facts that are only valid within a certain context and depend on human opinions and interpretation. Brute facts correspond to real world facts in Figure 4.1, whereas institutional facts are equivalent to the normative facts in the figure. According to Searle these human opinions are shaped by the normative setting the humans interact with, which also sets the context for the institutional facts (and consequently defines whether and to what extent they are valid). For this reason he calls them institutional facts (i.e. being dependant on an institution), with his notion of an "institution" being congruent to this dissertation's notion of normative frameworks. Based on this distinction, Searle defines *constitutive norms* as norms describing how institutional facts (i.e. facts in the normative framework) are created. In detail, he specifies them as norms describing what, when doing an action in one context, *counts-as* performing another action in a second context. If one interprets the physical world as the first context and defines when the presence of certain states, the execution of certain actions, or the occurrence of certain events leads to actions in the second context (i.e. the normative framework), one can explain the creation of institutional facts.

An extensive analysis of actions and events in the real world generating actions and events in a second context (e.g. the normative framework) was conducted by Goldman (1976), who distinguished four different cases of action generation: causal generation, simple generation, conventional generation and augmentation generation[2]. In respect to the normative frameworks and the

---

[1] Searle speaks of constitutive/regulative "rules" rather then constitutive/regulative "norms". Based on our definition of both terms in Chapter 2 we use the terms "norms" and "rules" synonymously. For reasons of uniformity with the dissertation's terminology we speak of constitutive/regulative norms.

[2] The causal generation takes a real world event and generates a corresponding action in the second context which interprets the results of the real world event. This interpretation of causal effects is not included in the simple generation. There, a real world event simply results in an action in the second context (possibly under the condition of a certain state of affairs in the first context). The conventional generation extends the simple generation by

counts-as principle, of these four different cases, the third one in particular (i.e. *the conventional generation*) has been attributed a high relevance (Cliffe, 2007).

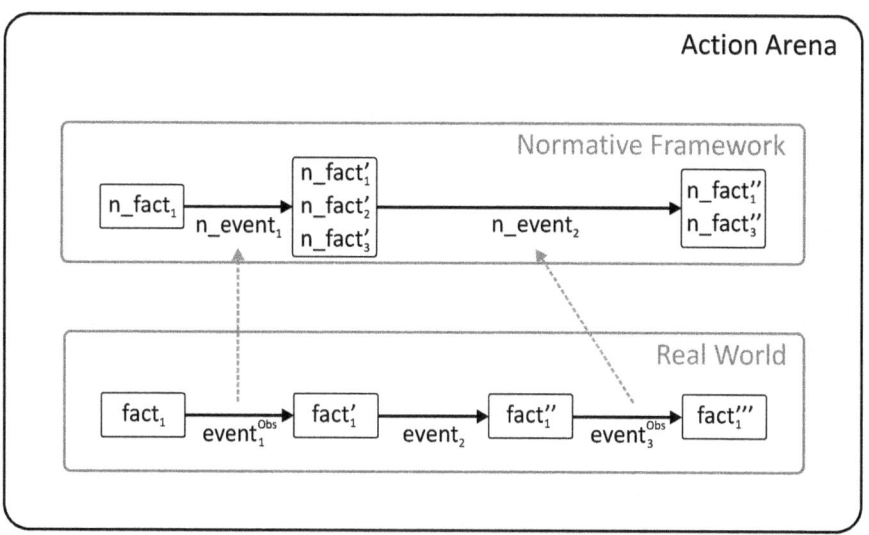

Figure 4.2: Conventional Generation turns Observations into Normative Events that lead to Normative Facts

Figure 4.2 shows conventional generation with regard to the normative framework, with the gray dashed lines representing the generation of normative events (and consequently normative facts) as result of the observation of real world events. Goldman's conventional generation concept describes how actions/events in a given context generate an action/event in the normative context within the bounds of certain conditions (i.e. he emphasizes the importance of the scope of the norms). To give an example of this: shooting a football into is considered as scoring a goal, but only in the context of football and if the game of football is still in progress.

What is also important for conventional generation is that real world events that should generate normative events which lead to normative facts need to be observed, i.e. they need to be perceived by the system participants. Returning to the football example again, the scoring of a goal for example needs to be observed by a person such as the referee, who has the relevant legal power to establish the normative fact that this is indeed a goal and counts for the score of the game. If the referee does not observe the scoring and as a consequence does not take any actions to bring about the change in the normative state, the goal does not count and therefore does not have any normative effect. This is

adding conditions for the generation of the action in the second context, that can be based on properties in the first and/or second context. Finally, augmentation generation refers to generations where the generated action in the second context elaborates on the event from which it is generated.

demonstrated in Figure 4.2 that refers to observed events ($event^{Obs}$) in the real world and where $event_2$ that is not observed does not bring about a normative event.

Another aspect worth noting in Figure 4.2 is that the conventional generation of one observed event in the real world can result in more than one normative fact in the normative framework. With regard to the football match example, one player kicking another player can result in both a foul being indicated as well as a card being given. Hence as a result of the first observed real world event, several normative facts (i.e. $n\_fact'_1$, $n\_fact'_2$ and $n\_fact'_3$) could be generated. As pointed out before, in the normative framework the transition from one state to the next state (which is based on the normative events generated) is defined with the help of constitutive norms. These norms include specifications of which normative facts result from which normative events in which context. Hence, they specify which of the normative facts are generated and initiated and terminated.

4.1.2   Normative Power and Permission

In Section 2.1 we outlined the difference between physical and legal power (with the latter also being referred to as normative power in the context of the normative framework) and distinguished the concept of power from permission. Applying these concepts to Goldman's conventional generation and Searle's counts-as principle the following relation can be established. Legal power acts as a constraint on conventional generation by specifying who has the legal capability to bring about action in the normative context. This legal power might be attributed to single persons only, or persons acting in a particular role, for example. Coming back to the football example: the game is finished by the final whistle. However, this whistle only ends the game (i.e. changing its normative state) when the referee blows it. Otherwise blowing the final whistle is considered an invalid action with regard to the normative framework and does not have any effect there (i.e. the game does not end). In contrast, the permission is independent of power and does not have an effect on the convention generation. The referee might for example not have the permission to blow the final whistle before 90 minutes are over. However if he does so, the game still ends (i.e. he triggers a valid action in the normative framework), he might however be punished by the board of referees afterwards because of it.

This notion of legal power was formalized by Jones and Sergot (1996) who use it to model conventional generations with the help of the logic expression $E_x A \Rightarrow_s E_x F$. This expression allows for the association of a person (x) whose real world actions E_x bring about a change in the state of the real world to state A. According to the expression, these actions might also bring about a change of state on the normative level (F) if the conditional statements ($\Rightarrow$) hold with regard to the normative framework s.

4.1.3   Events

In the preceding two sections we reviewed the notion of counts-as as well as that of conventional generation in order to answer the questions posed at the beginning of this chapter. The idea of conventional generation implied that an normative event can be generated in two ways: as a result of the interpretation of an event in the real world that acts as stimulus to the normative event, or as a result of a normative framework internal interpretation of rules. As a result, two types of events need to be distinguished for normative frameworks: normative events (i.e. events internal to the normative framework context) and external events (i.e. events from a context external to the normative framework, such as the real world). We further partition normative events into normative actions, that denote changes in normative state, and violation events, that signal the occurrence of violations. Violations may arise either from explicit generation, from the occurrence of a non-permitted event (e.g. the blowing of the final whistle in football before the 90 minutes of game time are over), or from the failure to fulfil an obligation.

4.2   Regulative Aspects of Normative Frameworks

The preceding sections dealt with different aspects of constitutive norms which are associated with the semantics of the normative framework. Although participants in the system might observe the resulting normative facts from their action, with constitutive norms alone the participants cannot determine which situations and actions a normative framework considers to be good or bad and therefore cannot account for that in their behaviour. That is why regulative norms that define what is perceived as normatively correct and incorrect behaviour by means of obligations, prohibitions and permissions are required.

4.2.1   Prohibition and Permission

The first aspect of regulative norms that this dissertation will consider are prohibitions and permissions. The latter indicate that actions are considered acceptable or desirable, whereas the former can be seen as their counterpart, because they indicate which actions are not permitted.

Although in legal theory permissions and prohibitions are not considered as being exact counterparts, in formal logic approaches this point of view is often taken (i.e. one is defined as the absence of the other). For the sake of simplicity in this dissertation we observe this simplification, allowing us to express one in terms of the other. Thus, this dissertation focuses on permissions only and defines prohibition as an absence of permission. This leads to a further simplification, namely that actions that are not permitted (including action about which no information is given about at all) are forbidden.

Having explained how the distinction between acceptable and not acceptable actions is made in a normative framework, another question that needs to be addressed is the question of the subject(s) of permissions. In the literature, two kinds of foci of permission can be found: permissions on normative facts and permissions on normative actions and events. Cliffe (2007), whose formalization of normative frameworks is used as foundation for this dissertation, focuses on the latter of the two. The reason for this decision was that the focus on events and actions allowed for an easier reasoning about violations. To illustrate this point, we observe that a focus on normative facts would only allow to determine that a violation had occurred. Reasoning about permission of actions and events allows for the determination of how the violations were brought about, which is helpful in trying to find ways to reduce violations.

4.2.2   Obligations

In contrast to permissions and prohibitions which serve as source of information about what actions/events are allowed/not allowed at a specific point in time, obligations are designed to trigger actions by a specific point of time.

As in the case of permission and prohibitions they can focus on normative facts that someone is supposed to bring about by a given time, or on actions someone needs to perform before a deadline or before another action/event. Using the same reasoning as with regard to permissions/prohibitions, in his formalization Cliffe prioritized the action/event view.

A further question that needs to be addressed it at whom an obligation is directed. Obligations can for example be simply associated with participants in the system or they can be associated with particular types of events which internally might be associated with specific participants or groups of participants. The former (i.e. specifying that a particular person X ought to perform an action A) is probably the more intuitive approach in contrast to the undirected obligation (i.e. X ought to be performed). However when groups of participants are involved (e.g. any member of the group can fulfill the obligation), it results in modelling problems. The reason for this being that the obligation then needs to be expressed for each member of the group and needs to include conditional statements on all group members' actions. To avoid these problems, Cliffe's formalization opts to associate obligations with particular type of events which in turn might be caused by any of several participants.

The final issue related to obligations is the treatment of pairs or even groups of obligations. As pointed out in Chapter 2, one important question with regard to normative frameworks is the consistency of the norms they include. What happens if pairs (or groups) of obligations are inconsistent? A common approach is just to eliminate one of the obligations by overriding it (Fasli, 2001). The problem with this approach is that it can lead to oversimplifications. In real world situations, normative conflicts are common and the decision makers need to choose between conflicting alternatives. Transferring the problem of normative conflicts to normative frameworks therefore seems a logical step. This does not necessarily mean that all conflicts need to be accepted, but allowing for conflicts provides the opportunity to analyse them and use the knowledge gained to refine the relevant obligations in order to arrive at a consistent normative framework in the long run. For the purpose of this dissertation it is assumed that normative frameworks are *internally consistent*, i.e. the process of analysis and adjustment has already taken place and the norms contained in the normative framework are not self-contradictory or mutually defeating. That does not mean that one cannot model inconsistent norms, but the models we present are internally consistent for ease of analysis. This assumption of internal consistency results in the requirement that the properties of the normative framework are verifiable at design-time.

4.2.3 Non-Compliance

By representing norms and giving the participants in the action arena a chance to learn and reason about them, one further question that needs to be asked is how to react if the actors do not comply with the norms (e.g. by performing non-permitted actions or by not fulfilling obligations). We have already discussed potential forms of enforcement in Section 2.3 and therefore will not go into detail about the different enforcement options at this point. Nevertheless we would like to point out two aspects that are of relevance for the implementation of normative frameworks.

The first is the practical consideration that if actors in the system are supposed to ensure norm compliance (e.g. by administrating a sanction) then they must be able to successfully perform the relevant actions. With regard to the normative frameworks this implies the requirement that legal power and permission to enforce compliance need to be ensured.

The second aspect is the question of whether or not the enforcement should be considered separate from the norms it aims to guarantee compliance with. One paper that considers the enforcement actions as distinct from the norms that were not complied with was presented by Vázquez-Salceda et al. (2005). The general problem of this approach, that is common to all stand-alone solutions, is that it requires a special treatment of the enforcement-related actions (and possibly a special way of expressing these enforcement actions). Furthermore it is more difficult to express the enforcement action in normative terms (e.g. as permissions or obligations) and link a possible decision on whether or not and to what extent to enforce sanctions as a result of the violation event. Finally, if the enforcement action is treated separately, cases where the enforcement actions are not executed despite a corresponding obligation (i.e. when a non-compliance to an enforcement obligation occurs) cannot be analysed in respect of the initial non-compliance which led to the enforcement obligation. This makes it harder to reason about the dependencies and events that were decisive for the problem. As a consequence of these limitations of the "stand-alone" approach we consider the enforcement action as part of the norms which they enforce and aim to express them in the same language as the norms themselves.

4.3   Further Aspects of Normative Frameworks

4.3.1   The Initiation and Termination of a Normative Framework

As pointed out in Definition 2, one important aspect of norms is their contextuality. Time, i.e. the temporal scope of a norm, is one such aspect. It defines when a norm or even the complete normative framework is in effect.

From a specification point of view it is therefore a requirement to define when a normative framework has an effect and when it does not. This can be done by specifying when, how and possibly with which parameters a normative framework is created and when and how it is destroyed. The time between those two points in time is the time-span where the normative framework is in effect, whereas for the remainder of the time, it is not.

4.3.2   Time

This leads us to the last important aspect to consider when trying to specify normative frameworks, namely how to deal with the problem of representing time in general. Allen (1991) presented a number of ways to represent time, ranging from explicit dating over intervals to temporal logics. Given the focus on events and states (and facts) of the normative frameworks, we pursue a similar approach with regard to time and focus on an event-driven specification. We assume time to consist of a number of ordered time instances and make no assumptions about the duration of time between these instances. At each of the time instances, one real world event occurs. This allows for a chronological analysis of events and of chains of events.

4.4   A Formal Definition of Normative Frameworks

Having presented the general idea of normative frameworks and their role in governing open distributed systems, as well as described considerations that need to be taken into account when modelling normative frameworks, the formal model for such frameworks to be used in this dissertation will now be presented. As noted before, the formal model is based on the work of Cliffe (2007), where the full specifications of the framework as well as the mathematical proofs for the internal correctness of the formalization are provided. To make this dissertation self-contained, based on the explanations given in this chapter we will outline the formal description of the normative framework components relevant to the questions of this dissertation.

Recalling the preceding sections, the essential elements of a normative framework are:

1. events e that are elements of the set of events $\mathcal{E}$ ($e \in \mathcal{E}$), and that bring about changes in state, and

2. a set of fluents $\mathcal{F}$, that characterize the state of a system at a given instant.

The function of the framework is to define the interplay between these concepts over time, in order to capture the evolution of a particular normative framework through the interaction of its participants. We distinguish two kinds of sets of events: normative events ($\mathcal{E}_{inst}$) that are the events defined by the framework and exogenous ($\mathcal{E}_{ex}$), that take place outside the context of the normative framework but whose occurrence triggers normative events in a direct reflection of the "counts-as" principle discussed earlier, with $\mathcal{E} = \mathcal{E}_{inst} \cup \mathcal{E}_{ex}$ and $\mathcal{E}_{inst} \cap \mathcal{E}_{ex} = \emptyset$. $\mathcal{E}_{ex}$ is sometimes also referred to as *observed event*, because in order to have an effect in the normative framework, the events first of all need to be observed and considered relevant. We use the terms observed event(s) and exogenous event(s) synonymously in this dissertation. We further partition normative events into normative actions ($\mathcal{E}_{act}$) that denote changes in normative state and violation events ($\mathcal{E}_{viol}$), that signal the occurrence of violations ($\mathcal{E}_{inst} = \mathcal{E}_{act} \cup \mathcal{E}_{viol}$, $\mathcal{E}_{act} \cap \mathcal{E}_{viol} = \emptyset$). As mentioned before, violations may arise either from explicit generation, from the occurrence of a non-permitted event, or from the failure to fulfil an obligation. In addition to the events just mentioned we also define a set of *creation events* $\mathcal{E}_+ \subseteq \mathcal{E}_{ex}$, that account for the creation of the normative framework and a set of *dissolution events* $\mathcal{E}_- \subseteq \mathcal{E}_{act}$ that account for its termination.

As we distinguish two main types of events, we also distinguish two main kinds of fluents: *normative fluents* that denote normative properties of the state such as permissions $\mathcal{P}$, powers $\mathcal{W}$ and obligations $\mathcal{O}$, and *domain fluents* $\mathcal{D}$ that correspond to properties specific to the normative framework itself. The set of all fluents is denoted as $\mathcal{F}$ ($\mathcal{F} = \mathcal{P} \cup \mathcal{W} \cup \mathcal{O} \cup \mathcal{D}$ and $\mathcal{P} \cap \mathcal{W} \cap \mathcal{O} \cap \mathcal{D} = \emptyset$). Looking at these sets of normative fluents in more detail the following definitions hold:

- $\mathcal{W}$ is composed of power fluents of the form $pow(e) : e \in \mathcal{E}_{act}$. The power proposition denotes the capability to bring about an action e in the normative framework.

- $\mathcal{P}$ is composed of permission fluents of the form $perm(e) : e \in \mathcal{E}_{act} \cup \mathcal{E}_{viol}$. Each permission proposition denotes the permission to bring about an action e in the normative framework.

- $\mathcal{O}$ is composed of obligation fluents of the form $obl(e, d, v) : e, d, v \in \mathcal{E}$, which specifies that an event e should be brought about before an event d to avoid a violation v.

A normative state is represented by the fluents that hold true in this state. Similar to the assumption made on missing permissions, fluents that are not presented are considered to be false. Conditions on a state are therefore expressed by a set of fluents that should be true or false. The set of possible conditions is referred to as $\mathcal{X} = 2^{\mathcal{F} \cup \neg \mathcal{F}}$.

In Section 4.1 we explained how changes in state are achieved through the definition of two relations: (i) the generation relation, which implements counts-as by specifying how the occurrence of one (exogenous or normative) event generates another (normative) event, subject to the empowerment of the actor and the conditions on the state, and (ii) the consequence relation. This specifies the initiation and termination of fluents subject to the performance of some action in a state matching some expression. The fluents to be initiated as a result of an event $\mathcal{E}$ are denoted as $\mathcal{C}^\uparrow(\phi, \mathcal{E})$ while the ones to be terminated are denoted as $\mathcal{C}^\downarrow(\phi, \mathcal{E})$. Formally, the generation relation can be expressed as $\mathcal{G} : \mathcal{X} \times \mathcal{E} \to 2^{\mathcal{E}_{inst}}$. Similarly the consequence relation can be described as follows: $\mathcal{C} : \mathcal{X} \times \mathcal{E} \to 2^{\mathcal{F}} \times 2^{\mathcal{F}}$.

The semantics of our normative framework is defined over a sequence, called a *trace*, of exogenous events $\mathcal{E}_{ex}$. Starting from the initial state, each exogenous event is responsible for a state change, through initiation and termination of fluents. This is achieved by a three-step process:

1. the transitive closure of $\mathcal{G}$ with respect to a given exogenous event determines all the generated (normative) events,

2. to this all violations from events not permitted and obligations not fulfilled are added, giving the set of all events whose consequences determine the new state,

3. the application of $\mathcal{C}$ to this set of events identifies all fluents that are initiated and terminated with respect to the current state so giving the next state.

So for each trace, a sequence of states is obtained constituting the model of the normative framework. Summing up, following Cliffe (2007) we formally define a normative framework as follows:

Definition 10:   Normative Framework (formal)

A *Normative Framework* is a 5-tuple $\mathcal{N} := \langle \mathcal{E}, \mathcal{F}, \mathcal{C}, \mathcal{G}, \Delta \rangle$ consisting of a set of events $\mathcal{E}$, a set of fluents $\mathcal{F}$, a set of causal rules $\mathcal{C}$, as set of generation rules $\mathcal{G}$ and an initial state Δ, where the respective sets have the following properties and relations between them:

1. $\mathcal{F} = \mathcal{W} \cup \mathcal{P} \cup \mathcal{O} \cup \mathcal{D}$

2. $\mathcal{G} : \mathcal{X} \times \mathcal{E} \to 2^{\mathcal{E}_{inst}}$

3. $\mathcal{C} : \mathcal{X} \times \mathcal{E} \to 2^{\mathcal{F}} \times 2^{\mathcal{F}}$ where $C(X, e) = (\mathcal{C}^\uparrow(\phi, e), \mathcal{C}^\downarrow(\phi, e))$ where
   (i) $\mathcal{C}^\uparrow(\phi, e)$ initiates a fluent
   (ii) $\mathcal{C}^\downarrow(\phi, e)$ terminates a fluent

4. $\mathcal{E} = \mathcal{E}_{ex} \cup \mathcal{E}_{inst}$ with $\mathcal{E}_{inst} = \mathcal{E}_{act} \cup \mathcal{E}_{viol}$

5. State Formula: $\mathcal{X} = 2^{\mathcal{F} \cup \neg \mathcal{F}}$

4.5   Summary

In this chapter we presented our view as well as the corresponding formal description of normative frameworks. We use normative frameworks to capture the relevant normative events in the action arena. Summarizing our normative framework represents both constitutive and regulative aspects of norms, as well as incorporating power and permission as well as obligations. It is furthermore able to track the evolution of normative frameworks by representing its consecutive states in form of traces. The normative framework consists of two abstraction layers: the normative layer as well as the real world layer. The general mechanism of our framework works as follows: observed events from the real world layer generate normative events on the basis of the count-as principle, and result in changes of the normative state.

Having presented our normative framework model, the next chapter will deal with information required to apply the model, by designing a normative framework for our WMG case study presented in Chapter 3. In this dissertation we use simulation as means of analysis. That is why, for being applicable in a simulation, our formal description of the normative frameworks needs to be realized as a computational model, which we present in the next chapter.

Chapter 5

Designing and Testing the Rules-in-Use

Section 4.4 presented the formal model of the normative framework underpinning this dissertation. We explained how it can help to address and analyse the effects of norms and the interaction with them by system participants. Although the formal framework has its merits (e.g. formally underpinning the concept of normative frameworks and thereby allowing us to reason about them), for being applicable in a simulation it needs to be realized as a computational model. Looking at the current literature on normative systems, a number of frameworks exist that present themselves as candidates for realizing the computational model. This chapter starts with an overview of the existing approaches for implementing the computational model of normative frameworks. As a result of the number of approaches presented for specifying computational models of normative frameworks in literature, we do not give a complete overview of all existing specification. Instead we have limited ourselves to only presenting a (in our view representative) selection of approaches that are of most interest to the presented normative framework definition and that are still supported and advanced by their authors.

5.1 Approaches for Specifying a Computational Model of Normative Frameworks

5.1.1 The e-Institutions Framework

The e-Institutions framework was first described in Esteva et al. (2000). The framework consists of three components: (i) a formal specification of a normative system (called institution by the authors (Esteva et al., 2001, 2002)), (ii) the ISLANDER tool for editing the specifications (Esteva et al., 2002), and

(iii) AMELI, a run-time environment for executing the system, based on the specifications (Esteva et al., 2004).

In regards to the normative framework defined in Chapter 4 (page 75) the e-Institutions are able to define a large part of it and translate it into a computational model. To be more precise, the framework includes a formal semantics for the notion of a normative system and its components (abstract and concrete norms, empowerment of agents, roles) and defines a formal relation between normative systems and organizational structures. Despite these clear advantages, the implementation of the framework poses a big problem, which renders it unsuitable for the normative systems as envisioned in this dissertation. This problem is shown in Figure 5.1.

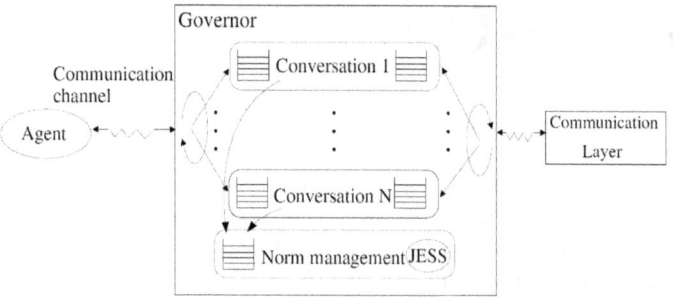

Figure 5.1: The Governor Function in the e-Institutions (Esteva et al., 2004)

In the e-Institution framework all actions are triggered with the help of messages. These messages trigger actions by the individual system participants are observed by so-called governors (that are control entities which are implemented in the middleware of the system). The governors check every message and do not forward it further in case it violates a norm. This way norm deviation is impossible for the system participants. This rigorous regimentation approach contradicts the underlying assumptions of the normative framework presented in Chapter 4, making the e-Institutions framework unsuitable for realizing the computational model.

5.1.2   OperA/OperettA

In contrast to e-Institutions, the OperA framework does account from non-regimentation and legitimizes the concept of autonomy between the goals of the system participants and society requirements (Dignum, 2003). The framework makes use of contracts that predefine the norms and their corresponding sanctions in case of violations. These contracts are implemented using the language LCR (Logic for Contract Representation) which is an extension of the temporal logic CTL* and includes deontic properties and a mechanism for reasoning about actions from the system's participants point of view (Dignum et al., 2003). Looking at the specifications of her framework, Dignum's work matches the idea of the normative frameworks presented in Chapter 4 to a large extent. Thus, Dignum's framework allows for a high degree of generality, both in the specification and enforcement of interactions, as well as in the composition and enactment of interactions. Dignum specifically focuses on roles for the latter purpose, distinguishing her work from the suggestion made in this dissertation. With regard to the implementation of a computational model, despite recent work on OperettA – a graphical environment for the specification and analysis of OperA models (see Aldewereld and Dignum (2010) for example) – so far no sufficient information exists. Although the formalization of the contract uses temporal logics that result in semantics which can be interpreted in a computational model of the running system, no specifications are given on how such a system should be build or how contracts can be enforced in it. That is the reason why Opera/OperettA is not suitable for building our computational model.

5.1.3   The $\mathcal{MOISE}$-Family

The $\mathcal{MOISE}$-family, consisting of $\mathcal{MOISE}$ (Hannoun et al., 2000), $\mathcal{MOISE}^{+}$ (Hübner et al., 2007), its extension $\mathcal{MOISE}^{Inst}$ (Gâteau et al., 2005) as well as $\mathcal{S} - \mathcal{MOISE}^{+}$ and its extension $\mathcal{SYNAI}$[1] (Boissier and Gâteau, 2007), is another set of frameworks that represent normative systems and provides a computational model for them. $\mathcal{MOISE}^{+}$ has the advantage that is not only has detailed specifications on normative concepts, but it has been linked to the simulation tool Jason (Hübner et al., 2009) and thus offers an existing well supported and tested computational implementation. $\mathcal{MOISE}^{Inst}$ is the component of the family that concentrates on norms. It offers definitions and implementations for permissions, obligations and prohibitions and deals with the questions of power and permission by assigning the actors in a system to roles and gives power/permission for certain actions to specific roles only. Despite this difference with respect to the focus of power and permission, by defining a large number of roles (possibly for every single agent) one would be able to use $\mathcal{MOISE}^{Inst}$ for the specifications given in Section 4.4 (page 84). Nevertheless, $\mathcal{MOISE}$ offers one problem similar to e-Institutions which renders it unsuitable for our purposes. It employs generic supervisor agents, aiming at controlling and enforcing the rights and duties of autonomous system entities operating under the normative framework. Whereas supervisor agents are dedicated to the control of the system at the normative and punishing level, the system entities implement the functionalities of the application level. As a consequence, the system entities can violate norms, but $\mathcal{MOISE}$ uses a rather harsh enforcement mechanisms that is able to detect and punish all events in the system. This kind of total observation is unsuitable for our model and does not conform to the ideas presented in Chapter 4.

---

[1] For reasons of simplicity we refer to $\mathcal{MOISE}$ when talking about the complete family.

5.1.4   Artikis et al.

One approach that focuses on the description of a computational model for normative frameworks was presented by Artikis (2003) and Artikis et al. (2007). The authors take a formal perspective based on the works of Sergot (2001) and Jones and Sergot (1996). Using event calculus (Kowalski and Sergot, 1986) in combination with the action language $\mathcal{C}+$ (Giunchiglia et al., 2004), Artikis gives examples on how the concepts of normative power, obligation, permissions and roles can be formalized as a computational model and how this computational model can be executed and analysed. Artikis et al. also consider observable real world events as the starting point of their framework. However, despite the notion of normative power, they do not include conventional generation in their framework. Instead they assume that every normative actions and observed event have a 1:1-mapping, i.e. that every normative action corresponds to exactly one real-world action which is constrained through normative power. This lack of the possibility to express conventional generation, rules out using Artikis et al. as base for the computational model of our normative framework.

5.2   Using Answer Set Programming for Design-Time Reasoning in the Action Arena

Having given a brief review of the most common existing frameworks and approaches that can be used to implement a computation model of normative frameworks and explained their weaknesses with regard to our definition given in Section 4, in this section we present the approach that is suitable for our purposes. It is based on Answer Set Programming (ASP). ASP is a *declarative* programming paradigm with a semantics known as the *answer set semantics* (Baral, 2003). The semantics is a model based semantics for normal logic programs. The declarative nature of ASP allows a programmer to only specify 'what' needs to be achieved instead of having to specify 'how' something should be achieve. Instead in ASP, the solution of the program corresponds to the 'how'-question. ASP therefore lends itself naturally to (normative) reasoning applications. *AnsProlog* is a logical language that forms the basis of ASP if being interpreted using the answer set semantics (Gelfond and Lifschitz, 1988). Cliffe et al. (2006) show that the formal model of a normative framework can be translated to an *AnsProlog* program such that the solution(s) of the program (the so-called answer sets of the program) correspond to the traces of the framework.

Based on the works by Cliffe et al. (see Cliffe et al. (2007a); Cliffe (2007); Cliffe et al. (2009)) this dissertation opts for ASP as the tool for representing normative frameworks in the form of a computational model.

In the next section we will therefore present the foundations of *AnsProlog* – the logic language underlying ASP – and describe how one can determine the answer sets from an *AnsProlog* program. Afterwards we outline how to specify normative frameworks using *AnsProlog*.

Cliffe et al. (2007b) provides a layer of abstraction on top of *AnsProlog* by introducing the domain-specific action language Inst*AL*. The goal of this additional abstraction layer was to make *AnsProlog* programs accessible to users without a background in logic programming. As this dissertation is oriented at such an audience, we use the action language Inst*AL* for the presentation of the normative specifications. We briefly describe Inst*AL* and its link to *AnsProlog*. Afterwards we present the formalization of the WMGs normative framework in Inst*AL*. The chapter ends with an analysis and a discussion of the results of this normative WMG model.

5.2.1 *AnsProlog*

AnsProlog is the short form of "*Programming in logic with Answer sets*" (Baral, 2003, p. 3). It is a declarative programming language mainly used for knowledge representation. It allows the programmer to describe a problem and the requirements that must be fulfilled by the solutions to this problem in an intuitive way, rather than the algorithm to find the solutions to the problem. The so-called *answer sets* of an *AnsProlog* program are the corresponding solutions of the problem. They are defined through (a variant or extension of) the stable model semantics (Gelfond and Lifschitz, 1988). The answer sets are computed using answer set solvers such as CLASP (Gebser et al., 2007) or SMODELS (Niemelä and Simons, 1997). *AnsProlog* is non-monotonic, i.e. it allows for conclusions to be retracted in the light of new knowledge. In contrast to standard monotonic logics this means that a set of conclusions drawn on the basis of a given knowledge base, given as a set of premises, does not necessarily increase if information is added to the knowledge base, but may actually shrink (Stanford Encyclopedia of Philosophy, 2010). In contrast to related techniques like the event calculus (Kowalski and Sadri, 1997) and C$^+$ (Giunchiglia et al., 2004), *AnsProlog* has the advantage to allow for the specification of the problem as well as the query for the answer set as an executable program. It therefore eliminates the gap between specification and verification language. In addition, as both the problem and the query are specified in the same language, a more straightforward verification and validation is possible.

Let us now look at the semantics and the syntax of *AnsProlog*. The basic components of the *AnsProlog* language are atoms (i.e. elements that can be assigned a truth value, i.e. either *true* or *false*). Atoms can be negated. A literal is an atom or its negation. We will only use one type of negation, namely negation-as-failure (denoted as *not*). This is different from classical negation ($\neg a$), where falsity needs to be proved[2]. Negation-as-failure indicates that something is considered false when it cannot proven to be true. Therefore *not a* (the negation-as-failure) can be different to the logical negation $\neg a$.

---

[2]Classical negation can be simulated by *AnsProlog* by the introduction of the atom *not_a* and a constraint that *a* and *not_a* cannot be true at the same point of time.

> ### Definition 11: *AnsProlog* program
>
> An *AnsProlog* program is a pair $\{\sigma, \Pi\}$, with σ being the *signature* and Π a finite set of statements (called rules) about σ (Gelfond, 2008). In this dissertation we will denote programs of *AnsProlog* by their second element Π. The corresponding signature will be denoted by $\sigma(\Pi)$. If $\sigma(\Pi)$ is not given explicitly we assume that it consists of symbols occurring in the program.
>
> Each of these rules in an *AnsProlog* program is comprised of a conjunct of literals (the body) giving the conditions for the rules to be applicable and a single literal (the head) giving the consequences of the rule. For example:
>
> $$a : - \ not \ b.$$
> $$b : - \ not \ a.$$
> $$c : - \ b.$$
> $$c : - \ a.$$
> $$d : - \ c, not \ b.$$

The head of an *AnsProlog* program is typically denoted $H(r)$ as a head and a body $B(r)$. They have the following form: $H : - \perp$, B, *not* C with H, $\perp$, B and C being sets of atoms and $\perp$ always having the truth value *false*. In a rule, the head can also be empty. This is called a *constraint*. For the *AnsProlog* program to be true, the constraints of this program need to be not true. A rule with an empty body is called a *fact*. Facts are often written as $H(r)$. for simplification purposes. The truth value of a fact must be true. The body can be divided into two parts, namely the set of all (not negated) atoms – denoted $B^{+}(r)$ – and the set of atoms that appear negated, denoted $B^{-}(r)$. Recalling the interpretation of negation-as-failure, a rule in an *AnsProlog* program can be read the following way: "if we know $B^{+}(r)$ and we do not know $B^{-}(r)$, then we can assume the head of the rule."

In *AnsProlog* it is possible to refer to a collections of rules that follow the same pattern, and also to abbreviate collections of atoms within the same rule. This is done using variables, which are indicated by capitalization. When an *AnsProlog* program is executed, the variables are replaced by all – according to the program – possible constants that suit the variable construct. This replacement of variables with constants is called *grounding*.

> ### Definition 12: Ground
>
> Terms, literals, and rules of program Π with signature σ are called *ground* if they contain no variables and no symbols for arithmetic functions. A program is called ground if all its rules are ground (Gelfond, 2008). It is denote as *ground*(Π).

For instance the following *AnsProlog* program

$$\Pi = \left\{ \begin{array}{l} p(a). \\ p(b). \\ q(X) : - p(X). \end{array} \right.$$

has the same meaning as

$$\Pi = \left\{ \begin{array}{l} p(a). \\ p(b). \\ q(a) : - p(a). \\ q(b) : - p(b). \end{array} \right.$$

As noted before, the goal of an *AnsProlog* program is to find the answer sets to the program, i.e. to answer 'how' the state for which the program has given specification can be achieved. If an *AnsProlog* program Π does not contain any negation-as-failure statements, its answer set is the minimal set of rules S, which satisfies the set of rules in Π in a consistent fashion. To give an example for this, let us have a look at the following *AnsProlog* program:

$$\Pi = \left\{ \begin{array}{l} a. \\ b. \\ c : - d, a. \\ d : - a. \\ e : - f. \end{array} \right.$$

Deductive closure is a widely accepted concept in the domain of logic of knowledge[3] It states that knowledge is preserved over known entailment. This mean that if S knows that p, and knows that p entails q, then S knows that q. Using deductive closure, one can determine the answer sets of an *AnsProlog* program.

Definition 13: Deductive Closure for Positive *AnsProlog* Programs

The general deductive closure to arrive at an answer set for a positive *AnsProlog* program (an *AnsProlog* program is said to be positive if Π does not contain any negation-as-failure statements) is defined as: (i) take all rules of which the bodies are true (ii) assume the heads to be true (iii) continue until you reach a fixpoint. This fixpoint is the unique answer set, $lm(\Pi)$ of the program.

Using this deductive closure on the example, the following can be concluded: a and b are facts and therefore true. From the third rule it can be concluded that d is also true. If d and a are true the third rule specifies that c is also true. At this point we have reached a fixpoint, with $lm(\Pi) = \{a, b, c, d\}$ being the answer set of the example *AnsProlog* program.

---

[3]For a discussion of the concept and its acceptance see White (1991).

If negation-of-failure is part of the *AnsProlog* program the search for answer sets gets more complex, as negation-as-failure gives no guarantee that something is indeed false and that information derived from it is actually correct.

The following example shall help to illustrate this problem:

$$\Pi = \left\{ \begin{array}{l} a : - \; not \; b. \\ b : - \; not \; a. \end{array} \right.$$

Using the deductive closure for positive *AnsProlog* programs, one reaches a dead end. The answer set to this program is found using the *Gelfond-Lifschitz transformation* or *reduct* (Gelfond and Lifschitz, 1991) on a set of ground literals.

Definition 14:   Gelfond-Lifschitz Transformation

Starting from an interpretation (which is an assumption on what might be an answer set), one proceeds step-wise and tries to verify that the interpretation is correct. This is done by reducing the original program to a simpler version without negation-of-failure. In order to arrive at the reduce program the interpretation is taken as a starting point and all rules that are considered *not x* that are considered false are removed while the remaining rules only retain their positive body elements. If the answer set obtained from that reduced program equals the interpretation the process was started with, the answer set is an answer set to the original *AnsProlog* program.

Looking at the example one arrives at two answer sets (i.e. $lm(\Pi) = \{a\}, \{b\}$) that are both acceptable solutions for the problem modelled.

Having explained *AnsProlog* as well as the determination of their solutions (i.e. answer sets), in the next section we will explain how it can be used to specify the normative frameworks presented in Chapter 4.

5.2.2   Representing Normative Frameworks in *AnsProlog*

In the last section we introduced *AnsProlog* and pointed out that if a normative framework is presented as an *AnsProlog* program then the answer sets to this program correspond to the traces of the normative framework. This statement will be explained in more detail in this section by using the translations presented in Cliffe (2007) and Cliffe et al. (2007a).

In order to achieve the goal of representing our normative framework definition in *AnsProlog* we need to map the formal definition presented in Section 4.4 (i.e. $\mathcal{N} := \langle \mathcal{E}, \mathcal{F}, \mathcal{C}, \mathcal{G}, \Delta \rangle$) and all its components into an *AnsProlog* program such that the program described by this translation models the semantics of the normative framework to be represented. This mapping consists of three parts that together generate the *AnsProlog* program $\Pi_{\mathcal{N}}$. These parts are:

1. a base component Π^{base} which is independent of the framework being modelled,

2. a time component Π^n, and

3. a framework specific component $\Pi^*_{\mathcal{N}}$.

Π^{base} consists of general rules that are true for any normative framework. These rules deal with the occurrence of observed events, inertia of the fluents (i.e. the transition of the states) and the generation of violation events resulting from not permitted actions and unsatisfied obligations. The time model Π^n defines the predicates for time and is responsible for generating a single observed event at every time instance. $\Pi^*_{\mathcal{N}}$ specifies all the components that are specific to the normative framework being modelled.

In general the mapping of the normative framework to *AnsProlog* uses the following atoms: `ifluent(P)` to identify fluents, `evtype(E,T)` to describe the type of an event, `event(E)` to denote the events, `instant(I)` for time instances, `final(I)` for the last instance, `next(I1,I2)` to establish time ordering, `occurred(E,I)` to indicate that the event happened at time I, `observed(E,I)` that the event was observed at that time, `holdsat(P,I)` to state that the normative fluent P holds at I, and finally `initiated(P,I)` and `terminated(P,I)` for fluents that are initiated and terminated at I (Cliffe et al., 2007a). Starting with Π^n and then continuing with Π^{base} and $\Pi^*_{\mathcal{N}}$, we now present these atoms in more detail by explaining their role and usage.

As described in Section 4.3.2, the general notion of time in a normative framework is based on the assumption that time consists of a number of ordered time instants, with no assumption about the duration of time between two of these instants being made. As a result we need to be able to model the semantics of a normative framework over a sequence of time instants of length n, such that $t_i : 0 \leq i \leq n$. Based on our definition of time, time instants represent the *state* of a normative framework at a given time, in such a way that each time instant corresponds to one possible state. Events are considered to occur between the time instants. For reasons of simplicity we do not explicitly specify any interval over which an event occurs, but refer to the time instance at the start of the interval when the event takes place. Figure 5.2 shows the *AnsProlog* definition of time.

$$
\begin{array}{rll}
0 < k < n & : & \texttt{instant}(i_k). \quad\quad\quad (1)\\
0 < k < n-1 & : & \texttt{next}(i_k,\ i_{k+1}). \quad\quad (2)\\
& & \texttt{final}(i_n). \quad\quad\quad\quad\ (3)
\end{array}
$$

Figure 5.2: The Time Translation Rules

Line (1) provides the definition for all time instants that need to be available in the normative framework. Line (2) specifies the transition between two time instants. This is required to go from one state to another. As it is impossible to have an observable event at the end of the last time instant, we also need to define a fact that represents the final state. This is done in line (3).

Having presented the interpretation of time, we now consider Π^{base}, which deals with the occurrence of observed events, inertia of the fluents, and the generation of violation events resulting from not permitted actions and unsatisfied obligations. The translation rules for Π^{base} are shown in Figure 5.3.

$$\text{occurred}(E, I) \quad :- \quad \text{observed}(E, I). \tag{4}$$

$$\text{holdsat}(P, I2) \quad :- \quad \text{holdsat}(P, I1), \tag{5}$$
$$\text{not terminated}(P, I1),$$
$$\text{next}(I1, I2), \text{instant}(I1, I2),$$
$$\text{ifluent}(P).$$

$$\text{holdsat}(P, I2) \quad :- \quad \text{initiated}(P, I1), \tag{6}$$
$$\text{next}(I1, I2), \text{instant}(I1, I2),$$
$$\text{ifluent}(P).$$

$$\text{occured}(\text{viol}(E), I) \quad :- \quad \text{occurred}(E, I), \tag{7}$$
$$\text{not holdsat}(\text{perm}(E), I),$$
$$\text{event}(E), \text{event}(\text{viol}(E)),$$
$$\text{instant}(I).$$

$$\text{occurred}(V, I) \quad :- \quad \text{holdsat}(\text{obl}(E, D, V), I), \tag{8}$$
$$\text{occurred}(D, I), \text{event}(E, D, V),$$
$$\text{instant}(I).$$

$$\text{terminated}(\text{obl}(E, D, V), I) \quad :- \quad \text{occurred}(E, I), \tag{9}$$
$$\text{holdsat}(\text{obl}(E, D, V), I),$$
$$\text{event}(E, D, V), \text{instant}(I).$$

$$\text{terminated}(\text{obl}(E, D, V), I) \quad :- \quad \text{occurred}(D, I), \tag{10}$$
$$\text{holdsat}(\text{obl}(E, D, V), I),$$
$$\text{event}(E, D, V), \text{instant}(I).$$

Figure 5.3: The Π^{base} Translation

The first rule (i.e. rule (4)) refers to an observed exogenous event (i.e. $\mathcal{E}_{ex}$). It ensures that each observed exogenous event (observed(E, I)) will be marked as occurred, as all observed events are valid events[4]. The next two rules (i.e. (5) and (6)) deal with standard inertia. Rule (5) specifies that fluents which are valid in the current state (holdsat(P, I1)) and which will not be terminated in the state (not terminated(P, I1)) are still valid in the next state holdsat(P, I2). next(I1, I2) and instant(I1, I2) denote the transition of time and its grounding as explained above. Rule (6) focuses on the initiation of rules. It specifies that a fluent that is being initiated in a state (initiated(P, I1)) becomes valid in the next state. Rule (7) deals with $\mathcal{E}_{ex}$ that are missing a permission. It specifies that for any occurrence (occurred(E, I)) of a non-permitted event (not holdsat(perm(E), I)) a violation (event(viol(E))) is generated. Finally, rules (8) – (10) cover the handling of obligations. Rule (8) deals with the case that not fulfilling an obligation holdsat(obl(E, D, V), I) before the deadline event occurred(D, I) results in the generation of a violation

---

[4]Validity in this context only refers to the fact that any $\mathcal{E}_{ex}$ can take place as a result of the autonomy of the actors in the normative framework. It does not specify anything about the permission involved, or whether this event might otherwise trigger a violation event.

event $\texttt{occurred(V, I)}^5$. An occurrence of an obligation deadline ($\texttt{occurred(D, I)}$) also leads to the termination of the obligation ($\texttt{terminated(obl(E, D, V), I)}$) (rule 10). Besides the expiration of the deadline, the fulfillment of an obligation can also result in its termination. The rule for this positive case of obligation-termination is encoded in rule (9). It specifies that the obligation is terminated ($\texttt{terminated(obl(E, D, V), I)}$) if the obliged event occurs ($\texttt{occurred(E, I)}$).

$$
\begin{array}{rll}
\{\texttt{observed(E, I)}\} & :- & \texttt{evtype(E, obs), event(E),} \hfill (11) \\
& & \texttt{instant(I), not final(I).} \\
\texttt{ev(E, I)} & :- & \texttt{observed(E, I), event(E),} \hfill (12) \\
& & \texttt{instant(I).} \\
& :- & \texttt{not ev(I), instant(I),} \hfill (13) \\
& & \texttt{not final(I)} \\
& :- & \texttt{observed(E1, I),} \hfill (14) \\
& & \texttt{observed(E2, I), E1! = E2,} \\
& & \texttt{instant(I), event(E1),} \\
& & \texttt{event(E2).}
\end{array}
$$

Figure 5.4: The *AnsProlog* Rules for ensuring Observable Traces

In Chapter 4 we described the semantics of the normative framework as a set of states that change as a result of a sequence of exogenous (real world) events ($\mathcal{E}_{ex}$) taking place and triggering normative events ($\mathcal{E}_{inst}$). When representing a normative framework in *AnsProlog* each of the sequences of $\mathcal{E}_{ex}$ corresponds to a single possible model for the interpretation of the normative framework. As a result, when specifying the computational model of a normative system, we are only interested in models that include trace information, i.e. that have had any $\mathcal{E}_{ex}$ taking place. For this reason we need to add rules to Π^{base} that constrain the answer set to those containing observable traces. Therefore we add the rules shown in Figure 5.4 to Π^{base}. Looking at these rules in more detail, rule (11) is used for the creation of $\texttt{observed(E, I)}$ atoms. It does so for each combination of event $\texttt{event(E)}$ of type $\texttt{evtype(E, obs)}$ (i.e. observable) that is not final $\texttt{not final(I)}$ at time instant $\texttt{instant(I)}$. $\{\texttt{observed(E, I)}\}$ is a choice rule. A choice rule states that if every positive atom in the body of the rule is applied, and every negated atom is not applicable then any subset of the head of the rule is applicable and can be chosen from. For each choice, an event $\texttt{ev(E, I)}$ is generated using rule (12). This event is then used by the restriction rule (13), which specifies that only answer sets that have $\mathcal{E}_{ex}$ at each time step are valid. Finally, rule (14) ensures that only answer sets that have one $\mathcal{E}_{ex}$ per time instant are valid.

So far, we have discussed the translation rules for Π^n and Π^{base}. The last missing component in the translation is $\Pi^*_{\mathcal{N}}$. This component will now be presented.

To start, we look at the condition statements that are modelled with the

---

[5] The atoms of rules (8) − (10) that were not described are required for grounding purposes.

help of the auxiliary function EX. For a given expression $\phi \in \mathcal{X}$ that is used at time instant t_i, Cliffe et al. (2009) suggest to use the term $EX(\phi, t_i)$ to denote the translation of ϕ into a set of *AnsProlog* atoms as shown in Figure 5.5:

$$EX(x_1 \wedge x_2 \wedge \ldots x_n, t_i) \overset{def}{\equiv} \begin{aligned} & EX(x_1, t_i), EX(x_2, t_i), \\ & \ldots EX(x_n, t_i) \end{aligned} \quad (15)$$

$$EX(\neg f, t_i) \overset{def}{\equiv} not\ EX(f, t_i) \quad (16)$$

$$EX(f, t_i) \overset{def}{\equiv} \mathtt{holdsat(f, t_i)} \quad (17)$$

Figure 5.5: The Condition Statement Translation

Using this translation the condition $\neg a, b, c$ at time instant t_i would for example be translated into the following sequence of extended literals in *Ans-Prolog*:

$$\mathtt{not\ holdsat(a, t_i),\ holdsat(b, t_i),\ holdsat(c, t_i).}$$

Keeping these translations in mind, we can look at the final translation components of $\Pi_{\mathcal{N}}^*$ shown in Figure 5.6.

$$
\begin{aligned}
p \in \mathcal{F} &\Leftrightarrow \mathtt{ifluent(p).} & (18) \\
e \in \mathcal{E} &\Leftrightarrow \mathtt{event(e).} & (19) \\
e \in \mathcal{E}_{ex} &\Leftrightarrow \mathtt{evtype(e, obs).} & (20) \\
e \in \mathcal{E}_{act} &\Leftrightarrow \mathtt{evtype(e, act).} & (21) \\
e \in \mathcal{E}_{viol} &\Leftrightarrow \mathtt{evtype(e, viol).} & (22) \\
\mathcal{C}^{\uparrow}(\phi, e) = P &\Leftrightarrow \forall p \in P \cdot \mathtt{initiated(p, I)\ :\text{-}} & (23) \\
& \qquad \mathtt{occurred(e, I), } EX(X, I). \\
\mathcal{C}^{\downarrow}(\phi, e) = P &\Leftrightarrow \forall p \in P \cdot \mathtt{terminated(p, I)\ :\text{-}} & (24) \\
& \qquad \mathtt{occurred(e, I), } EX(X, I). \\
\mathcal{G}(\phi, e) = E &\Leftrightarrow \forall g \in E \cdot \mathtt{occurred(g, I)\ :\text{-}} & (25) \\
& \qquad \mathtt{occurred(e, I),} \\
& \qquad \mathtt{holdsat(pow(e), I),} \\
& \qquad EX(X, I). \\
p \in \Delta &\Leftrightarrow \mathtt{holdsat(p, i_0).} & (26)
\end{aligned}
$$

Figure 5.6: The $\Pi_{\mathcal{N}}^*$ Translation

The first two rules in Figure 5.6, i.e. rule (18) and (19) deal with fluents and events respectively. Fluents are encoded as facts $\mathtt{ifluent(p)}$. Each event $e \in \mathcal{E}$ is translated with the help of a corresponding term $\mathtt{e}$. This term is applied to a number of facts that record the type of event $\mathtt{evtype(e, X)}$ $\mathtt{e}$ belongs to, such that $\mathtt{X} \in \mathtt{obs, act, viol}$ (rules (20) – (22)). Rules (23) and (24) are concerned with the consequence relation. They specify that whenever a fluent needs to be initiated or terminated a respective rule is generated. This rule has the event responsible for initiating/terminating the fluent `occurred(e, I)` together with conditional rules $EX(X, I)$ in the body and the respective initiation/termination atom in the head (`initiated(p, I)`/`terminated(p, I)`). Rule (25) deals with the event generation. Similar to before, in the body of the rule the event responsible for generation of the event `occurred(e, I)` together with conditional rules $EX(X, I)$ is used. In addition the power (`holdsat(pow(e), I)`) to execute the trigger event `e` is added. The head of the rule consists of the generated event (`occurred(g, I)`). Finally, the last rule (i.e. rule (26)) refers to the initial state Δ of the normative framework. It specifies that each fluent $p \in \Delta$ is translated into a fact `holdsat(p, i`$_0$`)`[6].

In this section we presented the translations for $\Pi^*_\mathcal{N}$, Π^{base} and Π^n, the three components that together compose $\Pi_\mathcal{N}$ (i.e. the corresponding *AnsProlog* program to $\mathcal{N} := \langle \mathcal{E}, \mathcal{F}, \mathcal{C}, \mathcal{G}, \Delta \rangle$). Using this translation, for any normative framework in accordance with the definition given in Section 4.4 we are able to generate a corresponding *AnsProlog* program that is able to provide ordered traces of length n in form of $lm(\Pi_\mathcal{N})$. Having such a translation, if required, we can add additional rules to the program that allow us to restrict the answer sets to a set which fulfills certain properties.

5.3   The Action Language Inst*AL*

In the last section we outlined how normative frameworks can be specified as computational model using *AnsProlog*. In order to make *AnsProlog* programs accessible to users with little or no experience in the *AnsProlog* formalization and syntax, Cliffe et al. (2006) provided a layer of abstraction on top of *Ans-Prolog* by introducing the domain-specific action language Inst*AL*. Inst*AL* uses semi-natural language to describe the various components of the normative framework.

An Inst*AL* programme generally consists of two components: the normative specification which is specified in the so-called ial-file as well as the domain information specified in the domain file that is used for grounding purposes. Figure 5.7 gives an overview of the components.

The domain file consists of general statements of the form `type` : `type – value`, such as `Handset : alice bob` (Figure 5.9, line 1 on page 108), which specifies that the `Handset` variable of the underlying *AnsProlog* program can be grounded with the values `alice` and `bob`. A domain file should always be complete, i.e. contain information for each type value that is to be grounded, but should never have more information than required either.

The normative specifications in the ial-file consist of four parts: template declaration, template rules, template initiation and concretisation. The template declaration part contains the general declaration of the normative framework, including the event and fluents as well as the name of the normative framework. In terms of the template declaration part, this means that events are defined

---

[6]It should be noted that $\Pi^*_\mathcal{N}$ is grounded except for the time instants. The grounding variables for these instants are provided by Π^n.

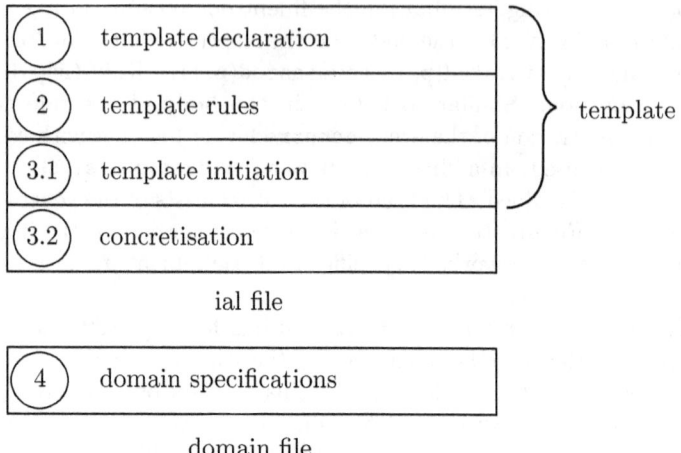

ial file

domain file

Figure 5.7: Normative Specification Components in Inst*AL*

by `typeOfEvent` event `nameOfEvent`; with `typeOfEvent` being one of `exogenous`, `create`, `violation` or `inst`, while fluents are defined by `fluent nameofFluent` `(ParameterType, ...);`.

The template rules part contains the actual rules of what follows from which actions, i.e. the description of how events cause the state of the normative framework to evolve. The generation of normative events from exogenous events is specified using the `generates` statement—in effect, adding a rule to the $\mathcal{G}$ relation – while `initiates` and `terminates` define the two parts ($\mathcal{C}^\uparrow$ and $\mathcal{C}^\downarrow$) of the consequence relation. Conditions on the state are expressed using `if`.

The two next parts of the normative specifications (i.e. template initiation and concretisation) contain information for the initial state of the normative framework with the help of an `initially` statement. This `initially` statement serves to specify the set of fluents that characterize the initial state after the normative framework is created. `initially` statements that are the same for each instance of the normative framework are part of the template initiation, whereas `initially` statements that might updated at run-time of the normative framework are part of the concretisation.

To explain in more detail how this looks in Inst*AL*, this dissertation now explains the language features of Inst*AL* by looking at the Inst*AL* fragments that will be used later when modelling the case study:

- `institution` *name* declares the name of the normative framework, such as `institution grid` in line 1 of Figure 5.10, page 109.

- `type` *identifier* declares a type, such as `type Handset` (line 3 in Figure 5.10). Type declarations establish a disjoint set of monomorphic types[7]. The types are specified in a domain file providing the acceptable values

---

[7]Monomorphic types are types having parts that exist in only one form.

for each declared type. An example of a domain file is given in Figure 5.9 on page 108. The types in the domain file are specified type-identifier: type-value(s) with the values being separated by spaces. An example of such a specification of grounding values in the domain file is Handset: alice bob (Figure 5.9, line 1). Inst*AL* will then substitute those values (e.g. alice and bob) whenever the respective type (e.g. Handset) is specified for the events and fluents.

- create event *event-name*(*type*⁺) declares a new creation event, such as create event creategrid in line 14 in Figure 5.10.

- dest event *event-name*(*type*⁺) declares a new destruction event. In our model, we do not use a destruction event. If being used an example of such an event would be dest event destroygrid.

- exogenous event *event-name*(*type*⁺) declares a new physical world event and the types of its parameters, such as exogenous event download(Handset, Chunk, Channel) (line 9 in Figure 5.10).

- inst event *event-name*(*type*⁺) declares a new normative event and the types of its parameters, such as inst event intDownload(Handset, Chunk, Channel) in line 17 in Figure 5.10.

- violation event *event-name*(*type*⁺) declares a new violation event, such as violation event misuse(Handset) in line 24 in Figure 5.10.

- fluent *fluent-name*(*type*⁺) declares a new normative fact—that is, an object that can be an element of the normative framework state, such as fluent downloadChunk(Handset, Chunk) in line 27 in Figure 5.10.

- noninertial fluent *fluent-name*(*type*⁺) declares a fluent that is non-inertial, i.e. is not automatically persistent between states without termination. An example could be noninertial fluent busyHSending(Handset) in line 39 in Figure 5.10.

- *event-name* generates *normative-event*⁺ [*condition*] adds a new pair to the generation relation with domain event (exogenous or normative framework) and rangeinstitutional world event, subject to an optional condition. For example: send (A, X) generates intSend(A) if hasChunk(A, X), ...; (Figure 5.14, line 99), where the condition is the presence of the fluent hasChunk, with the corresponding A and X (these variables are unified) in the normative framework state.

- *event-name* initiates *institutional-fluent*⁺ [*condition*] adds a new pair to the consequence (addition) relation, with domain event (physical or institutional) and range fluent. Thus, intDownload(A, X, C) initiates hasChunk(A, X) in Figure 5.13 line 74 adds the corresponding fluent to the normative framework state.

- *event-name* `terminates` *institutional-fluent*$^+$ [*condition*] adds a new pair to the consequence (deletion) relation, with domain event (physical or institutional) and range fluent. Thus, `intDownload(A, X, C)` `terminates` `pow(intDownload(A, X1, C1)` (Figure 5.13 line 80) deletes the corresponding fluent from the normative framework state.

- *non-inertial-fluent-name* `when` [*condition*] adds the conditions in which a non-inertial fluent should be true in a given state, like for example *busyHReceiving(Handset)* `when` *areceive(Handset, Time)* in line 52 (Figure 5.11).

- `perm(`*event*`)` is a special fluent whose presence indicates that the event is permitted, such as `perm(intDownload(A, B, C))`, and is typically the subject of an `initiates` or `terminates` rule (e.g. in line 139, Figure 5.15).

- `pow(`*event*`)` is a special fluent whose presence indicates that the event is empowered, such as `pow(intDownload(A, X1, C1))` (as above), and is typically the subject of an `initiates` or `terminates` rule.

- `obl(`*event, event, event*`)` is a special fluent whose presence indicates the existence of an obligation, such as `obl(send(A, x1), intDeadline, misuse(A))` (Figure 5.15, lines 154–154). This fluent can be terminated by the occurrence of the first specified event – that is the obligation that needs to be satisfied. If not, and the second event (typically a deadline) occurs, then the third event is triggered, which is typically a violation event, leading to the addition of a violation fluent to the normative framwork state.

Inst*AL* can be translated into an equivalent *AnsProlog* program as the framework specific part of the formal model. This process is shown in Figure 5.8.

Using the normative framework definition (i.e. the normative specification) which is specified in Inst*AL* as well as grounding information derived from a domain file as input, a translator generates a set of answer set programs which describe the semantics of the input normative framework. The grounding information in the domain file consists of values for all the variables that are present in the *AnsProlog* program. This includes the variables for the time instants. The translator used in the research of this dissertation is called *instal2asp* and was developed by Nicholas Robert Jones as part of his Bachelor's thesis at the University of Bath (Jones, 2011). These resulting normative framework programs are then grounded along with a trace program which defines which set of traces should be analysed and a query program which describes the desired properties that should be validated by the Inst*AL* reasoning tool[8]. The derived grounded problem description is then used as input for an answer set solver. In this dissertation we use *CLINGO*, developed by the University of Potsdam (Gebser et al., 2008) for both the grounding as well as the answer set

---

[8]The trace program and the query can be empty except for the specification of the time steps the program should reason about. These time steps are specified in the trace program.

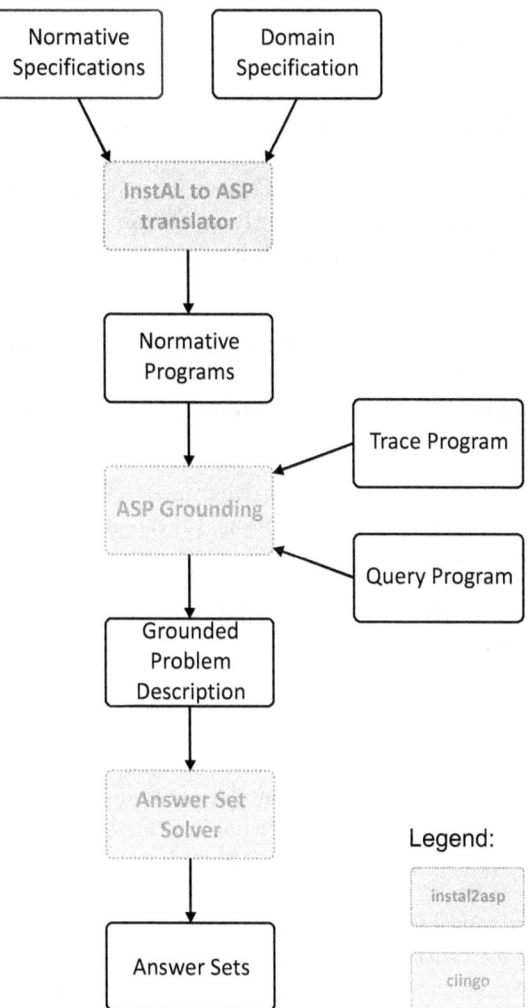

Figure 5.8: Overview of the Inst*AL* Translation Process (Cliffe, 2007)

solving. The result of this process can be zero or more answer sets, of which each corresponds to a possible model of the described normative framework for the given trace and query specified.

5.4 Formalizing the Wireless Mobile Grid Scenario

Now that we have described the WMG scenario in some detail from the technological and stakeholder perspective (Chapter 3, page 53) as well as our approach for modelling normative frameworks (Sections 5.2–5.3) we combine these two threads and present our (computational) normative model of the WMG Scenario.

Based on the three groups of underlying factors distinguished in the IAD framework we distinguish three perspectives to present our normative model of the WMG:

- The technological components including the quantification of battery costs for a given action as well as the technical sending restriction both by the mobile phones and the base-station (in the biophysical characteristics perspective),

- the users that populate the normative frameworks and choose which action to take, based on utility considerations (in the attributes of the community perspective), and

- the actions that users may take, as prescribed by the normative framework (in the rules-in-use perspective).

In this section, we focus on the (normative) actions and hence we present the implementation of the normative framework in InstAL. The purpose of this model is to provide a means for a first design-time analysis of the system and to serve as a reference point for the run-time simulation model – that is, the integration of utility consideration through the autonomous users that participate in the normative framework. Implementing a simulation environment is a complex and time-consuming task. Before starting the process it is best to verify that the protocol is indeed suitable for the task at hand. The normative model provides the protocol designer with a tool to verify, from the theoretical perspective, that the protocol is correct and the methods of enforcing collaboration between the users will indeed benefit all participants in terms of reduced communication costs and that none of the participants can take unfair advantage without risking sanction.

The model is intentionally limited in that it focuses on the *essential* interactions and the communication costs that arise from those interactions. Although a more elaborate model may be desirable for the sake of greater realism, we believe that more details would also distract and unnecessarily complicate the model.

The features of the prototypical scenario are:

- $1 \times$ base-station: B,

- $m \times$ users with handsets: $\mathcal{A} = \{a_1, \ldots, a_m\}$,

- $1 \times$ digital good: G divided into,

- $n \times$ chunks: $\{g_1, \ldots, g_n\}$.

We further assume that $n|m$, which is to say the number of chunks, is a multiple of the number of users.

In the design-time model, we focus on the actual interaction between participants, neglecting the preceding phase of searching for cooperation partners. We identify three phases to the interactions for handset to base-station and handset to handset[9]. These phases are the *negotiation* phase between the handsets on who is to download which chunk, the *download* of the chunks from the base-station and the *sharing* phase afterward where the handsets can decide on whether to cooperate and send chunks to its collaborators or not. Expressing this in more formal terms, the three phases are as follows:

- Negotiation: assign g_i to a_j s.t. $f : G \to \mathcal{A}$ and

$$f^{-1} : A \to G \text{ where } f^{-1}(a_i) = \{g_j, \ldots\} \text{ and } f(g_j) = a_i$$

- Downloading: handset a_i receives chunks $f^{-1}(a_i)$ from base-station B

- Sharing: handset a_i sends chunks $f^{-1}(a_i)$ to and receives chunks $G \backslash f^{-1}(a_i)$ from other handsets.

While these three phases are independent, they need not be sequential. Of course, the negotiation phase has to precede both the downloading and sending/receiving phases. Sharing a chunk is possible as soon as its downloading has commenced; thus the two can be interleaved. In the following paragraphs we discuss each phase in more detail and how each is encoded in Inst*AL*.

Each Inst*AL* specification starts with the name given to the normative framework, followed by the different types of variables it uses. This together composes the template declaration component of the normative specifications. The values of the variables are supplied by the domain specifications in the separate domain file.

Our scenario, which is kept simple for illustration purposes, consists of two handsets, alice and bob, four chunks, x1 to x4, one base-station with two channels, c1 and c2. Furthermore, we have a type `time` to represent transmission times. The complete domain file of the design time model is displayed in figure 5.9.

After the types are specified the Inst*AL* specification continues with the description of the fluents and events it recognises using the defined types. The full definition of the template declaration can be seen in Figure 5.10. The meaning of the various elements of the specification should become clear as we discuss the different phases.

---

[9] As far as the model is concerned, users and their mobile phones are a single entity and

```
1    Handset: alice bob
2    Chunk: x1 x2 x3 x4
3    Channel: c1 c2
4    Time: 1 2 3 4 5 6 7 8 9 10 11 12 13 14 15 16 17 18 19 20 21
```

Figure 5.9: The Domain File of the Design Time WMG Model

5.4.1 Negotiation Phase

Given this dissertation's focus on enforcement, we are not particularly concerned with the technicalities of the negotiation phase. We assumed that any off-the-shelf protocol could be employed as long as the post-condition is satisfied: that each chunk is assigned to exactly one handset and that each handset is assigned the same number of chunks. These conditions can readily be relaxed at the cost of a lengthier specification[10].

Once the download tasks are allocated to the handsets based on the above-mentioned conditions, the allocation needs to be translated for the normative model. This is done by modelling the allocation as part of the initial state of the model (see Figure 5.15, lines 142–143) via the `downloadChunk` fluents indicating which handsets are tasked with downloading these specific chunks from the base-station. Together with their chunk assignment the handsets receive the necessary permission download (lines 140–141).

5.4.2 Downloading Phase

In the download phase each handset downloads its assigned chunks from the base-station. This process should result in each handset holding $n|m$ distinct chunks. Because the base-station uses several different frequencies (frequency division multiplexing), many users may download chunks simultaneously. We refer to a frequency division in the model as a *channel*. Of course, there is a physical limit to the number of frequency divisions and hence the number of simultaneous user connections but we view this as a feature for future incorporation, rather than an essential property to model. So we only work with two channels (c1 and c2) in the example.

The full specification of this phase can be seen in Figure 5.13. Each handset can only obtain one chunk at a time from the base station, while each channel can only be used to download a single chunk. This is modelled by the non-inertial fluents `busyBReceiving` and `busyChannel` (lines 43–46) which are implied on the basis of the handset downloading and the base-station transmitting (lines 54–55).

The first Inst*AL* rule (lines 70–72) indicates that a request to download a chunk is granted (`intDownload`) whenever the handset does not have the requested chunk already, there is an available channel and the handset is not currently receiving from the base-station. When a chunk is downloaded, the handset and

---

there is no need to make a distinction between the two. For convenience in the Inst*AL* model we use the term "handset" to refer to this user/mobile-phone entity.

[10]In Section 5.5 we will discuss the issue of less restrictive approaches with regard to the negotiation as well as resilience issues.

```
 1    institution grid;
 2
 3    type Handset;
 4    type Chunk;
 5    type Channel;
 6    type Time;
 7
 8    %% exogenous events %%
 9    exogenous event download(Handset,Chunk,Channel);
10    exogenous event send(Handset,Chunk);
11    exogenous event clock;
12
13    %% creation event %%
14    create event creategrid;
15
16    %% normative events %%
17    inst event intDownload(Handset,Chunk,Channel);
18    inst event intSend(Handset,Chunk);
19    inst event intReceive(Handset,Chunk);
20    inst event transition;
21    inst event intDeadline;
22
23    %% violation event %%
24    violation event misuse(Handset);
25
26    %% fluents %%
27    fluent downloadChunk(Handset,Chunk);
28    fluent hasChunk(Handset,Chunk);
29    fluent areceive(Handset,Time); % receiving from handset
30    fluent asend(Handset,Time); % sending by handset
31    fluent creceive(Handset,Time); % receiving from basestation
32    fluent transmit(Channel,Time); % sending by basestation
33
34    %% fluents for time-related aspects %%
35    fluent previous(Time,Time);
36    fluent countdown(Time);
37
38    %% non-inertial fluents %%
39    fluent busyHSending(Handset);
40    % indicates that the handset is sending to a peer
41    fluent busyHReceiving(Handset);
42    % indicates that the handset is receiving from a peer
43    fluent busyBReceiving(Handset);
44    % indicates that the handset is receiving from the base
45    fluent busyChannel(Channel);
46    % indicates that the channel is busy
```

Figure 5.10: Declaration of Types and Events in the Design Time WMG Model

```
48    %-----------------------------------------------------------------------
49    % noninertial rules
50    %-----------------------------------------------------------------------
51
52    busyHSending(Handset) when asend(Handset,Time);
53    busyHReceiving(Handset) when areceive(Handset,Time);
54    busyBReceiving(Handset) when creceive(Handset,Time);
55    busyChannel(Channel) when transmit(Channel,Time);
```

Figure 5.11: Specification of the Noninertial Rules in the Design Time WMG Model

```
57  %-------------------------------------------------------------------
58  % countdown rules
59  %-------------------------------------------------------------------
60
61  transition initiates countdown(T2) if countdown(T1), previous(T1,T2);
62  transition generates intDeadline if countdown(1);
63  misuse(A) terminates pow(intReceive(A,X));
64  misuse(A) terminates perm(intReceive(A,X));
```

Figure 5.12: Generation and Consequence Relations for Deadline-Countdown in the Design Time WMG Model

```
66  %-------------------------------------------------------------------
67  % rules for downloading
68  %-------------------------------------------------------------------
69
70  % handset A requests a block from the base station on channel C
71  download(A,X,C) generates intDownload(A,X,C) if not hasChunk(A,X),
72          not busyChannel(C), not busyBReceiving(A);
73
74  intDownload(A,X,C) initiates hasChunk(A,X);
75  intDownload(A,X,C) initiates perm(send(A,X));
76  intDownload(A,X,C) initiates creceive(A,4), transmit(C,4);
77  % handset and channel are busy for 4 time-units when a chunk
78  % is downloaded from the base station
79
80  intDownload(A,X,C) terminates pow(intDownload(A,X1,C1));
81  intDownload(A,X,C) terminates pow(intDownload(B,X1,C));
82  intDownload(A,X,C) terminates downloadChunk(A,X);
83  intDownload(A,X,C) terminates perm(download(A,X,C1));
84
85  download(A,X,C) generates transition;
86  clock generates transition;
87
88  transition initiates transmit(C,T2) if transmit(C,T1), previous(T1,T2);
89  transition initiates creceive(A,T2) if creceive(A,T1), previous(T1,T2);
90  transition initiates pow(intDownload(A,X,C)) if creceive(A,1);
91
92  transition terminates creceive(A,Time);
93  transition terminates transmit(C,Time);
```

Figure 5.13: Generation and Consequence Relations for Downloading in the Design Time WMG Model

```
 95   %------------------------------------------------------------------
 96   % rules for sharing
 97   %------------------------------------------------------------------
 98
 99   send(A,X) generates intSend(A,X) if hasChunk(A,X), not busyHSending(A),
100          not busyHReceiving(A);
101
102   send(A,X) generates intReceive(B,X) if not hasChunk(B,X),
103          not busyHSending(B), not busyHReceiving(B), hasChunk(A,X),
104          not busyHSending(A), not busyHReceiving(A);
105
106   intSend(A,X) initiates asend(A,2);
107   intSend(A,X) terminates pow(intSend(A,X));
108   intSend(A,X) terminates perm(intSend(A,X));
109
110   intReceive(A,X) initiates hasChunk(A,X);
111   intReceive(A,X) initiates areceive(A,2);
112   intReceive(A,X) terminates pow(intReceive(A,X));
113   intReceive(A,X) terminates perm(intReceive(A,X));
114
115   send(A,X) generates transition;
116   clock generates transition;
117
118   transition initiates asend(A,T2) if asend(A,T1), previous(T1,T2);
119   transition initiates areceive(A,T2) if areceive(A,T1), previous(T1,T2);
120   transition initiates pow(intReceive(A,X)) if areceive(A,1);
121   transition initiates pow(intSend(A,X)) if asend(A,1);
122
123   transition terminates asend(A,Time);
124   transition terminates areceive(A,Time);
```

Figure 5.14: Generation and Consequence Relations for Sharing in the Design Time WMG Model

```
126   %------------------------------------------------------------------
127   % countdown
128   %------------------------------------------------------------------
129
130   initially countdown(20), pow(transition), perm(transition),
131           perm(clock), pow(intDeadline), perm(intDeadline);
132
133
134   %------------------------------------------------------------------
135   % downloading
136   %------------------------------------------------------------------
137
138   initially pow(transition), perm(transition), perm(clock),
139           pow(intDownload(A,B,C)), perm(intDownload(A,B,C)),
140           perm(download(alice,x1,C)), perm(download(alice,x3,C)),
141           perm(download(bob,x2,C)), perm(download(bob,x4,C)),
142           downloadChunk(alice,x1), downloadChunk(alice,x3),
143           downloadChunk(bob,x2), downloadChunk(bob,x4);
144
145   %------------------------------------------------------------------
146   % sharing
147   %------------------------------------------------------------------
148
149   initially pow(transition), perm(transition),
150           perm(clock), pow(intReceive(Handset,Chunk)),
151           perm(intReceive(Handset,Chunk)),
152           pow(intSend(Handset,Chunk)),
153           perm(intSend(Handset,Chunk)),
154           obl(send(alice,x1), intDeadline, misuse(alice)),
155           obl(send(alice,x3), intDeadline, misuse(alice)),
156           obl(send(bob,x2), intDeadline, misuse(bob)),
157           obl(send(bob,x4), intDeadline, misuse(bob));
158
159   %------------------------------------------------------------------
160   % time
161   %------------------------------------------------------------------
162
163   initially previous(20,19);
164   initially previous(19,18);
165   initially previous(18,17);
166   initially previous(17,16);
167   initially previous(16,15);
168   initially previous(15,14);
169   initially previous(14,13);
170   initially previous(13,12);
171   initially previous(12,11);
172   initially previous(11,10);
173   initially previous(10,9);
174   initially previous(9,8);
175   initially previous(8,7);
176   initially previous(7,6);
177   initially previous(6,5);
178   initially previous(5,4);
179   initially previous(4,3);
180   initially previous(3,2);
181   initially previous(2,1);
```

Figure 5.15: Initial State of the Design Time WMG Model, Post-Negotiation Phase

the channel are busy for a fixed amount of time — 4 time steps in this case (line 76). In the case of a successful interaction with the base-station (the normative event `intDownload` was generated), the handset is considered to have downloaded the chunk, so it can be shared (line 74). As soon as a channel and a handset are engaged, the framework (i) removes the power from the handset (line 80) and from the channel (line 81) to engage in any other interactions[11], (ii) stops the handset from needing the chunk (line 82) and (iii) cancels the permission to download the chunk again later on (lines 83).

As this example presents the code for a design time analysis of the WMG and its energy consumption, we need a mechanism to mark the transition of time. This is done with the help of the exogenous `clock` event (Figure 5.10, line 11) and the normative `transition` event (line 20). Each exogenous event generates a transition to mark the passing of time (lines 85–86). The `clock` event indicates that no handset was interacting with the normative framework. The `transition` event reduces the duration of the interaction between the channel and handset (line 88–89) using the initial (predecessor,successor)-time information specified in Figure 5.15 in lines 163–181. When the interaction comes to an end, `transition` restores the power for handsets to download chunks via the channel and for the handset to download more chunks (line 90). The event also terminates any busy fluents that are no longer needed (line 92–93) which in turn results in satisfying of the `not busyBreiceiving(A)` condition for downloading in line 72. Consequently afterwards the handset fulfills the requirements to download again (with his newly restored power and permission).

5.4.3   Sharing Phase

Once the handsets possess the chunks they are supposed to share, the sharing phase can commence. In this phase each handset needs to send chunks to or receive chunks from other handsets, with the goal that at the end of the process, each handset has a complete set of the chunks, which is the entire digital good. The full specification can be found in Figure 5.14. The idea behind the model is more or less the same as with downloading chunks from the base-station, only that we build in a very basic mechanism to encourage handsets to share their chunks with others rather than just downloading them.

Just like for downloading we introduce two non-inertial fluents `busyHReceiving` and `busyHSending` that keep track of the activity between handsets (lines 39–41 in Figure 5.10). These fluents are true whenever a handset is sending or receiving (Figure 5.11, lines 52–53).

A handset sharing a chunk can possibly trigger two normative events, `intSending` and `intReceiving`, depending on the availability of the handsets

---

[11]The code in line 81 in detail specifies that `intDownload` terminates the power of any handset other then the one engaged in the interaction to download any chunk from the channel, that has is involved in the download. Together with line 80 which removed the permission from the downloading handset to download any other chunk from any channel, this blocks both the involved channel and the involved handset from any further download interaction until this restriction is removed.

and whether they have possession of the chunk or not (lines 99–104). The duration for sending or receiving a chunk is set to 2 time steps (lines 106 and 111 respectively)[12].

The initial sharing assignment is modelled with the help of obligations (Figure 5.15, lines 154–157) which specify which handset has to share which chunk before the a specific deadline[13]. The deadline chosen in the example is a temporal deadline which comes into effect after 20 time-steps (line 130)[14]. An non-fulfillment of the obligation (i.e. not sending the chunk before the deadline) results in a violation event named misuse (represented by the third event in the respective obligation specifications). The simple penalty we chose to implement in our model is that the violating handset permanently loses the power and permission to intReceiving (Figure 5.12, lines 63–64) which they were initially given (Figure 5.15 lines 150 and 151). The result of this sanction is that for all intents and purposes the punished handset has been expelled from the collaboration group.

5.4.4   Trace and Query

Figures 5.9 to 5.15 give the complete specification (and domain information) of our WMG scenario. As explained earlier, when these two components are translated to *AnsProlog* and combined with the framework-independent program components, we obtain all the possible traces over a specified number of time instants. A successful trace makes sure that at the end all handsets have all chunks and are no longer engaged.

This success criteria can be expressed in *AnsProlog* as shown in Figure 5.16. From the traces we are only interested in the observed events that have taken place, violations and the state of the non-inertial fluents busyHSending, busyHReceiving and busyBReceiving to monitor communication costs. Figure 5.17 shows how this can be requested.

Figure 5.18 contains the output of a successful trace for a scenario with two handsets (bob and alice), four chunks (x1, x2, x3 and x4) and a base-station with two channels (c1 and c2). Figure 5.19 shows the graphical representation of this trace. The small circles indicate the time steps. A light grey circle means the handset is busyBReceiving (or for the channel busyChannel) while darker grey indicates busyHReceiving and black indicates busyHSending. Circles with two colours indicate that the handset is both receiving from the base-station and another handset. The arrows indicate which chunk goes to which handset. The left-hand side labels indicate the exogenous event and the current distribution

---

[12]The time steps chosen for downloading and sending/receiving have been chosen arbitrarily to demonstrate the difference between the two communication mechanisms. They were kept relatively small to reduce the length of the traces while still showing possible simultaneous receiving and downloading.

[13]The chunks the handsets are obliged to share are the ones they were asked to download beforehand, i.e. by the point of the sharing phase the handsets should be in possession of the chunks they should share.

[14]The time is again counted down using the **transition** system and the **previous** operator (Figure 5.12, lines 61–62).

```
1   busy :- holdsat(transmit(C,T),F),final(F).
2   busy :- holdsat(creceive(A,T),F),final(F).
3   busy :- holdsat(asend(A,T),F),final(F).
4   busy :- holdsat(areceive(A,T),F),final(F).
5
6   % success criteria
7   success :- holdsat(hasChunk(alice,x1),F),
8              holdsat(hasChunk(alice,x2),F),
9              holdsat(hasChunk(alice,x3),F),
10             holdsat(hasChunk(alice,x4),F),
11             holdsat(hasChunk(bob,x1),F),
12             holdsat(hasChunk(bob,x2),F),
13             holdsat(hasChunk(bob,x3),F),
14             holdsat(hasChunk(bob,x4),F),
15             not busy, final(F).
```

Figure 5.16: The Success Criteria for the Design Time WMG Model

```
27   busyHReceiving(A,I) :- holdsat(busyHReceiving(A),I).
28   busyHSending(A,I) :- holdsat(busyHSending(A),I).
29   busyChannel(C,I) :- holdsat(busyChannel(C),I).
30   busyBReceiving(A,I) :- holdsat(busyBReceiving(A),I).
31
32   violSh :- occured(viol(X),I).
33
34   #hide.
35
36   #show observed(E,I).
37
38   #show busyHReceiving(A,I).
39   #show busyHSending(A,I).
40   #show busyChannel(C,I).
41   #show busyBReceiving(A,I).
```

Figure 5.17: Defining the Trace Output of the Design Time WMG Model

of chunks. The observed events `clock` and `transition` are not displayed to avoid cluttering the diagram.

```
observed(creategrid,i00) observed(download(alice,x1,c1),i01) observed(download(bob,x2,c2),i02) observed(transition,i03)

observed(transition,i04) observed(transition,i05) observed(send(alice,x1),i06) observed(download(bob,x4,c1),i07)

observed(transition,i08) observed(download(alice,x3,c2),i09) observed(transition,i10) observed(transition,i11)

observed(send(bob,x2),i12) observed(transition,i13) observed(transition,i14) observed(send(alice,x3),i15)

observed(transition,i16) observed(transition,i17) observed(send(bob,x4),i18) observed(transition,i19) observed(transition,i20)

observed(clock,i21) observed(transition,i20) busyHReceiving(alice,i14) busyHReceiving(alice,i19) busyHReceiving(alice,i20)

busyHReceiving(bob,i16) busyHReceiving(bob,i17) busyHReceiving(alice,i13) busyHReceiving(bob,i07) busyHReceiving(bob,i08)

busyHSending(alice,i16) busyHSending(alice,i17) busyHSending(bob,i14) busyHSending(bob,i19) busyHSending(bob,i20)

busyHSending(alice,i07) busyHSending(alice,i08) busyHSending(bob,i13) busyChannel(c1,i02) busyChannel(c1,i03) busyChannel(c1,i04)

busyChannel(c1,i05) busyChannel(c1,i08) busyChannel(c1,i09) busyChannel(c1,i10) busyChannel(c1,i11) busyChannel(c2,i03)

busyChannel(c2,i04) busyChannel(c2,i05) busyChannel(c2,i06) busyChannel(c2,i10) busyChannel(c2,i11) busyChannel(c2,i12)

busyChannel(c2,i13) busyBReceiving(alice,i02) busyBReceiving(alice,i03) busyBReceiving(alice,i04) busyBReceiving(alice,i05)

busyBReceiving(alice,i10) busyBReceiving(alice,i11) busyBReceiving(alice,i12) busyBReceiving(alice,i13) busyBReceiving(bob,i03)

busyBReceiving(bob,i04) busyBReceiving(bob,i05) busyBReceiving(bob,i06) busyBReceiving(bob,i08) busyBReceiving(bob,i09)

busyBReceiving(bob,i10) busyBReceiving(bob,i11)
```

Figure 5.18: Ouput of a Successful Trace

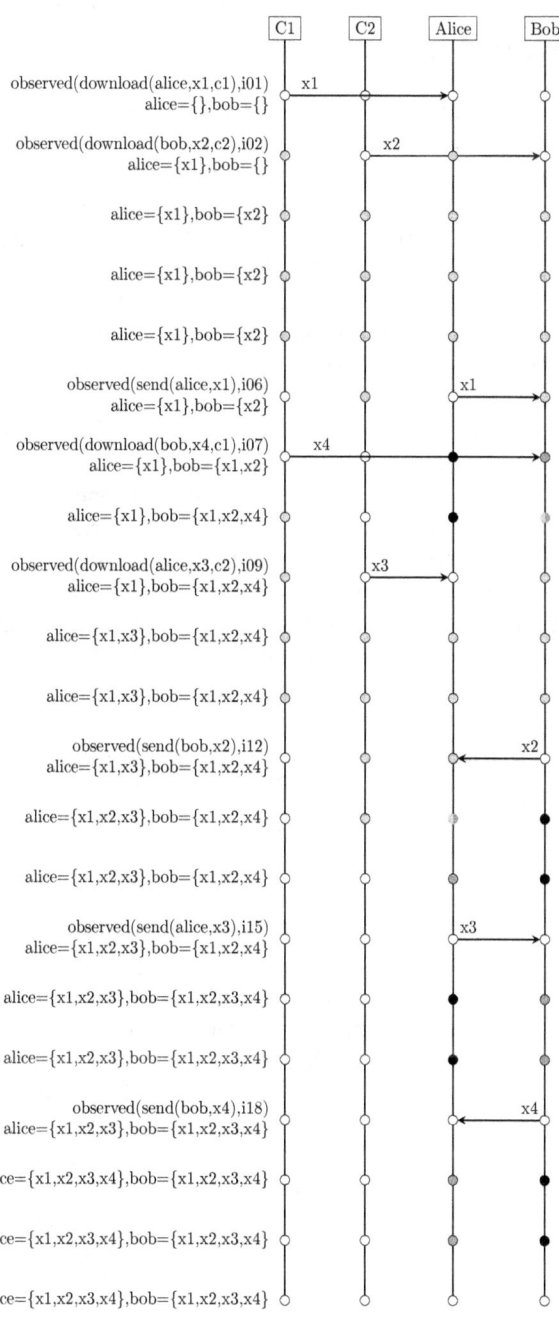

Figure 5.19: One Trace of the Interaction between Alice, Bob and the Channels of the Base-station

5.5   Discussion of the Results of the Design-Time Model

Now that we have set out the normative framework and presented the energy data to quantify communication costs for the particular situation of a 3G structured network and an WMG (see Tables 3.1 and 3.2), we can use the model to examine the traces for expected, but also unexpected behaviour. For our immediate goal, simply by counting the number of `busyBReceiving`, `busyHReceiving` and `busyHSending` in the trace for each handset and afterwards connecting them to the energy values determined by Perrucci et al. (see Tables 3.1 and 3.2), we can obtain an estimate for communication cost under different circumstances. When we investigate the traces, it is straightforward to notice that the traces in which handsets collaborate as much as possible result in better outcomes than those traces where handsets resort to the more expensive 3G links to obtain all the chunks if no free-riding occurred in an interaction.

The model is at this stage a design time tool and generates all possible trace for the encoded normative framework. As mentioned earlier, when analysing the traces, we focused in particular on the ones that that lead to success (i.e. resulted in all handsets having the complete set of chunks after a successful interaction). The specifications for this success-query, which furthermore specifies that no traces are shown in which exogenous event do not lead to an normative event, are shown in Figure 5.20.

```
19   good :- occured(creategrid,i00).
20   good(T) :- occured(download(A,X,C),T),occured(intDownload(A,X,C),T).
21   good(T) :- occured(send(A,X),T), occured(intSend(A,X),T).
22   good(T) :- final(T).
23
24   :- not good.
25   :- not good(T), instant(T).
```

Figure 5.20: The Success Criteria for our Design Time WMG Model

Even with this restriction the model still results in an enormous number of acceptable traces. When examining these traces more closely, we notice that the likelihood of a high proportion of these traces occurring in practice depends on the relative intelligence and (bounded) rationality of the handset participating in the normative framework. Our design-time model purposely avoids modelling the reasoning of the handsets. The reason for this is our objective to design the space in which the handsets interact and to verify that when the handsets follow the norms the entire community benefits, i.e. to verify the possible gains achievable in ideal WMG situations. One further interesting aspect is that enforcement for norms that are breached at the end of the trace will no longer have an effect in the particular interaction and could possibly only be applied on the respective next encounters.

In conclusion, the design-time model is relatively easy to construct and allows for an easy first assessment of the impact of norms on WMGs in an early prototyping stage. It can be used to determine whether a particular state of affairs is reachable or not from a given set of initial conditions and as such it is an effective tool to design and verify properties of protocols and the effectiveness of sanctions. In our WMG example, the design-time model was used as a prototype to demonstrate that normative reasoning can be applied to the domain and to evaluate whether cooperation between the handsets is beneficial to the individual agents.

However, despite the usefulness of the model for quick assessment purposes in an early prototyping stage of the development process, it has several limitations which need to be addressed:

Autonomy Restrictions The design-time model is an abstraction of a possible running system and cannot take into account participants' reasoning capabilities as some of the participants might not be norm-aware or even be irrational. In the design-time model, it was explicitly specified that a handset was only allowed to download a chunk if it did not own this chunk already. In a real running system, keeping track of such information should not be necessary and may not be possible. Furthermore, as described in Chapter 2.3.1 it is impossible to regiment actors in open distributed systems in such a way. This restriction of the autonomy of the system participants is therefore one limitation of the design-time model.

Artificial Artifacts The design-time model also does not have access to information available in a running system. As a consequence it needs to manufacture some information for itself. In the WMG example this means that the design-time model has to keep track of which channels are in use at any given time in order to prevent simultaneous downloads on the channel. This also implies it has to monitor the duration of the download. The same is true for the sending and receiving of the chunks. In a running system this is taken care of by the system and its components (such as the base-stations) or the physical limitations of the devices. The modelling of such system artifacts in the design-time model forces the designer to be very precise about his or her intentions, and results in a possible impairment of significance of the acquired results.

Enforcement The enforcement mechanism employed in the design-time model is very rudimentary and part of the normative system, as discussed in Chapter 4. In general, norm enforcement in a running system is the responsibility of the participants rather than the normative model. The only exception is the granting or removal of permissions to actions (the exogenous events) of the people which is part of the normative specification. In a running system, participants' actions can have side effects, limiting the removal of power from normative events, limiting these in the design-time model might be sensible or useful in that context but cannot be enforced in a running system. An example of this is removing the power of participants to receive chunks in the design-time model when they violate the sharing norm. In the running system this impossible because the data is broadcasted and asking a participant to penalise himself is hardly sensible. For a more realistic model more advanced enforcement concepts need to be analysed.

5.6   Summary

The design-time model is a useful, easy and fast to implement and therefore financially cheap way of generating a first assessment of the possible impact of norm changes in open distributed systems at an early prototyping stage. In the particular WMG example we modelled it showed that energy advantages that can be achieved of norms to ensure cooperation are in place and followed by the system participants. Assured by the results of the design-time model, we use it as a foundation to develop a run-time model that addresses the issues mentioned above. The general concept of this run-time model is shown in Figure 5.21.

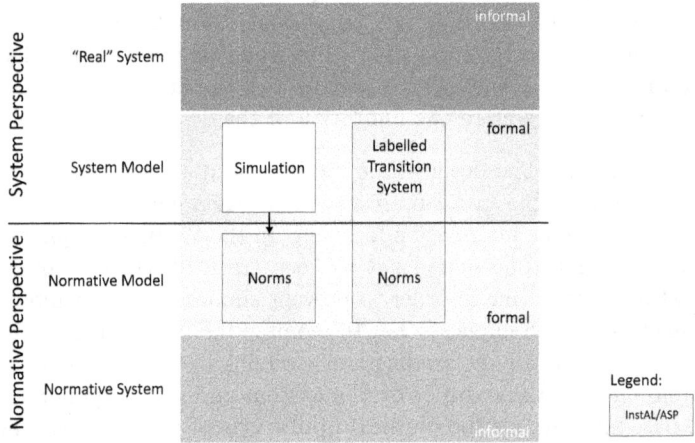

Figure 5.21: The General Run-Time Model Design

The right hand side of Figure 5.21 shows the design-time model in Inst*AL* / *AnsProlog*. It covers both the formal description of the normative framework as well as the system level (i.e. what is referred to as *real-world* in Figure 4.1). For a given normative system, both the design-time and run-time model should have the same normative intentions, making the design-time model a good starting point for the development of the run-time one. For this reason, the normative basis from the design-time model is transferred to the run-time model with some modifications that will be explained in detail in Section 6.1.

This design-time model can be understood in terms of a labelled transition system where all possible traces and the transitions between them are determined by the answer sets. In contrast to the design-time model, where all possible events and the resulting transitions are determined – no matter whether the traces are sensible or not – in the run-time model the real world events are not generated by solving the logic program, but with the help of a *social simulation*. This approach is illustrated by the left hand side of Figure 5.21. As a result, in contrast to the design-time model, the complete possible transition system is not generated. By generating events the social simulation determines which paths are taken through the transition system and allows for a statistical analysis of the paths chosen as well as the normative results of the chosen paths.

Part III

Analysing Enforcement in Wireless Mobile Grids using Normative Multi-Agent System Simulations

Chapter 6

Simulation Design

6.1 Multi-Agent Systems Simulations for Analysing the Action Arena

In the last Chapter we presented a design-time model of our WMG case study. This model is very useful for quickly assessing the possible energy advantages of the WMG concept. Despite its merits, due to its focus, it has some limitations (e.g. the restriction of the autonomy of users) which were outlined in Section 5.5. As a consequence of these restrictions with design-time reasoning, this dissertation uses the design-time model as foundation to develop a run-time model. The run-time model is derived from the design-time model by stripping the design-time model of everything that results in unrealistic restrictions on autonomy as well as artificial artifacts, so that after this process it will only contain normative information and domain facts. Furthermore and most importantly we will change the creation of the exogenous events. Thus, the normative information is still encoded using answer set semantics, however on the real world level, the events that trigger normative events are created using a particular kind of simulation that is called a *Multi-Agent System (MAS) Simulation*. This chapter focuses on the method of simulation for generating exogenous events for the normative framework[1]. In the following we will briefly outline the general idea of simulations – as well as the specific concept of MAS simulations – and explain which features make them particularly suitable for the research conducted in this dissertation. Afterwards we will outline the general simulation research process and then explain the general design of the simulation experiments of this dissertation.

---

[1]This chapter focuses on simulations only. A detailed description on how the simulation is linked to the normative framework components follows in section 7.2.

> ### Definition 15:   Simulation
>
> A simulation is the representation of the behaviour or key characteristics of a physical or abstract system through the use of another system.

In the context of this dissertation, the latter system is an abstract and simplified computational representation of the real system (the future WMG), with the purpose of analysing a particular problem of the real system (the effect of enforcement mechanisms). By abstracting from the real system, simulations reduce complexity of the problem being analysed and allow for problem-oriented experimentation, which might otherwise not be possible in the real system (e.g. due to time or – such as in the case of WMGs – financial constraints in the early prototyping phase). The goal of the simulation process is to transfer the results from these experiments back to the real system. The advantages of simulation are manifold (Shannon, 1998):

- The simulator can test new potential problem solutions / designs, etc. and analyse consequence relationships (i.e. to answer "what if" questions) without having to commit resources to the implementation of the real system. With regard to the WMG idea one can test cooperation concepts and their impact on the stakeholders of the WMG at an early design stage without having to invest in the financially risky implementation and deployment beforehand.

- Simulations allow possible bottlenecks in the real system to be identified.

- They allow hypotheses about whether, why and how the represented system behaves in certain circumstances to be tested and also may lead to theory discovery. According to Axelrod (1997) simulations takes a model, composed of a structure and rules that govern that structure and produce output (observed behavior) in form of data and behaviour of the model. By comparing different outputs obtained via different structures and governing rules, researchers can predict what might happen in the real situation if such changes in input parameters were to occur. As for theory discovery, unexpected emergent phenomena may occur in a simulation, leading to new theories about the interplay of the simulation units (as well as the entities they represent) (Dooley, 2002).

- Simulations grant insight into how a represented system works and thus contribute to the understanding of the respective system. By doing so they may indicate the parameters that are likely to be most important with regard to the performance of the represented system.

- Simulations allow the control of time, either by speeding up processes which in reality might have taken several months or even years; or by slowing down phenomena for in-depth analysis.

Looking at simulations more closely a large number of different simulation methods can be found, hence one needs to decide which of these to employ for a particular problem. The simulation methods found in literature according to Davidsson (2001) and Siebers and Aickelin (2007) can broadly be classified into three categories: Discrete Event Simulation (DES), System Dynamics (SYD) and MAS.

DES models a system as a set of entities being processed and evolving over time according to the availability of resources and the triggering of events (Dooley, 2002). The simulator maintains an ordered queue of events. DES is widely used for decision support in manufacturing (batch and process) and service industries. SYD are models in which key system variables and their interactions with one another are explicitly (mathematically) defined as differential equations. They follow a top-down modelling approach and thus require a good knowledge by the system designer about how the state variables and the system interact with one another. As a result neither DES or SYD are seen as an appropriate paradigm for representing systems where behaviour is best defined through the actions of the constituent entities, as it is the case in collaborative situations like in WMGs. Instead, in settings with cooperation situations like in the WMG, MAS are the favoured simulation method.

Definition 16:   Multi-Agent System (MAS)

A Multi-Agent System (MAS) is the representation of a system as a collection of autonomous self-directed decision-making entities – so-called agents. These agents engage in complex, often non-linear interactions. They act locally and individually towards their own goal, based on their assessment of their current situation with the help of a basic set of agent-individual rules.

By using MAS as computational experiments, one may test in a systematic way different hypotheses related to attributes of the agents, their behavioural rules, and the types of interactions, and their effect on macro level stylized facts of the system. Simulations based on MAS are often also referred to as Agent-based simulations, complex agent systems simulations or agent-based model simulations (Gulyás, 2005) – all referring to the same idea just described. This dissertation uses the term MAS simulation.

To understand better what is understood by the term agent and as a consequence by MAS, we will now look more closely at the definition of the term *agent* in the computer science context. In this context agents are referred to as reactive systems (e.g. pieces of software) that exhibit some degree of *autonomy* in the sense that if being delegated a task or goal, the system determines how to achieve this goal. An important difference to other modelling approaches is that, rather than being given low-level detail on how to fulfill the task, the agents pursue the goals actively and decide themselves how best to accomplish their goals with the (possibly limited) amount of resources they possess. When looking at the properties of agents, as shown in Figure 6.1 agents are systems that are considered to be situated in some environment.

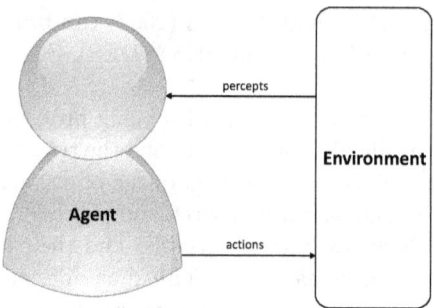

Figure 6.1: Agent and Environment

Agents are capable of sensing their environment via sensors and have a number of possible actions that, based on their internal reasoning and decision making (with regard to their multiple and possible conflicting goal(s)), that they can decide to perform in order to affect the environment. This environment the agents populate and interact with can be physical (e.g. robots that inhabit a physical world) or a software environment such as a computer simulation.

Besides being situated in and interacting with an environment, further properties are attributed to agents (Wooldridge and Jennings, 1995) that are advantageous with regard to representing the cooperation problem in WMGs:

Autonomy As mentioned earlier, agents operate independently in order to achieve goals being delegated to them by their principals. Thereby, they make independent decisions on how to achieve these delegated goals.

Proactiveness Proactiveness is very closely linked to goal delegation. It implies that agents exhibit goal-directed behaviour. Hence, given a goal, an agent is expected to actively work to achieve this goal.

Reactivity Agents respond to changes in the environment. This also implies that if conditions on which they based their earlier decision change, they can adapt to these changes and change their plans for how to achieve their goals accordingly.

Social Ability This refers to the property of agents to be able to cooperate and coordinate activities with other agents (including a communication at knowledge level where agents are able to communicate their goals, beliefs and plans).

Keeping these properties of agents in mind, one can clearly see that they exhibit properties that make them suitable for representing mobile users together with their mobile phones[2]. Users with their phones act in the environment and interact with other users, each being driven by their own objectives (file download and energy saving) as well as being constrained by resource limitations (battery capacity). The actions that the users perform are on the one hand based on their perception of their environment, (e.g. of the other users and their actions), and on the other hand on their resources (e.g. files they have and need) as well as utility considerations (battery costs for different actions).

So far in this chapter we have mainly discussed single agents and their properties. However, what makes agents and MAS particularly relevant for this dissertation is their *social ability*, which allows us to analyse situations in which agents interact with each other. As shown in Figure 6.2 an MAS describes a system from the bottom-up, i.e. from the perspective of its constituent (possibly heterogeneous) units. The macro result on the global systems level is perceived as a result of the interaction of the constituent entities on the micro level. The overall aim is to observe emergent global system behaviour resulting from the sum of the individual actions of the units (Holland, 1992).

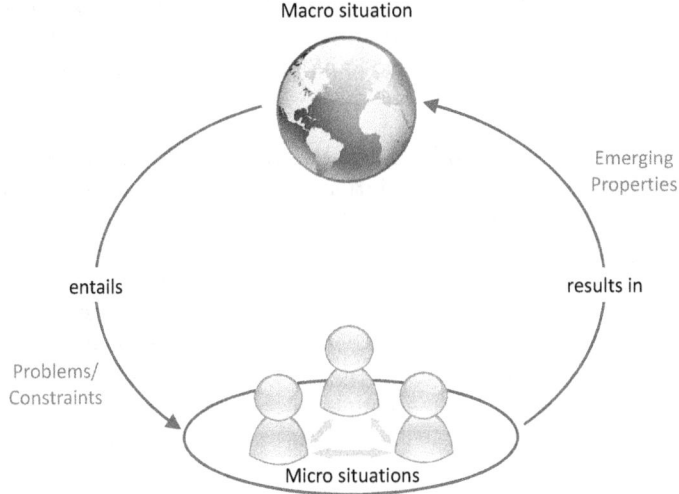

Figure 6.2: The actions of agents situated on the micro level of a system produce macro situations which in turn generate constraints on the micro level (Ferber, 1999)

This property of MAS is of particular interest for the research conducted in this dissertation, because the WMG case study has certain features that make MAS simulations especially suitable for analysing the problems described in Chapter 3. Concerning the initial problem description of cooperation in WMGs, the main feature of a WMG is that *humans* interact with one another with the help of their mobile phones. This is important, as in contrast to the assumptions made when modelling their behaviour with analytical approaches, humans (and

---

[2]We see a user with his mobile phone as one entity and model them as such.

their behaviour) exhibit certain features that are difficult or even impossible to
express (Bonabeau, 2002):

- Human behaviour tends to be complex and non-linear. For example, it
  exhibits memory effects, forms of learning and adaption, path dependence,
  temporal correlations and non-Markovian components. As a result, for
  modelling purposes it is difficult to capture human behaviour in a purely
  analytical form. Of course, almost everything can be defined with the
  help of some form of equation, in principle, however the complexity of the
  required differential equations or – as seen before – logic programs increases
  exponentially as the complexity of behaviour increases. Thus, one would
  either have to make restrictive assumptions such as in the design-time
  ASP-model described in section 5.4[3], or face the problem that describing
  human behaviour with analytical means becomes intractable and difficult
  to compute.

- Human behaviour is characterised by stochasticity. In contrast to analytical
  approaches where, typically, a "noise term" is added more or less arbitrarily to an aggregate equation, MAS simulations allow to add sources of
  randomness at very specific and appropriate points in the agent's reasoning
  process.

- Humans tend to act and base decisions on local information and their limited knowledge. Analytical solutions very often assume global knowledge,
  whereas MAS can represent this local focus easily.

- Finally, humans and their interactions are heterogeneous. The heterogeneous interaction can generate network effects that may deviate a lot from
  predicted aggregate behaviour, based on the emergent network topology
  of the individuals interacting. Representing a system from the agent's
  perspective takes this into account. In contrast, purely mathematical systems typically assume global homogeneous mixing which mainly portrays
  aggregate behaviour and does not account for any network topology, etc.

In addition to these advantages, when trying to represent humans and their
decision making process, MAS simulations offer further advantages with regard
to the flexibility of the analysis. In a MAS simulation it is easy to add more (or
take away) agents, i.e. scaling the system up (or down) as required. But not
only is the number of agents easy to scale, the complexity of the agents (e.g.
their degree of rationality or their ability to learn and evolve) as well as the
level of description and aggregation of the system are also flexible. Thus, one
can easily analyse a system with certain groups of agents or single agents, and
work with different facets of the system description. This is particularly useful
if – as in the questions being asked in this dissertation – the appropriate level of
description or complexity are not known beforehand.

---

[3]A detailed discussion of these restrictive assumptions is presented in section 5.5.

One final research approach that has been overlooked so far and will not be used in this dissertation, is empirical research. Despite being very valuable when analysing humans and their behaviour, empirical research seems inappropriate for the kind of research explored here. The reason for this – as pointed out before – is that this dissertation focuses on an early prototyping stage. WMGs are still very much a conceptual idea. Although a first series of mobile phones that are capable of performing WMG tasks are already being designed and developed, no large scale testing has been performed and no empirical data about mobile user behaviour in the WMG context exist so far. This is particularly due to the time and costs involved in such studies, making them only applicable with difficulty for testing business ideas at an early stage. As a consequence simulation seems the best approach for analysing the cooperation problems in WMGs before their deployment.

6.2   The Simulation Research Process

It is important to keep in mind when using the scientific method of simulations that simulations are gross simplifications of reality because they include only a few of the real-world factors, and as a consequence are only as good as their underlying assumptions, i.e. the problem-specific model of the real system. Amid all the improvements of computer hardware and software, one thing has not changed: the need to develop sensible models. It is important that models are fit for purpose. This means, in most cases, that they support someone's understanding or improve the system being simulated. This in turn implies that the models must correctly represent the features of the simulated system that are relevant to the problem being studied. They furthermore should be developed in an appropriate timescale and have a form that makes analysis as straightforward as possible (Pidd, 2008; Siebers et al., 2010).

In order to give a guidance on how to derive good simulation models Gilbert and Troitzsch (2005) have proposed a generic simulation process similar to the one shown in Figure 6.3. They suggest that this process should be followed to avoid the design problems mentioned above. Their process was largely used in the development of the MAS simulation for this dissertation and therefore will now be explained in more detail.

The start of each simulation experiment is the observation of a problem / question whose answer is not known and which it will be the aim of the simulation to solve. Typically these are problems / questions derived from existing phenomena in the real world which cannot be answered reasonably with the help of analytical or empirical methods. In the case of WMGs or for example nuclear power plant stress test simulations (see Hsueh and Mosleh (1996) for example) – the problems / questions can also focus on future problems that the researcher wants to analyse before their occurrence. Resulting from the questions / problems to analyse, the next step is the representation of the target, which the simulation should help to reason about in form of a model. This stage is referred to as *model design*. For reasons already outlined in the earlier section,

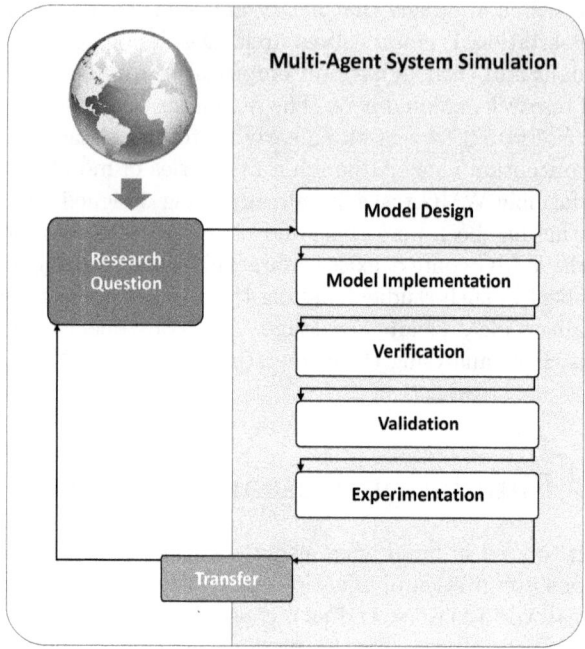

Figure 6.3: The Simulation Research Process

every model will necessarily be a simplification of the target being modelled. As a result one essential and very important step in designing the model is to decide which simplifications can be made in the model and which ones are not allowed because they might distort the results of the simulation experiments with respect to their relevance for solving the initial question / problem. Thus, the more simplifications are made in the simulation model, the greater the conceptual leap required between the conclusions drawn from the simulation experiments and their interpretation with respect to the target. However, the more factors are included in the simulation model, the more precisely parameters for the factors have to be measured or assumed and the more complex the simulation gets, as each of the factors can have an impact on the validity of the conclusions obtained. Unfortunately no general guidelines on how to arrive at a good model exist, but a rigour and and a focus on relevance are needed when developing the simulation. In this dissertation, we aim to pursue both and point out, as well as explain, the design choices made at each step of the conceptualization of the simulation.

Once the simulation model has been conceptualized the next step that follows is the *implementation* of the simulation model, i.e. encoding of the simulation as a computational program. In general two approaches for this step can be distinguished: the utilization of existing toolkits and simulation environments, or the development of the simulation from scratch. Toolkits have the advantage that they offer a wide range of predefined and frequently required components, such as visualization tools or data handlers. Furthermore, if the toolkits are widely used, most of the bugs in the code of packages will have been found by the developers and subsequent users, and communities that can provide advice and support often exist. Developing a simulation from scratch, in contrast, often allows for greater flexibility , as toolkit-based restrictions can be avoided. For this dissertation we opt for a middle-way between the two described approaches. The main agent-reasoning component of the simulation is developed with the help of the Jason MAS simulation environment but we have extented it to allow for interaction with our normative framework. Looking at the simulation literature a large number of MAS simulation environments exist. Gilbert (2008); Railsback et al. (2006) give an overview of existing MAS simulation environments and discuss advantages and disadvantages of each. We have opted for the Jason simulation environment for the following reasons:

- Jason was designed with cooperation in mind (Bordini et al., 2007, p. xiii). It comes with predefined components that enable agents to communicate and coordinate with one another in a high-level way, i.e. focusing on knowledge level communication rather than the technicalities of the communication transmission. This is particularly suitable when economic issues – such as the cooperation in WMGs – are the focus of the simulation experiment and not technical protocols. In addition, the predefined communication components of Jason include single- and broadcast message transmission protocols that are typical to the telecommunication domain and therefore make it well suited for the WMG case study.

- Jason uses AgentSpeak as the programming language for the agents and their reasoning. AgentSpeak is an agent-oriented programming language. It was inspired by and based on a model of human behaviour that was developed by philosophers, the so-called *Belief-Desire-Intention* (BDI) model (Rao and Georgeff, 1995; Rao, 1996). This allows for the reasoning of agents in a human-like fashion based on goals and responses to the environment they interact in. AgentSpeak also has the advantages of having an easy-to-read syntax (Machado and Bordini, 2001), making it accessible for a non-technical audience and facilitating the publication of the specifications and results of the Jason simulations to a wider community.

- Besides the usage of AgentSpeak for programming the agents and their reasoning, Jason employs the Java language to code the agents' environment in the simulation. This usage of Java allows for an easy customization of the simulation (e.g. the inclusion of the normative framework components in the simulation with the help of Java system calls) and thereby makes it very flexible.

- The execution of a Jason simulation is event-driven. This focus on events goes hand-in-hand with the event-focused definition of normative frameworks given in Chapter 4 and facilitates an holistic analysis of the agent interactions and the normative framework.

- Finally, Jason has extensive related literature, simulation experiment examples and an active user community, which provides advice and support and help to reduce bugs as well as programming errors.

Once the simulation has been implemented, the next step in the simulation process is to check that it is behaving as expected (Balci, 1994). This is done by verifying as well as validating it. *Validation* is ensuring that the model, within its domain of applicability, behaves with satisfactory accuracy consistent with the studied target. *Verification*, in contrast, is substantiating that the computational software implementation correctly represents the model, and is in effect the debugging step of the simulation implementation (Ormerod and Rosewell, 2009). The MAS simulation presented in this dissertation was verified by including a large number of output diagnostics in the implementation phase, by observing the simulation step-by-step with the built-in Jason debugging tool, by adding assertions to the simulation, by using unit tests for separate parts of the simulation, as well as by testing the simulation with edge cases.

Besides the verification of the simulation model, it is obvious that its validity needs to be checked, i.e. whether the model portrays the target to be represented. This is a difficult process in MAS simulations, for several reasons: (i) The big advantage of MAS, i.e. the ability to describe micro level behavior and observe the macro level result (i.e. a directly encoded missing link between the two), can also present significant difficulty in verifying their validity. Not only are direct cause-effect mappings not always obvious, but this often tends to be the motivation for the use of simulation experiments. (ii) Other aspects that cause difficulties in the validation of MAS simulations are the inherent non-linearity and the resulting brittleness of outcomes, as well as the mere number of simulated entities (MAS can incorporate several thousands of agents). (iii) Finally, because of the early prototyping stage of WMG's the lack of appropriate data hinders statistical validation (Klügl, 2008).

To validate the model used in this dissertation, we present and justify all the assumptions made concerning the model and the structure of the simulation and discuss them with domain experts such as Frank Fitzek. Furthermore we make an immersive assessment. This is composed of analysing the simulation through the eyes of single agents (the Jason debugging mode offers this possibility) and checking that what the agent perceives and how it reacts, is a valid representation.

Having verified and validated the correctness of the simulation model, the actual execution of the simulation (i.e. the running of the simulation) experiments follows. We execute the simulation experiments in a multi-threaded fashion as this allows for the simulation of a larger number of agents[4]. A detailed description of the design of the simulation experiments is given in Section 6.4.

---

[4]A detailed explanation of this choice as well as the description of the experiments results to ensure that single- and multi-threading experiments yield the same results can be found in Appendix D.

Once the simulation experiments have been carried out, the final step is the analysis of the results as well as their interpretation with regard to the research formulated at the beginning of the simulation process.

6.3 Result Evaluation in the Context of Multiple Stakeholders

Recalling the description of the WMG case study in Section 3.1.3 we identified four main stakeholders with different interests in the system. These are the mobile phone users, mobile phone manufacturers, the manufacturers of infrastructure components (such as base stations, etc.) as well as network and service providers that are subsumed as infrastructure providers (iPr) in this dissertation. This is important when aiming to evaluate the simulation experiment results with respect to the initial research question(s) asked. The reason for this is that depending on the stakeholders the analysis focuses on, it can yield different results.

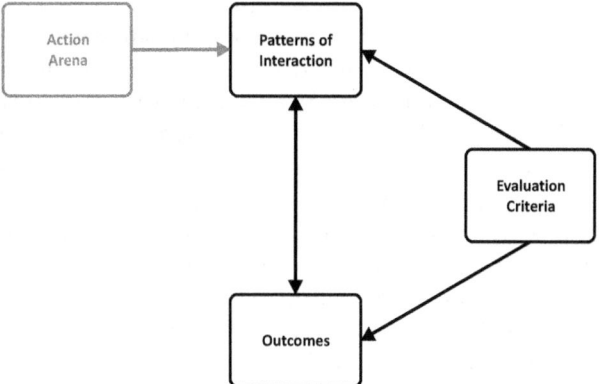

Figure 6.4: Outcomes in the IAD context

Figure 6.4 recalls this problem in the light of the IAD framework, by showing the outcomes components of the framework[5]. As a result of the interaction in the action arena and consequently as a result of our simulation experiments we obtain so-called patterns of interaction, which are some form of emergent macro behaviour. These patterns of interaction are different to outcomes but can be turned into them by evaluating the patterns with the help of evaluation criteria. Depending on the evaluation criteria chosen, consequently the outcomes (i.e. the evaluated patterns of interaction) can be different. Therefore choosing appropriate evaluation criteria is highly significant. In this dissertation what should be measured is the *success* of governance decisions in an early prototyping

---

[5]The full framework is shown in Figure 2.3, on page 46.

stage of development of business ideas for open distributed systems, to enhance cooperation in these systems. Success, unfortunately is a very wide term that can be interpreted in a number of ways according to the domain and context the evaluation is done in. In business terms success is often judged by the *going concern* principle, i.e. to what extent the business can function without the threat of liquidation in the foreseeable future (which typically at least covers a period of one year). For evaluating new business ideas in business ventures, the contribution of the venture to the parent company's going concern as well as the overall monetary success of the product or service being developed are often chosen as criteria. As such, these metrics appear to be appropriate to select for this dissertation. In the WMG case study setting, the driving entities behind the WMG idea are the iPr together with the mobile phone manufacturers. One major problem in determining the monetary success of WMGs for them is that this success depends on the adoption of the WMG by other stakeholders such as the mobile phone users. These stakeholders, however, are not interested in the monetary success of the WMG, but have other priorities by which they judge the success of a WMG (and decide about its adoption) such as the saved battery capacity, or the fairness of the system. Focusing on the monetary success only and neglecting the interests of the users therefore might result in a distortion of the evaluation, which in the worst case can be completely inappropriate. If the users do not adopt a WMG because it does not offer any benefits to them, then predictions of financial are obsolete. As a result, with regard to the evaluation criteria in the IAD framework, one needs to incorporate these inter-dependencies and account for criteria that reflect the interests of the various stakeholders.

In order to reflect this issue in this dissertation, in the next paragraphs of this section we will present evaluation criteria for the two stakeholder groups, namely the iPr and the users and outline the relevance as well as the dependencies of these criteria. Afterwards we will present the models from literature that deal with the issue of multi-criteria decision situations and discuss which model best suits our purposes. In Chapters 7–9 we use these criteria to evaluate the patterns of interaction resulting from the simulation experiments. It is important to note that – despite reviewing possible models for multi-criteria analysis – this dissertation will not do any complete analysis, but evaluate the criteria separately. The reason for this decision is that for the presented analysis various assumptions need to be made. These include assumptions concerning the values of the dependencies of different criteria as well as assumption which importance the stakeholders assign to the different criteria. Given the current early prototyping stage of WMGs, information about values for these assumptions are not available. They would need to be determined with the help of quantitative or qualitative empirical surveys. Without these empirical surveys, scientific rigour is at risk. These studies however go beyond the scope of this dissertation. For this reason we cannot conduct any multi-criteria analysis without endangering scientific rigour. For this reason, we do not perform any full multi-criteria analysis, but instead focus on the energy consumption only when presenting the simulation results. In Chapter 10.1 we outline what impact different evaluation criteria could have for the simulation results. As mentioned above, we will only focus on evaluation criteria for two stakeholder groups – the iPr and the mobile phone users – and neglect the infrastructure and phone manufacturers. The reason for this decision is that both manufacturer groups are typically contracted by the iPr to execute the iPr's hardware and software requirements and are not active themselves in the WMG. They are only involved indirectly through their contractual relation with the iPr, and can be swapped by the iPr, if they do not fulfill their contractual responsibilities. As a consequence their criteria for the WMG success count for little in terms of the overall WMG success criteria and are therefore neglected in this evaluation criteria analysis.

Looking at the evaluation criteria that can be found in literature, generally speaking four different groups of evaluation criteria can be distinguished: technical, economic/financial, social and legal criteria (McKnight et al., 2007). We exclude the last because of current vagueness of the legal status of WMGs (especially in cross-national scenarios) and focus on the first three groups only.

We start by looking at the technical components, which are important to the iPr, as they measure the general suitability and technical functioning of an enforcement mechanism. This suitability/functioning is important, as in case of a non-functioning enforcement mechanism – as explained in Section 3.2 – the whole WMG idea would be at risk. With regard to evaluating enforcement mechanisms the following criteria can be found in the literature:

Scalability The first technical criterion concerns the scalability of the enforcement mechanism. Scalability refers to the fact, that in a WMG any number of mobile phone users $[2, \infty]$ can participate. The enforcement mechanism therefore should be able to cope with any number of WMG participants.

Robustness This criterion measures to what extent an enforcement mechanism can cope with malicious participants. Malicious participants are those whose aim is to sabotage the enforcement mechanism, even at personal cost. They can be compared to malicious nodes in relay routing (see Section 2.4.4.1). Robustness indicates with which number or percentage of malicious participants (of the total number of system participants) a mechanism can cope.

Looking next at the economic/financial criteria, we mention above that monetary success is of high relevance to the iPr. This monetary success is often determined by calculating the profitability of an investment or a technology, etc. A typical metric for measuring profitability in business contexts is the so-called *EBIT*.

Definition 17:   EBIT

EBIT is an approximate measure for a company's financial performance that excludes interest and income tax expenses. It is calculated as revenue minus the expenses before the deduction of tax, interest (Weber and Schäffer, 2005).

Infrastructural Costs Savings For the iPr one aspect of the EBIT is of
particular interests with respect to the WMG. This aspect is related to the
expenses. Thus, as pointed out in Section 3.1.3, one huge advantage of the
WMG for iPr is its lower transmission error rates and the resulting reduced
exposure of the network infrastructure. This in turn could result in reduced
investments in the infrastructure required for mobile communications and
consequently increase the EBIT of the iPr. Buck (2010) determined that
approximately 31.11% of the costs per user for an iPr are related to
infrastructural costs. According to Gruber (2005) base stations account for
more than 50% of these infrastructural costs. Any saving on base station
infrastructure expenses as a result of a transfer of the transmission from
the base stations to the WMG is a big incentive for iPr. In order to get an
idea about these savings we compare the amount of data downloaded in
the WMG case to the amount of data downloaded that would be required
without a WMG. With regard to the WMG enforcement mechanisms it
is therefore interesting to determine to what extent they foster WMG
cooperation (and to what extent they reduce the downloaded data from
the base station) and at what costs.

Enforcement Costs and their Stakeholder Group Distribution
Assuming that enforcement does not come for free, but has associated
costs, another important financial aspect is the level of these costs (as well
as their distribution between the stakeholder groups). In the simulation
we therefore determine whether the enforcement mechanisms produce any
direct costs (e.g. in form of additional infrastructure needed) and also
look at indirect costs such as the communication overhead caused by the
enforcement mechanism, as well as which stakeholder group has to carry
these enforcement costs. In fact the iPr and the users have conflicting
interests in this respect, as neither of them wants to carry the enforcement
costs.

Finally, looking at the social criteria, distribution aspects as well as false
positive and false negative information are of particular relevance:

Cost and Benefit Distribution between Users As well as measuring
which stakeholder group has to carry the costs of enforcement, for the
individual users it is also of considerable relevance how much of the user
group share they each have to carry, as well as what share of the overall
benefits of the system they receive. The relevance of these criteria for
the iPr directly follows from their relevance to the users, as a high user
satisfaction is in the interest of the iPr. A measure typically employed for
determining the inequality of a distribution is the so-called Gini coefficient
(Gini, 1912). The Gini coefficient can take values between 0 and 1 and
is commonly used as a measure of inequality of income or wealth. It
can however be used for more general measurements of the equality of
distributions.

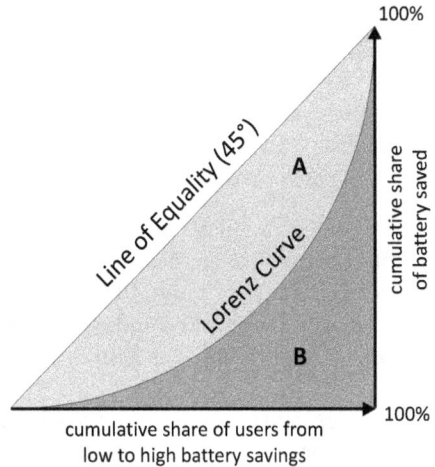

Figure 6.5: The Gini Coefficient as Ratio of the Areas of the Lorenz Curve Diagram

The Gini coefficient is usually defined mathematically as ratio of the areas of the Lorenz curve diagram, as shown in Figure 6.5. The Lorenz curve is a graphical representation of the cumulative distribution function showing the distribution of an independent variable such as population against dependent variables such as income or saved battery capacity. In a perfectly equal distribution of the dependant variable, the Lorenz curve is a straight $45\,^\circ$ line, whereas unequal distributions will yield the curve. The Gini coefficient is calculated as the quotient of the area between the Lorenz curve and the perfect equality line (labelled A in Figure 6.5) and the complete area under the equality line (i.e. $A + B$). The closer the Gini coefficient (i.e. $\frac{A}{A+B}$) gets to 0, the more equally distributed the dependant variable is. In the WMG case study the independent variable is the group of mobile phone users in the system, whereas the dependent variables are the energy consumption, the cost of enforcement or the energy gains per user.

False Positive and False Negative Rates False positive and false negative rates refer to misdetections in the system. Whereas a false negative specifies how often defective actions have not been detected and sanctions not applied, the false positive rate refers to events where sanctions are applied despite no violation having taken place. The false negative rate thus determines the ability of an enforcement mechanism to detect non-cooperative behaviour in terms of detected non-cooperation events over total non-cooperation events. It is especially important for the iPr, because the fundamental purpose of the simulation is to investigate whether a WMG has a general chance of success or whether the idea is at risk of failure because of the cooperation problems outlined in Section 3.2. The false positive rate is equally important for both the iPr as well as the users. Users who receive sanctions despite having cooperated might consider the system to be "unfair" and abstain from using it in the future. This in turn affects the success of the WMG and consequently directly the iPr. It is therefore in the interest of the users as well as the iPr to keep both the false positive and the false negative rate low.

Looking at possible indicators for determining the success of a governance decision, it is apparent that a number of indicators can be considered. These indicators might be viewed differently by the different stakeholder groups and therefore this multi-criteria and multi-stakeholder perspective needs to be accounted for by any mechanism determining the overall success of an enforcement mechanism.

In the decision theory literature, a number of mechanisms to evaluate decision alternatives / different mechanisms in multi-criteria settings exist. Multi-attributive settings are settings where one decision maker (or one group of decision makers with homogeneous preferences) has to evaluate alternatives whose utility / success is determined by two or more possibly conflicting attributes / decision criteria. The most prominent of these multi-attribute decision mechanisms are Multiple Criteria Decision Making (MCDM) (Buchanan and Daellenbach, 1987), the Multiattribute Utility Theory (MAUT) (Keeney and Raiffa, 1976) and the Analytic Hierarchy Process (AHP) (Saaty, 1977). All of these mechanisms allow for decision making under uncertainty, both for multiple criteria discrete alternative problems and multiple criteria optimization problems[6]. They are all supported by computational algorithms and tools that help to apply the mechanisms. To some extent they allow for dependencies between the different parameters / decision criteria. Despite these features fitting well with the evaluation problem in the WMG scenario, all three mechanisms (i.e. MAUT, MCDM and AHP) have the same limitation: they do not account for multiple stakeholders with different interests but assume one decision maker or one group of decision makers with homogeneous preferences. As a consequence the classical multi-attribute decision mechanisms do not fit the case study invesitgated in this dissertation.

Unfortunately, so far the literature on mechanisms accounting for multi-stakeholder situations is scarce with only a few papers attempting to advance the multi-attribute decision mechanisms to incorporate multiple stakeholders. Looking at the papers addressing the issue (see Bertsch and Geldermann (2008); Malczewski et al. (1997); Hämäläinen (2004); Schneck et al. (2004); Merino et al. (2003); André et al. (2007) for example) two aspects can be noted. The first one is that most of these papers deal with practical decision-making situations (typically in the ecological domain) and the second aspect is that most of the works base

---

[6]Multiple criteria discrete alternative problems are decision problems such as the planning of a new airport location where the sets of alternatives that one has to compare and decide between is typically modestly-sized. In contrast, in multiple criteria optimization problems, feasible sets of alternatives usually consist of a extremely large or even an infinite number of alternatives which are defined by equation and inequality systems that identify the bounds for the decision variables (Wallenius et al., 2008).

their calculations on some form of the so-called Compromise Programming or its extension the so-called Composite Programming (CP) (Bardossy et al., 1985). The CP method uses functional groups (with each group for example representing a stakeholder group) and establishes a hierarchical set of decision criteria for each of the groups. In this process single decision criteria are not limited to one functional group only, but can appear in multiple functional groups. In the next step, for each functional group separately the decision criteria are weighted according to their relevance for the respective functional group. The final aggregation of the results of the individual functional groups to a global system-wide decision parameter is done with the help of compensation factors, which determine to which extent the particular decision criteria within a functional group affect decision criteria within other functional groups. The literature (see Schneck et al. (2004) for example) suggest that the compensation factors for the CP are best determined with the help of empirical studies. In the case of WMGs these empirical studies problematic due to the WMG's early prototyping stage of development and go beyond the scope of this dissertation. That is why currently most compensation factors would have to be guessed or derived from other studies and literature and thereby endanger the rigour of any analysis based on them. For this reason, in the following chapters presenting the simulation experiments of dissertation, we do not attempt to conduct any full CP analysis or establish a CP decision structure with functional groups. Instead we focus on the analysis of the above derived single decision criteria only. A short general analysis of the impact the multi-stakeholder scenario on the results found in the simulation experiments is reported in the evaluation part of this dissertation (Section 10.1).

6.4 Experimental Design

Having presented the general idea and process of MAS simulation as a research tool, this section presents the experimental design that has been used to test the different hypotheses in this dissertation by means of simulation experiments. The reason for dedicating this section to the description of the general experimental design, is to make the chapters in which the actual experiments are described (i) more focused on the experiments, (ii) less repetitive and thus (iii) more readable. This section serves as a reference for the reader remembering how the general experimental design is set up and how the experiments were conducted in general.

In general the dissertation is structured in such a way that we start with a simple simulation model of the WMG concept and its inherent cooperation problems in Chapter 7 and then extent this model incrementally in the following two chapters, first with enforcement mechanisms (Chapter 8) and then movement patterns (Chapter 9), conducting experiments at each stage.

The experiments consist of MAS simulations of the models defined in Chapters 7–9. All the models described in these chapters have several variables each, which can take many possible values. Nonetheless – depending on the hypotheses

– each experiment deals with different ranges of these variables. Table 6.1 gives an overview of the types as well as ranges of the main variables used in this dissertation. These will be explained in detail in Section 6.4.2. These variables are taken as input parameters for the executions of the simulation experiments. The goal of the simulation is to record measurements during and at the end of the simulation experiments to allow analysis of the recorded data afterwards. The general assumption behind this process is that the input parameters and the data extracted are correlated. As a result the extracted data for different input parameter settings are compared and analysed in order to support or discard the dissertation's hypotheses.

As a result of the high number of variables in each simulation model, factorial experiments are the best way to test the dissertation hypotheses. They are especially well suited when – as in the case of this dissertation – the experiments focus on two or more *factors* (i.e. important variables) that are analysed with regard to their impact on simulations and the underlying hypotheses (Freedman et al., 2007). For factorial experiments a subset of the discrete set of input variables available for each factor is selected. The general idea of factorial experiments is to run simulations with all possible combinations of the values of the subsets of chosen input variables across all factors. Factorial experiments allow the user to study the effect of each factor on the simulation output data, as well as the effects of interactions between factors, while cancelling out influences of other factors on a particular setting.

6.4.1   Simulation Parametrisation and Statistical Analysis

The MAS simulation was implemented using the MAS simulation tool Jason (Bordini et al., 2007) in combination with a variation of the *AnsProlog*-based normative systems specifications introduced in Chapter 5[7]. The Jason components were specified in AgentSpeak (Rao, 1996; Machado and Bordini, 2001) and Java and the simulation was run in a multi-threaded fashion[8]. The simulation was parameterised using comma-separated values from a text file that were passed to the simulation with the help of command line arguments at its initialization. The usage of Jason as the simulation tool resulted in some random effects – such as the order of interaction or the randomization of some agent parameters. We do not consider these as part of the experimental factors, in order to keep the number of experimental factors low. To avoid the impact of these random effects on the simulation results and the analysis thereof, as well as to be able to apply mean comparison methods, for each set of input variables 50 simulation runs were conducted (Field, 2009).

After running all simulation experiments and gathering the output data, the

---

[7]The nature of the changes in the normative systems specifications will be explained in section 7.2.

[8]Detailed information on the execution of the simulation in a multi-threading mode, the suitability of this execution mode as well as the resulting impact of the mode choice are given in Appendix D.

statistical analysis of the data was conducted. For this purpose we used the analysis of variance (ANOVA) (Fisher, 1918) to test whether specific input parameters (or parameter combinations) have effects on the output measurements. In order to be able to apply the ANOVA test effectively, its underlying assumptions about the experiment need to be fulfilled. These assumptions are independence, normal distribution and homoscedasticity of the input parameters as well as any combination of them (Scariano and Davenport, 1987). Independence refers to the concept that values of one factor do not influence the probabilities of the values of another factor. Normal distribution means that the graph of the associated probability density function of each variable is "bell"-shaped and follows the function $f(x) = \frac{1}{\sqrt{2\pi}} e^{-\frac{(x)^2}{2}}$. Finally, homoscedasticity assumes that the residuals at each level of the input parameters have similar variances. ANOVA tends to be robust to non-normality as well as non-homoscedasticity to a certain extent (Lindman, 1974). Nonetheless we run the Shapiro-Wilk test (Shapiro and Wilk, 1965) through which we test the experimental data for normality as well as Levene's test (Levene, 1960) which verifies homoscedasticity. The independence of the variables is ensured through rigorous experimental design and simulation modelling.

In addition the general analysis resulting of the ANOVA tests, for the experiments where we want to evaluate the specific effects that input variables had on the specific output data, we run a Tukey's test and – if only two means are to be compared – the two-sample t-test for the respective significant factors. Tukey's test is a single-step multiple comparison procedure and statistical test that is used to find means that are significantly different from one another. It can be used to compare the means for the different input values of the given factor that is being checked. The test determines which input values have significantly smaller or larger output measures than the standard error predicts. The t-test is used to compare two sample means, in order to determine whether these are significantly different from each other.

All the statistical analysis tests in this dissertation were conducted using MATLAB software[9].

6.4.2 Input Factors

After having outlined the general experimental design used for all simulation experiments described in chapters 7–9, this section focuses on the input factors that are used in the simulations. As pointed out in the previous section, the simulation models can be parameterised with a large number of different input parameters that need to be taken into account for the statistical analysis. In this section we describe the main factors of the simulation experiments, and explain how they are used in the simulations.

An overview of these main factors is given in Table 6.1. The main variables are the number of agents, the relative number and ratio of the type of agents

---

[9]More information on MATLAB can be found on the following website: http://www.mathworks.com/products/matlab/.

Table 6.1: The Main Simulation Input Factors

Name	Type/Range
Number of Agents	$[2, \infty]$
Agent Types as % of $\mid \mathcal{A} \mid$	Utility-Maximizing Agents as % of $\mid \mathcal{A} \mid$, Honest Agents as % of $\mid \mathcal{A} \mid$, Malicious Agents as % of $\mid \mathcal{A} \mid$
Number of Interactions per Agent	$[1, \infty]$
Partner Search Interval	$(0, 0.5]$
Enforcement Mechanism	None, Normative Empowered Agents, Image Information, Reputation
Movement Pattern	Simple/Parameterised Boundless Random Waypoint, Reference Point (Nomadic) Group, Sedentary Movement
Average Neighbourhood Density ($\rho_{neighbourhood}$)	$[0, \mid \mathcal{A} \mid)$

represented in the simulation, the number of interactions, the partner search interval and the enforcement model as well as the movement pattern employed.

The number of agents is a natural number that must be at least 2. It defines the total number of agents that are represented in the simulation. In order to test different values for this variable we use numbers from the following power distribution: 100×2^1, 100×2^2, 100×2^3,... [10]. Within this set of the total number of agents, we distinguish three general groups of agents: utility maximizing agents that – as the name indicates – try to maximize their own utility in the transaction, honest agents that try to cooperate in their transactions as well as malicious agents that try to sabotage the system by not contributing at all (even if the utility of cooperation is higher than the utility of defection). In the simulation experiments we will analyse the effects of different proportion sizes of these groups (in respect to the total number of agents and in terms of ratios to one another) have on the simulation results.

The number of interactions refers to the number of transactions each agent is participating in. This, together with the number of agents determines the length of the simulation runs. It is a natural number larger than or equal to 1.

As outlined earlier (see Section 3.2 for example), we assume bounded rationality of the actors. As explained in Section 2.5.1, one of the results of this assumption is that the actors in the system cannot perceive the whole system, but only their local neighbourhood (e.g. every other handheld that is in the radius of their WLAN reception range). The search interval size specifies the size of this local neighbourhood, i.e. how far an agent can perceive other agents. It is specified as a real number between 0 and 0.5. For cooperation to take place, an agent must be in the proximity of other agents (i.e. they must be in their local

---

[10]Smaller numbers than 100×2^1 have not been considered as real world WMG are expected to have large number of participants.

neighbourhood). As we will explain in Section 7.1.3.3, an accurate Cartesian model of location is not actually necessary, so in the simulation experiments we represent an agent's location one-dimensionally as a number between 0 and 1 and proximity is determined by the visibility radius $[location - x, location + x]$ with x being the radius in which the agents can perceive other agents. It can be thought of as the WLAN signal radius, which limits the number of other mobile phones the agents can detect. We refer to the number of other agents an agent can on average perceive as a result of them being within its proximity as neighbourhood density (denoted as $\rho_{neighbourhood}$)[11].

One factor which determines the location of the agents (which however is independent of the proximity) is the movement pattern of each agent. One obvious, but important aspect of mobile phones is that they and their users are mobile. As real world movement traces of mobile phone users are not available, in the simulation experiments we need to model the movement of mobile phone users and analyse whether different movement patterns result in different simulation outputs. In the experiments, we analyse different movement patterns for the agents, including a simple boundless random waypoint pattern, a parameterised boundless random waypoint pattern, a reference point (nomadic) group movement pattern and a sedentary movement pattern. All these patterns will be explained in detail in Section 9.

Given this dissertation's focus on the reciprocity issues inherent in open distributed systems, the final main factor used in the simulation experiments is the enforcement concept employed. We start the simulation experiments with no enforcement mechanism and then consecutively include three different enforcement mechanisms, namely normative empowered agents, image information and then finally a reputation mechanism. These mechanisms will be explained in detail in Chapter 8.

6.4.3 Measurements

During, as well as at the end of, each simulation experiment we take measurements of the simulation output. By default the measurements are taken at every event that corresponds to an exogenous event ($\mathcal{E}_{ex}$) of the normative framework and then aggregated per interaction round. With regard to the interaction "rounds" it should be noted that the simulation experiments do not use synchronized rounds, but instead the simulation uses asynchronous events. Thus, an agent x might perform two interactions (e.g. ac_{x1} and ac_{x2}) whilst another agent Y only performs one interaction (e.g. ac_{y1}). A "round" aggregates the respective n-th interaction per agent, e.g. all agents' first interaction events, or all agents' second interaction events by all agents, etc. For each agent the total number of interactions is fixed and the same. This ensures that in each round, one interaction for each agent is analysed.

Table 6.2 gives an overview of the different main output measurements taken for the different simulation experiments. These are measurements that cannot

---

[11]A more detailed explanation of the location related aspects in the MAS simulation is given in Chapter 7 on page 153.

Table 6.2: The Main Simulation Measurements

Name	Type
Round	$\mathbb{N}$
Number of Defection Actions	$\mathbb{N}$
Number of Detected Defection Actions	$\mathbb{N}$
Number of Punished Actions	$\mathbb{N}$
Number of Correctly Punished Malicious Actions	$\mathbb{N}$
Download-related Data Transferred per Agent per Interaction	$[\mathbb{R}, \mathbb{R}, \ldots]$
Sending-related Data Transferred per Agent per Interaction	$[\mathbb{R}, \mathbb{R}, \ldots]$
Reception-related Data Transferred per Agent per Interaction	$[\mathbb{R}, \mathbb{R}, \ldots]$
Communication-related Data Transferred per Agent per Interaction	$[\mathbb{R}, \mathbb{R}, \ldots]$

be directly derived from other metrics. All derived metrics will be explained in detail in the respective experiment section.

In each interaction round[12] the agents can decide to obtain the required files themselves, or to cooperate with other agents. In the latter case they can furthermore decide to contribute to the WMG by sending their chunk / share or to defect. For the simulation experiments the total number of defection actions (i.e. not contributing to the common interaction pool) is determined. As every agent performs one interaction per round, based on the input parameters the total number of interactions can be determined. This number, in combination with the number of defection actions, can be used to determine the number of norm-compliant actions. The number of punished actions specifies the number of times agents were punished for their behaviour. This does not automatically imply that all of these actions where correct or justified, as false positive or false negative punishment situations can arise. By determining the correctly punished defection actions (i.e. the instances where a sanction was correctly performed as a result of a defection action by the punished agent), we are able to determine other economic parameters such as the false positive and false negative rates. Comparing the number of defection actions with the number of detections we are also able to calculate the detection rates of defective behaviour in the simulation experiments.

The next four measurements concern the battery consumption and the amount of data transmitted per interaction round. By measuring the data transmitted as for each interaction protocol (i.e. $E_{3G,rx}$, $E_{WLAN,rx}$ and $E_{WLAN,tx}$) and using Tables 3.1 and 3.2, we are able to determine the relative energy per bit consumption rates in each round. With regard to these values we distinguish the battery consumption / bits transmitted for downloading, sending and receiving chunks as well as for the additional communication involved in the interactions.

---

[12]When speaking of interaction rounds, we refer to situations in which the agent is involved in an interaction situation. Agents also do have the option of doing nothing, these situations are however not considered as interaction rounds.

6.4.4   Experimental Design and Results

In the sections where experiments are presented, a table is given to represent the experimental setup. An example of such a table in shown in form of Table 6.3.

Table 6.3: Example of an Experimental Setup Description Table

Input Factors	Values
Agents	200, 400, 800
Number of Interactions per Agent	5, 25, 50
Utility Agents as % of $\mid \mathcal{A} \mid$	0, 25, 50, 100
Malicious Agents as % of $\mid \mathcal{A} \mid$	0, 25, 50, 100
Honest Agents as % of $\mid \mathcal{A} \mid$	0, 25, 50, 100
Measurements	**Type**
Number of Defection Actions	$\mathbb{N}$
Number of Detected Defection Actions	$\mathbb{N}$
Download-related Data transferred per Agent per Interaction	$\mathbb{R}$
Integrity Constraints	**Formula**
Full Partitioning	Utility Agents % + Honest Agents % + Malicious Agents % = 100

This table consists of three parts: the first part contains the input factors of the simulations with the values for the simulation experiments, the second one the measurements taken throughout the experiments and the third one specifies the integrity constraints required for the input factor values, i.e. constraints that concern the coherence of the input factors. It is important to note that as a result of the integrity constraints, not all input factors of the first part can be used as variables for the ANOVA. The percentage of honest agents for example results from the percentages of utility maximizing as well as malicious agents in the system. The same applies for determining the percentage of utility maximizing agents from the percentages of honest malicious agents. As a result of this possibility of deriving one percentage from the other percentages, only the ones used for deriving the third are considered in the ANOVA.

After the execution of the experiments, the results are presented in the respective sections.

Table 6.4 gives an example of how the results are summarized in this dissertation. In this table, each row in the table represents an potential cause-effect relationship. This relationship is specified with reference to information that is arranged into four columns. The first column gives the input factor that can possibly have an effect, the second column provides the measured output variables that the input factor might exhibit an effect on. The third column shows the significance value for the respective relationship. If this value is 95% or higher, a significant correlation between the input factor and the measurement exists. The last column gives specifics on the correlation if any was found.

Table 6.4: Example of an Experimental Results Description Table

Input Factors	Measurement	α	Relationship
% Malicious Agents	Defection Actions	0.98	A higher percentage of malicious agents results in a higher number of defection actions.
% Honest Agents	Energy Consumption	0.95	A higher percentage of honest agents results in an overall decrease in energy consumption.

6.5 Summary

In this chapter we have presented the simulation design that we employ to answer the research questions posed in Chapter 1. For reasons given in Section 6.1 we have opted for MAS simulation experiments. The general research process of simulations was described in Section 6.2. Following this description as well as the illustration of particular challenges in the process (as well as answering how this dissertation addresses the challenges), we presented the general experimental design of this dissertation in order to give the reader a reference to the experimental design set up and explain how the experiments are conducted. Having presented the simulation design, we move on to present the actual simulation experiments used to analyse compliance in WMG. For this purpose, we start by presenting the WMG without enforcement in Chapter 7, extent the simulation with enforcement mechanisms (Chapter 8) and then analyse the effects of movement patterns on the enforcement mechanisms in Chapter 9.

Chapter 7

The Basic Wireless Mobile Grid Scenario

In parts two and three of this dissertation we presented the motivation, foundations, and the related work of our research and introduced the idea of combining normative frameworks and MAS simulations to reason about the run-time interaction of actors in a system. Chapters 7–9 present the simulation experiments conducted as part of the dissertation research. Chapter 7 deals with the basic WMG scenario without any enforcement mechanisms. It presents the basic WMG setting and explains the general representation of mobile phone users and their actions in a MAS simulation. Furthermore, the results of the simulation experiments without any enforcement mechanism serve as "worst case" reference point for the later experiments where different enforcement mechanisms are used. Worst case implies that with an enforcement mechanism, the WMG should not perform worse than without any mechanism. In addition – in comparison to the simulation experiments without any enforcement – all cooperation gained and battery savings as a result of the utilization of an enforcement mechanism needs to be evaluated against the additional costs due to the enforcement mechanism. The simulation experiments with enforcement mechanisms are presented in Chapter 8. Finally, Chapter 9 repeats the experiments conducted in Chapter 8, but changes the movement pattern of the agents in the simulation in order to investigate the effects of different movement behaviours of mobile phone users on the results of the enforcement mechanisms from Chapter 8.

7.1 Representing the Basic Wireless Mobile Grid Scenario as MAS Simulation

7.1.1 General Formalizations

In Section 6.1, we explained that we will use a MAS simulation in order to represent the simulated world view of the interactions that are taking place in the action arena. After giving a abstract definition of MAS in Definition 16 (page 127), we now present the formal definition. This definition links the normative frameworks with the idea of MAS and is used to underpin our simulation experiments. The definition is based on the works of Centeno et al. (2009) who developed a formal framework for combining MAS and norms which suits the ideas presented in this dissertation very well. Throughout the formalisation as well as the following Chapters presenting the simulation and its results, the following types of symbols will be used: Latin calligraphic capital letters refer to sets (e.g. $\mathcal{A}$), lower case Latin letters refer to elements of sets (e.g. $a \in \mathcal{A}$) and lower case Greek letters refer to functions (e.g. γ).

Definition 18:   Multi-Agent System (formal)

A MAS is a tuple $\langle \mathcal{A}, \mathcal{S}, \gamma, \kappa, s_0, \vartheta \rangle$ where:

- $\mathcal{A}$ is a set of agents where $|\mathcal{A}|$ is the size of the population;

- $\mathcal{S}$ is the state space of the system that is being modelled (in the normative frameworks this system corresponds to the real world state);

- $\gamma : \mathcal{S} \times Ac^{\mathcal{A}} \times \mathcal{S} \to [0,1]$ is the MAS transition probability distribution, describing how the system evolves as a result of agents' actions (Ac). Thus, as in the normative frameworks, we assume time to be discrete. At each time step the agents in the MAS perform one action (this includes a "wait"-action of doing nothing). The new state of the system is determined as a result of the joint actions of all agents.

- $\kappa : \mathcal{S} \times Ac^{\mathcal{A}} \to \mathcal{E}_{ex}$ is the MAS event function, associating an external event to a change in the system's state as a result of agent's actions, i.e. it specifies under which conditions $\mathcal{E}_{ex}$ for the normative framework are being generated;

- $s_0 \in \mathcal{S}$ stands for the initial state of the MAS; and

- $\vartheta : \mathcal{A} \times Ac \times \mathcal{X} \to 0,1$ is the agents' capability function describing the actions agents are able to perform in a given state of the environment. $\vartheta(a, ac, s) = 1 \, (= 0)$ means that agent a is able (not able) to perform action ac in the state s.

Within this context, in formal terms, an agent is defined as follows:

Definition 19:   Agent (formal)

An agent is a tuple $\langle \mathcal{I}, Ac, \mathcal{O}, \nu, \xi, \tau, s_0, \eta \rangle$ where:

- $\mathcal{I}$ is the set of internal states of the agent;

Definition 19:   Agent (formal) (cont.)

- Ac is a possibly infinite action space that includes all possible actions that can be performed by an agent. Ac includes an action ac_{wait}; the action of doing nothing.

- Ob is the observation space of the agent; i.e. the set of possible observations the agent is able to perceive from the MAS;

- $\nu : Ob \times \mathcal{I} \to \mathcal{I}$ is the agent's state transition function;

- $\xi : \mathcal{I} \to Ac$ is the agent's decision function describing the action it will choose given an internal state. It follows the principle of maximizing the expected utility.

- $\tau : \mathcal{S} \to Ob$ is a perception function assigning an observation to an environmental state;

- s_0 is the agent's initial internal state;

- $\eta : \mathcal{I} \to \mathbb{R}$ is an utility function that assigns a value to each possible internal state of the agent.

Keeping these definitions in mind, we now turn to the explanation of the simulation components. Using the IAD framework as the underlying concept of our analysis of the impact of governance decisions, we have structured the following sections according to the underlying factors of the IAD framework (i.e. the biophysical characteristics, the attributes of the community as well as the rules-in-use) and then turn our focus to additional relevant aspects.

7.1.2   The Biophysical Characteristics

The first part of the underlying factors we discuss are the biophysical characteristics. In the WMG scenario, these biophysical characteristics are given by the underlying communication network and technologies. The design of the communication network and technologies in the simulation experiments of this dissertation is based on a assumptions that are commonly known as the *Flat Earth model* (Kotz et al., 2004).

Definition 20:   Flat Earth Model

The Flat Earth Model is a simplified model of real world mobile phone communication systems often used in mobile phone simulations, which is based on Cartesian $X - Y$ proximity. In detail, the model makes the following assumptions / simplifications:

> **Definition 20:   Flat Earth Model (cont.)**
>
> - The world is flat.
>
> - A mobile phone's or base station's transmission area is circular.
>
> - The communication is based on a Cartesian $X - Y$ proximity, that is, nodes a_1 and a_2 communicate if and only if node a_1 is within some horizontal distance of node a_2.
>
> - All mobile phones have equal range.
>
> - If node a_1 can hear node a_2, a_2 can hear a_1 (symmetry).
>
> - If node a_1 can hear node a_2 at all, it can hear it perfectly.
>
> - Signal strength is a simple function of distance.
>
> - There is no external interference.
>
> - There are no obstacles.

Besides following the Flat Earth model assumptions, for reasons of simplicity we furthermore assume that all WMG-capable mobile phones have the same specifications, i.e. that they will have equal battery consumption rates for equal actions performed. For the simulation the consumption rates used are taken from a Nokia N95 mobile phone, for which detailed measurements exist (Perrucci et al., 2009). In detail, the power [in W] and data rate [in Mbps] values given in Tables 3.1 and 3.2 are being used for determining the respective energy consumptions[1], i.e.: (i) 1.314 W and 0.193 Mbps for receiving data via the cellular link, (ii) 1.629 W and 5.623 Mbps for sending data via a IEEE802.11 WLAN broadcast/multicast, and (iii) 1.213 W and 5.115 Mbps for receiving data via the IEEE802.11 WLAN connection (at a distance of 30m between the devices).

Based on this data (especially the assumed fixed transmission speed for each transmission type), the energy consumption required for the sending and receiving of data, as well as communicating in the WMG simulation is determined by the size of the transmitted data packages (the data packages are referred to as chunks in the simulation). Hence, checking which data packages have been sent / received in which transmission mode allows us to determine the overall energy consumption. For the transmission themselves we distinguish between transmission between base stations and mobile phones (cellular link) as well as an IEEE802.11 WLAN transmission between phones. We model the IEEE802.11 WLAN sending as a multicast transmission. This means that a single message by one mobile phone can be sent simultaneously to a group of recipients in a

---

[1]The energy consumption is determined by the power used over time. The time is a result of the data rate and the amount of data being processed.

single transmission action. If this group of recipients corresponds to all handsets in the transmission range of the sender, multicasting equals the broadcast idea, whereas if it is sent to only one recipient, it is equal to the unicast transmission.

The base stations in the WMG infrastructure are modelled with the help of special base station agents. In a real WMG, base stations are stationary. To reflect this, we model the base station agents as having a fixed location in the simulation and serve as download points for the agents representing the mobile phone users. In the simulation they are located in such a way that full coverage is ensured. This means that mobile phone agents have access to a base station from any location in the simulation area. One particular feature of the base station agents is that they have a limited capacity for providing downloads. This limited capacity corresponds to the channel (i.e. frequency division multiplexing) restrictions described in Sections 5.4.2. When mobile phone users with their handsets want to download a chunk from the base station, they have to check that this base station still has a channel available. For this purpose, the base station agent keeps track of its own capacity and can be queried with questions on its availability.

As stated above, we use the Flat Earth Model assumptions in our WMG simulation model. The location of the mobile phones (and hence their distance between each other), determines which can interact. As a consequence, for our simulation, the proximity of the different mobile phones is highly relevant. In order to keep the simulation model as simple as possible without losing any relevant information – as already mentioned in Section 6.4.2 – we model a location as a number $l \in \mathbb{R}$ from the interval $[0, 1]$. The proximity of two agents in two locations is determined with the help of a *visibility radius* ($r_v \in (0, 0.5]$). This visibility radius corresponds to the transmission radius of a mobile phone IEEE802.11 WLAN adapter. Using the transmission radius as well as the location of a mobile phone, the area in which a mobile phone user can find transaction partners is determined by the interval $[l - r_v, l + r_v]$. As we assume that all mobile phones in the simulation have the same technical specifications, we can conclude that the visibility radius for all mobile phones is the same. As a result, if one mobile phone A is in the visibility radius of another mobile phone B, then B is also in A's visibility radius. This also corresponds to the assumptions of the Flat Earth Model. Concerning the border areas of our location interval $[0, 1]$, we assume a continuous space (or *ring* form) of our simulation area. This means leaving the location interval at one border automatically results in entering it at the other border. To give an example of this: if an mobile phone user at location 0.9 and moves a distance of 0.2 in direction of the 1 border[2] it arrives at location 0.1. Using the visibility radius we are able to determine the so-called average neighbourhood density $\rho_{Neighbourhood}$ in the simulation, i.e. the number of mobile phone users (i.e. agents) an agent can perceive in its transmission range[3],

---

[2]Mobile phone users can make moves of sizes $[0, 1]$. The direction of the move is indicated by attributing a $+/-$ indicator to the move. The $-$ indicator thereby refers to a move in the direction of the 0 border, whereas the $+$ indicator is used for moves in the direction of the 1 border.

[3]This number includes the agent itself.

by multiplying r_v with 2 and $|\mathcal{A}|$, i.e. $\rho_{Neighbourhood} = r_v \times 2 \times |\mathcal{A}|$[4]. In the simulation experiments we use $\rho_{neighbourhood}$ as one input parameter and analyse the effects it has in different WMG settings with and without enforcement.

7.1.3 Representing the Community

Having presented the biophysical characteristics of our WMG case study and explained how we intend to model them in the MAS simulations, this section turns its focus to the next component of the underlying factors of the IAD framework – the attributes of the community, i.e. the modelling of the actors in the system.

As in the case of the base station, we model the actors in the system using agents. The simulation itself is constructed in such a way that at each internal individual reasoning step, agents receive an observation from the environment, change their internal state and take an action – which may be doing nothing – that is finally executed. The internal state of an agent possibly encodes its history of actions and observations, its beliefs about the state of the environment, as well as its own preferences. The internal state evolves by integrating observations from its environment. The agent's decision function reflects its behaviour or policy and determines which action it will take in the next step. Using κ, the actions the agents perform can generate $\mathcal{E}_{ex}$ for the normative framework.

As indicated earlier, one of the most important feature of MAS is that agents act locally, but their actions result in a global system behaviour. For this reason, in order to explain the simulation in more detail, we first look at the actions an agent can perform.

In the simulation, for reasons of simplicity, we have limited the agents' actions to the actions relevant for analysing the cooperation problem described earlier. This is why, in the simulation, $\mathcal{A}c$ includes the following main actions for interaction purposes:

- $ac_{download}$: downloading chunks of a file from a base station;

- ac_{search}: searching for cooperation partners;

- $ac_{receive}$: receiving chunks of a file via the IEEE802.11 WLAN link;

- ac_{send}: sending file chunks via the IEEE802.11 WLAN link;

- ac_{cheat}: not sending chunks of a file via the IEEE802.11 WLAN link, despite having promised cooperation partners to do so[5];

- ac_{move}: changing the location of the agent within the environment;

---

[4]The multiplication with 2 is required, because r_v is a radius. The mobile phone transmission we model however is circular and not directed into one direction only. Therefore we need to use the diameter (i.e. $2 \times r_v$) when calculating $\rho_{neighbourhood}$.

[5]The lack of performing the sending action is modelled as action, because the agent makes a conscious decision not to send their chunks.

- ac_{enter}: entering the WMG (this includes entering a WMG by switching the mobile phone on);

- ac_{leave}: leaving the WMG (this includes switching off the mobile phone); and

- ac_{wait}: doing nothing.

What is important with regard to these actions is that we explicitly allow for cheating by the agents, i.e. leave it as an action option to them. What kind of actions an agent chooses to perform in an interaction depends on its type, i.e. its attitude to cooperation and particular utility considerations.

7.1.3.1   User Types

To represent the possible variety in utility considerations by mobile phone users, we opted to model three different kinds of agents representing mobile phone users in our simulation. These modeled types were chosen to model a broad range of agent behaviour, without increasing complexity unnecessarily. For all three agent types presented in this section we assume the agents to have bounded rationality (see Section 2.5.1) and only able to interact as well as directly communicate with agents that are within their visibility radius. The three agent types we model in this dissertation are:

Utility Maximizing Agent In an interaction a utility maximizing agent (or utility agent for short) will try to maximize its utility in terms of saving as much battery as possible. Hence a utility agent tends to cheat if it does expect the costs of a potential sanction to be lower than the gains from cheating and will cooperate if it perceives the situation to be the opposite way around (i.e. the possible punishment cost higher then the cheating gains). Given their characteristics, the agents are based on the typical *homo economicus* considerations in economics.

Honest Agents In contrast to the utility agents, honest agents prefer to co-operate and always send their chunks.

Malicious Agents The last agent type we model are malicious agents which are based on the malicious/faulty node idea in relay routing (see Section 2.4.4.1, page 41). Malicious agents behave in the opposite way to honest agents and pursue the goal to harm the system to the greatest extent possible (e.g. because the want to demonstrate the vulnerability of the system, or because they favour ideas that are competitive to the WMG one). This means they will always defect, even if the costs of a possible punishment are higher then the gains from defection.

7.1.3.2   The Basic Agent Decision Process

Having presented the different agent types and explained their behaviour in cooperation situations, this section serves to illustrate the general agent decision

process in the simulation. The general process is visualized in the activity
diagram in Figure 7.1; we explain it here in more detail. For easier reference, we
have numbered the decision nodes in Figure 7.1 from 1 to 6 and will refer to
these node numbers when discussing the respective decision situations.

Throughout the simulation experiments, the agents are given the task of
obtaining different files. Once an agent has been given such a task and wants to
download a file, it sequentially goes through the nodes 1 to 6 and every time
decides how to act in the respective decision situation.

Node 1: Wanting to download a file, the agent starts in node 1 and considers
the opportunity costs of the download. These are the battery costs
accumulating from searching for and negotiating with potential cooperation partners. In case these costs are higher than the gain from
a cooperative download (i.e. higher than the saved energy if the cooperation is successful), the agent will download the file itself. Since
the energy saving from a cooperative download is dependant on file
size, we compare this information (and the resulting minimum energy
gains in case of a successful cooperation) against the minimum energy
needed for sending 1 cooperation request and 1 negotiation message.
Hence, if the file is large enough for the cooperative download energy
saving to outweigh the opportunity costs, the agent will move to node
2 (otherwise it will simply download the file itself).

Node 2: Being at node 2, the next consideration for the agent is whether it has
any neighbours (i.e. other agents in its visibility radius) that it could
possibly cooperate with (action ac_{search}). If this number is too low
(e.g. if it cannot find many other agents in its vicinity) then sending a
cooperation request has little point, since it only reduces the battery
life and will not result in enough responses; so the agent downloads
the file itself. If the number of neighbours is high enough, it sends out
a collaboration request to its neighbours via an IEEE802.11 WLAN
broadcast message and waits for responses[6]. To encourage interaction
we have set the threshold of "enough neighbours" to one agent. However,
the simulation can be modified to give each agent a separate threshold.

Node 3: Decision node 3 is similar to node 2. However, instead of checking
whether enough neighbours are in its vicinity, the agent checks whether
it has received enough positive responses to its collaboration request.
If this is not the case, it will download the file itself.

Node 4: Having received enough responses in decision step 3, the agent checks
who has responded and decides whether it wants to collaborate with
these agents. TIn reality a user might get a response from someone
who has betrayed him before or who for other reasons does not seem

---

[6]To reduce the number of messages and increase cooperation, agents will only send out
requests for collaboration if they have not received matching requests themselves and have
agreed to join a group.

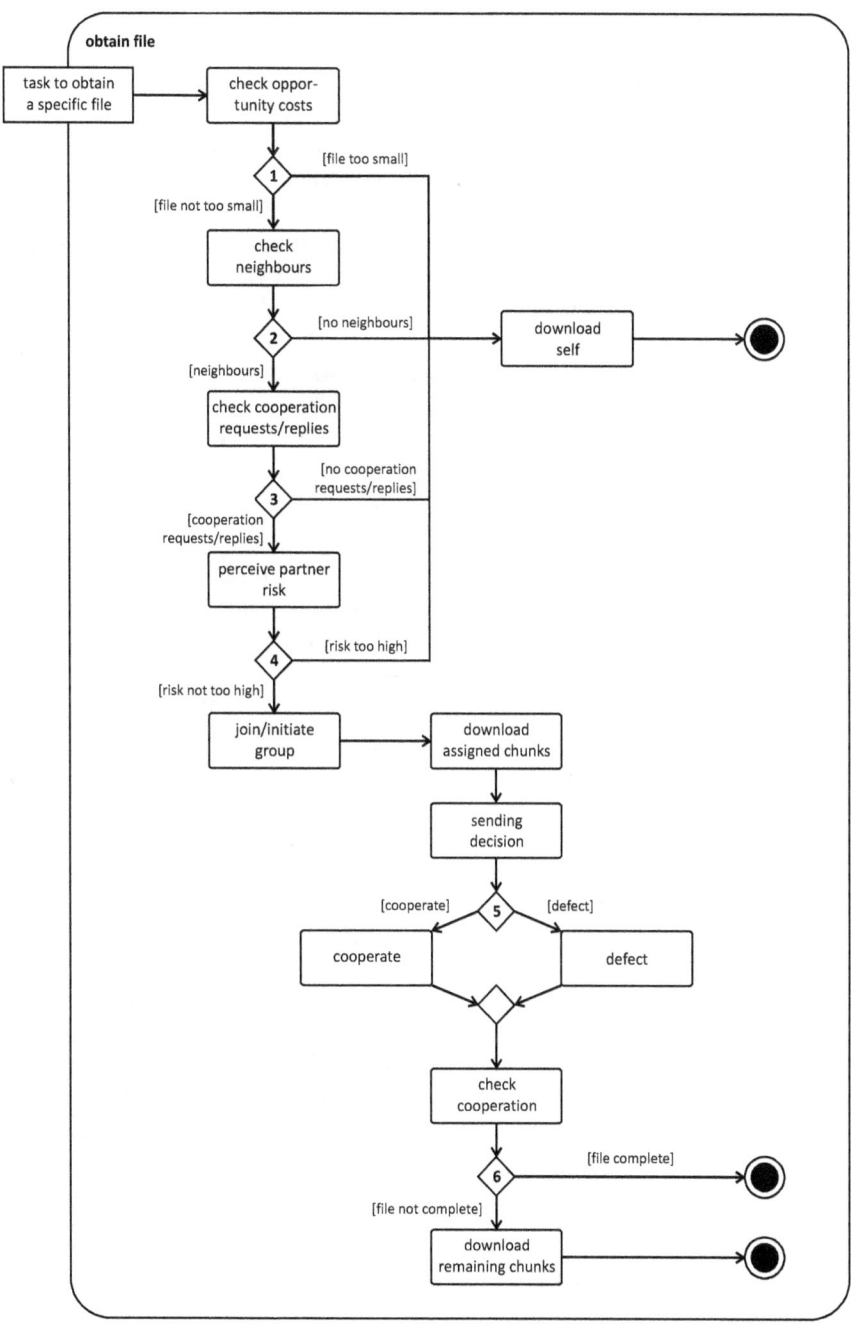

Figure 7.1: Activity Diagram of Agent Download Considerations

trustworthy in his eyes and therefore could decide not to cooperate with this user. If agents decide to collaborate and form a group this group elects a group leader (we set the agent initially sending out the cooperation request to be the leader of the group). This leader then initialises a protocol that assigns the chunks the group members should download and share equally to all agents in the group. This protocol furthermore determines a communication key for the interaction between the group members. In our simulation experiments we do not focus on this interaction, but assume the protocol assigning the chunks to be functioning. For reasons of simplicity in the simulation we assign one chunk to each agent of the group, with all chunks assigned having the same size. Our simulation can be extended to incorporate several chunks, adding a method to account for chunk numbers in order to identify specific chunks per agent. Once the agents have been assigned their chunk, in a next step they have to download the assigned chunk from the base station (action $ac_{download}$) in order to be able to send it to other agents afterwards[7].

Node 5: Having agreed to join or initialise a cooperation group and having downloaded the chunk(s) it is supposed to contribute, finally, in decision step 5, the agent has to decide whether to cooperate (i.e. send its share to its cooperation partners) or to defect (i.e. not send its share). This decision is based on the agent types described earlier. In case of the utility agents the decision depends on the assumed costs of sending and punishment by the agents.

Node 6: The last node in Figure 7.1 does not refer to an actual decision situation for the agents, but instead focuses on a control situation. When arriving at this node, the agents have made their sending decision (in the case they decided to send their share to the group, the sending has taken place). They now need to consider whether they have received the chunks from the other group members. In the case an agent has received chunks from all other group members, it will have obtained the complete file it needed. In the case it did not receive all missing chunks from its interaction partners it will have to obtain these chunks another way, i.e. by downloading the missing chunks itself or by searching for other collaboration partners.

Earlier in this dissertation we emphasised the aspect of movement and location in the WMG scenario. In the next section, we will therefore outline the movement pattern used in the basic WMG simulation experiments.

---

[7]In the simulation the assumption is being made that all group members initially do not have any chunks of the file they want to obtain. That is why every agent has to perform the download action at least once.

7.1.3.3   The Basic Movement Pattern

Despite its importance to various applications (e.g. urban planning, traffic forecasting as well as the spread of biological or software viruses), so far no models that represent human motion in a realistic way exist. The main reasons for this are a lack of tools and legal problems for monitoring time-resolved movement of individuals (González et al., 2008). This lack of large scale real movement traces results in the necessity to use abstract models for determining the movement of the mobile phone users. In the literature, a number of models to represent movement patterns of humans have been proposed. Camp et al. (2002) give an overview of the most common movement models used in ad-hoc scenarios similar to the WMG one[8]. We start these experiments by using an adaptation of one of the most basic models, which is used widely in literature: the random waypoint movement model[9].

The random waypoint movement model is a so-called entity movement model, meaning it assumes the movements of the entities in the system to be independent of each other[10]. In this model, the entities (i.e. in the MAS simulation the agents representing the human users with their mobile phones) choose random locations every time they move and pause in between two movement steps. In our simulation this means that the agents choose a random location in our boundless $[0,1]$-space every time they move. They find this new location by choosing a random movement distance $r_d \in [-1,1]$ (with the numbers in the interval being uniformly distributed[11])[12] and changing the current location by r_d. If r_d is negative the agents move $\mid r_d \mid$ into the 0-direction, otherwise they move $\mid r_d \mid$ into the 1-direction. As a result of the boundless movement area, if $r_d \in \{-1,0,1\}$, the agent will stay in its current position. We assume the agents to move at infinite speed between two locations. When having arrived at a new location agents the agents pause their movement and can choose to perform several actions. These actions include ac_{search}, i.e. searching for cooperation partners if wanting to download a file, as well as ac_{move} i.e. moving again immediately. Figure 7.2 shows the frequency by with which agents have chosen locations in the boundless $[0,1]$-space in the experiments conducted with the

---

[8]A detailed discussion of the different movement models is given in Chapter 9 when the effects of different movement models on the simulation results are discussed.

[9]Our adjustments to the classical random waypoint movement model are made as a result of our notion of location. Thus, we view location as a one-dimensional value only ($l \in \mathbb{R}$ and $l \in [0,1]$), whereas the classical random waypoint movement model assumes two dimensional movement. We therefore adjust the model to reflect one-dimension movement.

[10]The second kind of movement models are group movement models. Both kinds of models will be explained in more detail in Chapter 9.

[11]We use the Java random number generator as well as the random number generator implemented in Jason for our simulation experiments. These algorithmic random number generators are not truly random, they are algorithms that generate a fixed but random-looking sequence of numbers. As a consequence in the simulation experiments the random numbers generated will not be perfectly uniformly distributed. For reasons of analysis whenever using the two random number generators, we assume the algorithms to work perfectly and generating a perfectly uniform distribution.

[12]The absolute value $\mid r_d \mid$ of the movement distance is referred to as step length. The step length in a movement model will be of importance in Section 9.1.2.

basic WMG simulations. As expected from a random movement pattern, the locations are uniformly distributed[13].

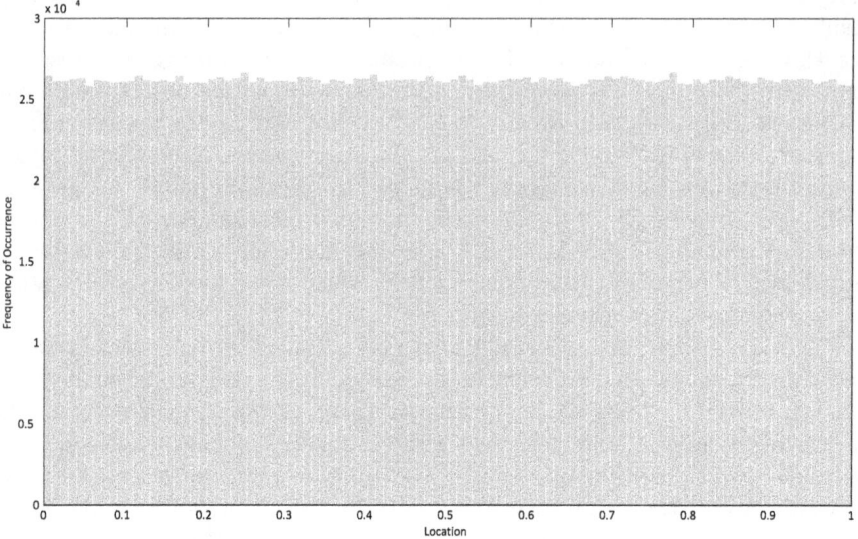

Figure 7.2: Distribution of the Locations chosen by Agents in the Basic WMG Experiments

Having explained how the simulation is designed with regard to the first two underlying factors of the IAD framework, namely the biophysical characteristics and the attributes of the community, we now turn our attention to the final underlying factor, i.e. the rules-in-use.

7.1.4 Rules-in-Use

In the earlier chapters of this dissertation we explained that the rules-in-use are the central element of our analysis, as one main goal of this dissertation is to analyse how governance decisions can effect a future system that is currently still in an early development stage. Governance decisions are understood as decisions on the changes of the rules-in-use. Using the simulation experiments we analyse several governance decisions (i.e. the integration of different enforcement mechanisms) in future WMGs. In particular, we focus on three particular enforcement mechanisms in the simulation experiments and analyse the possible impact they have on the predicted cooperation problems in WMGs.

Concerning the rules-in-use, in the basic WMG scenario, we start with simulation experiments that do not include any rules-in-use, to have a "worst case" reference point for the enforcement experiments. In Chapter 8 this assumption is

---

[13]The locations in the figure are rounded to the fifth decimal place.

lifted and specific rules-in-use are employed (in the form of different enforcement mechanisms).

7.1.5   Implementing the Basic MAS Simulation of the Wireless Mobile Grid Scenario

The implementation of our MAS simulation experiments was realised using the Jason Simulation Platform (Bordini et al., 2007). Jason uses an extension of the AgentSpeak agent-oriented programming language (Rao, 1996) to program the behaviour of individual agents (including $\mathcal{S}, \mathcal{O}b, \xi$ and τ) and thereby follows the Belief – Desire – Intention (BDI) model of agency (Rao and Georgeff, 1995).

BDI architectures originated in the work of the Rational Agency project at Stanford Research Institute in the mid-1980s. The main idea of BDI is that computer programs (i.e. the agents in the simulation) are viewed as if they have a "mental state". Thus, when programming the agents, computational analogues of beliefs, desires and intentions are used, enabling a form of reasoning by agents about their goals and the different options to achieve them.

Beliefs are information the agent has about the world. This information could possibly be out of date or inaccurate, however it represents the agent's view of the world. *Desires* are all the possible states of affairs that the agent might like to accomplish. Having a desire, however, does not imply that an agent acts upon it: they only potentially influence an agent's actions. Finally, *intentions* are the states of affairs that the agent has decided to work towards. Intentions may be goals that are delegated to the agent, or may result from considering options: we think of an agent looking at its options and choosing between them. Options that are selected in this way become intentions. Therefore, one can imagine an agent starting with some delegated goal, and then considering the possible options that are compatible with this delegated goal; the options that it chooses are then intentions, to which the agent is committed (Bordini et al., 2007). Jason makes use of the BDI concept by repeatedly executing the following control loop:

Step 1:   the individual agents look at the world, perceive their environment and other agents, and update their individual beliefs on this basis (it is important to note that not all agents perceive the same, but have individual percepts that can be different between all agents);

Step 2:   as a result, they deliberate to decide which intention to achieve;

Step 3:   and use means-ends reasoning to find a plan (a sequence of actions) to achieve this intention;

Step 4:   in the last step the agents then execute the plan in order to fulfill the intention.

Whereas the agent reasoning in Jason is written in AgentSpeak, Jason itself is developed in Java and allows the customisation of most aspects of an agent or the

MAS (Bordini et al., 2005). One of the customisations we made is the inclusion of a normative framework component. This inclusion will be explained in detail in Chapter 7.2. Besides the normative framework connection, in our simulation all environmental related aspects as well as the mathematical calculation of utilities by the individual agents and the logging of the simulation data (except for the normative states, which are recorded in the normative framework) are programmed in Java.

To not distract from the main reasoning of this dissertation, we do not include the details of the implementation at this point, but refer the interested reader to Appendices A and B. Appendix A contains the complete code of all Jason agents (including the code of the agents of the later experiments). In Appendix B we give a detailed explanation of the basic WMG agent code.

Having presented the agent reasoning, we now turn our attention to the question of how the Jason agents and the normative framework described earlier can be connected in order to allow for normative reasoning by the agents at the run time of the simulation.

7.2 Connecting Normative Frameworks and the MAS Simulation in the Action Arena

As discussed in Section 5.5, as a result of the special features of the design-time model, when developing it, a system designer could not take into account participants' reasoning capabilities and is required to incorporate artificial artifacts in the model. In case of our WMG these artifacts included the explicit tracking of whether a channel or a handset is busy or not. This further required adjusting the power and permissions to perform certain normative framework actions accordingly, or the necessity to explicitly model time as well as its transition. With the agent simulation in place, we can avoid these problems and concentrate on the modelling of the normative aspects only. In this section we will therefore explain how the design-time model presented in Section 5.4 (page 106) can be cleared of all artifacts. Afterwards we highlight how the agent simulation and hence the participants with reasoning capabilities can be connected to the normative framework.

7.2.1   The Run-Time Model

In starting to develop the run-time model, the first important thing to keep in mind is that for a given normative system, both the design-time and run-time model should have the same normative intentions, making the design-time model a good starting point for the development of the run-time one. A sensible next step is to remove rules and conditions that deal with simulating a running system. Figure 7.3 gives one example of design-time specifications that are being removed because of not being required in the run-time case[14]. All the

---

[14]The complete run-time specifications as well as their comparison to the design-time specifications are provided in Appendix B.

specifications that are printed in bold are kept in the run-time model, whereas all other specifications will be removed.

```
 1   institution grid;
 2
 3   type Handset;
 4   type Chunk;
 5   type Channel;
 6   type Time;
 7
 8   %% exogenous events %%
 9   exogenous event download(Handset,Chunk,Channel);
10   exogenous event send(Handset,Chunk);
11   exogenous event deadline
12   exogenous event clock;
```

Figure 7.3: Declaration of Types and Events in the Run Time WMG Model – Example

The purpose of the run-time model is solely to observe and keep track of normative behaviour, not the system's behaviour. Thus, it only monitors the external events resulting from agents' actions and does not predetermine all agent behaviour. As a consequence of moving to a run-time model, we no longer need to be concerned with modelling system data.

Concretely for our example, this means that the model does not have to track whether a channel is being used at a given moment or that a particular handset is incapable, from a technical perspective, of sending or receiving chunks. These are properties of the agents, but not normative information as such.

This means starting from the original design-time specifications, we can remove all rules that deal with these issues.

One implication of not having to model busy states of handsets and the base-station any more is that the exogenous event `clock` (Figure 7.3, line 12) and the normative framework event `transition` are no longer required. By the same reasoning, we longer need fluents to indicate that a handset or channel is engaged or to indicate elapsed time. With all of these events and fluents gone, the type `Time` is no longer needed either (line 6).

Removing the rules involving these events and removing the fluents from the remaining rules as well as the concretisation statements or the time transition we are almost left with that part of the run-time specification that is printed in bold print. One component that in the design-time model is explicitly controlled by the time transition, is the deadline event, for which an normative framework event was automatically generated when the time transition had reached point 1. In the run-time model this deadline is not pre-defined in length by the normative framework, but can be decided upon by the agents. At the point of the deadline an exogenous (and no longer normative) event is generated by the agents and the obligation statements will be checked as a consequence. That is the reason why in the run-time model we need to introduce the exogenous event `deadline` and can delete the respective normative event as well as the statements relating to it. However, modelling the deadline as an exogenous event means that we need to add the permission to perform this exogenous deadline event to the normative framework specifications. To highlight their addition to the run-time model we have printed the respective definitions in italics (Line 11 in Figure 7.3).

In the design-time model we penalise misbehaviour by taking away the power of a handset to receive chunks. While this may be a reasonable simplification in a design-time model for verification purposes, it cannot be enforced in a running system unless one expects agents to penalise themselves. Instead, the system notes the violation and agents may use this information in future interactions with the offending agent. Thus, we keep the violation event `misuse` that still can be generated as a result of the deadline passing for the obligations that the agents have agreed to fulfill, but delete the consequence rules that take away power from the agents as a result of the misuse, as well as any rules that assume the ability to terminate the power of agents.

In contrast to the design-time model, in which the chunk attribution to agents (i.e. the initial configuration of the agent/chunk combinations indicated by the 'initially' identifier in the normative specifications) is pre-determined, in the run-time model this changes. Thus, one cannot pre-determine the assignment configuration in advance, but it is the agents who after meeting and deciding to cooperate and negotiating which agent is to download and share which chunk. Hence, the concretisation part (i.e. part 3.2 in Figure 5.7 on page 102) of the normative specification file is determined at run-time and added to the template specification, whenever agents inform the normative framework (see sections 7.2.2 for details) about their negotiation results[15].

7.2.2   Monitoring Dynamic State

The UML component diagram in Figure 7.4 shows the components of the run-time MAS simulation model. The two main components of the run-time architecture are the Jason MAS simulation at the bottom of the figure as well as the normative framework (with the normative template specifications) at the top of the figure. We link Jason with the normative model and answer set solver using system calls.

In order to maintain the normative state in our running system we introduce a special type of agent or entity: the Governor, which when created is given the template part[16] of the normative framework specifications. The governor is a well-known concept in the agent community and has been used in earlier works such as, for example, the PhD dissertation of Noriega (1997).

The governor component – which is implemented in Java – is conceptually located between the Jason MAS simulation and the normative framework and (i) handles all the instantiations of the normative framework, (ii) stores the normative states and the respective domain files, and (iii) helps to decouple the normative framework and the agents. Thus, using the governor entity, the agents do not need to know any specifications of the internal structure, the

---

[15]As well as the concretisation component of the normative specifications, the domain file is also created dynamically in the run-time case.

[16]See Figure 5.7 on page 102 as well as the description of the figure for more information.

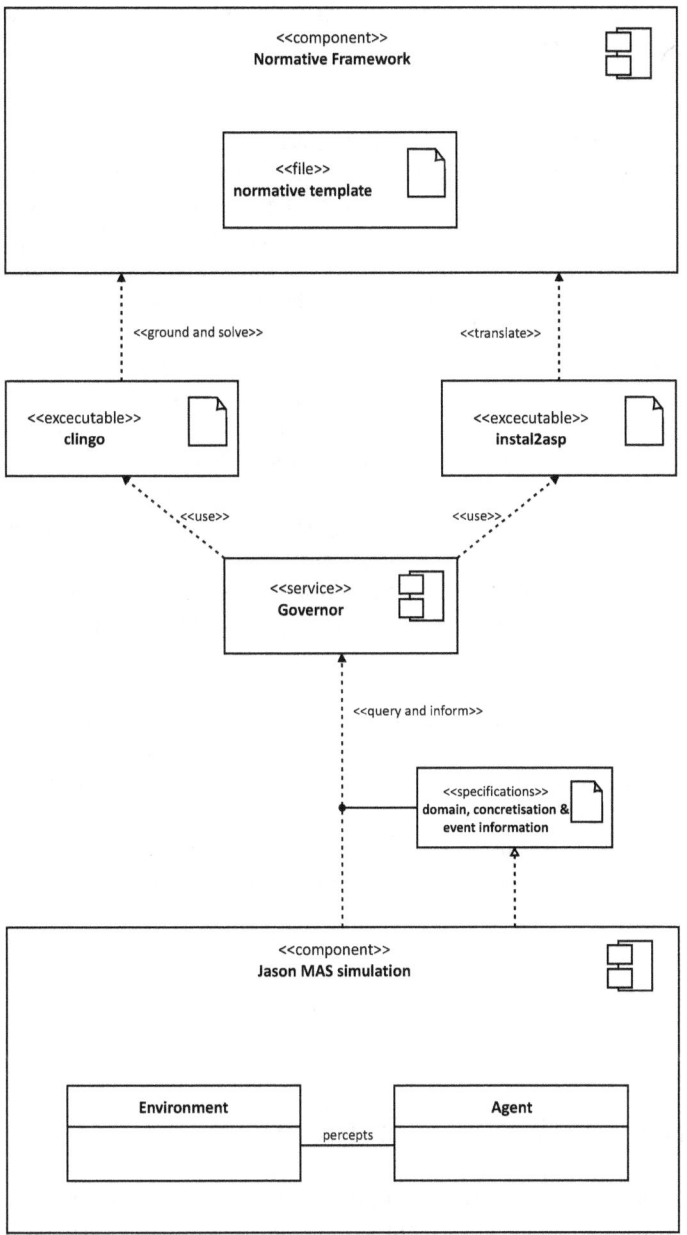

Figure 7.4: The Components of the Run-Time Normative MAS simulation

semantics or the syntax of the normative framework, but can pass on simple information to the governor who then translates this information to be usable in the normative framework component. In detail, the interaction between the three components just described works as follows: When agents agree to collaborate, one agent of the group contacts the governor with the specifications of the collaboration to establish a *contract* through a new instantiation of the normative framework. The specifications that are passed on to the governor include a unique group ID (that is used as an identifier for the group) as well as the names of all group members and information about which chunk which group member has been assigned to the governor. The governor uses this collaboration information to create the domain file as well as the concretisation part of the InstAL specifications and combines these with the template part. These complete specifications are then used to determine the first state of the agents' contract. When the state needs updating (a new exogenous event takes place)[17], the governor combines the template part of the InstAL specification with the current state and solves the correpoding *AnsProlog* program with a time-step of length 2. The resulting answer set is then parsed to obtain the new normative state, which the governor stores in respective data structures.

Having the information for the initial contract as well as tracking the normative state of each contract by analysing the respective exogenous events, the governor can act as a normative query processor for the agents. Contracting agents can query the current state and obtain consequences of potential actions.

At this point, we have implemented three kinds of queries that are useful for agents:

- queries about the current state, including the norms applying to that state (e.g. "What norms affect my current situation?" or "Given the current situation, following the norms, am I allowed to execute action ac_Y?"),

- queries about the possible impact of the agent's own actions (e.g. "What is going to happen if I take action ac_Y (e.g. download chunk x1 from channel 1)?"), and

- general queries on what might happen in the future (e.g. "What would happen if a series of actions (e.g. events e_A, e_B, e_C and e_D) take place?").

7.3   Simulation Experiments

Having discussed the structure and implementation of the MAS simulation, in this section we now focus on the simulation experiments. We start by introducing our simulation hypotheses and in a second step derive the simulation setup from these hypotheses. The results of the simulation experiments are presented in Section 7.3.2.

---

[17]We have implemented the MAS simulation in such a way that every exogenous event will have a reference to the contract (i.e. agent interaction group) it refers to. This reference is the unique group ID. Using this ID, the governor is able to distinguish between several contracts and can attribute the exogenous event to the correct normative framework instance.

7.3.1   Simulation Hypotheses and Setup

Recalling Chapter 3, one of the main drivers behind the WMG idea was that in WMG cooperation has the potential to reduce energy consumption of mobile phones.

Hypothesis 1: Higher cooperation levels in the WMG reduce the energy consumption in the WMG.

Despite this generally assumed energy advantage of WMGs, as expressed in Section 3.2, without any enforcement the cooperation (and thus the WMG advantages) is at risk if the participants try to cheat in order to save their own battery, which is required for sending. We also explained that not all users necessarily follow the same decision rationales, but can react differently in cooperation situations based on their personal attitude to cooperation and the WMG. Because cooperation is very much dependent on the mobile phone users' decisions, and that these decisions are guided by their cooperation attitudes (which we summarized as behavioural types earlier), we hypothesize the following:

Hypothesis 2: The composition of the agent population (in terms of the numbers of each behavioural types present in the population as well as the percentage of each type) has an effect on the energy consumption.

One huge advantage of MAS is that it can incorporate entities (i.e. agents) with very different decision concepts. In theory, one can think of an infinite number of behavioural types, however in order to keep the simulation as simple as possible and still allow for very different decision processes, we use the three different behavioural groups presented in Section 7.1.3. Thinking about these different groups and their expected interaction behaviour – assuming Hypothesis 2 to be correct – we can extend our analysis to check the impact that particular groups in the population have on the energy consumption. To give an example of this detailed analysis, we assume that, as honest agents will always cooperate if given the chance, an increase in the percentage of honest agents in the total population should increase cooperation and reduce defection, and therefore reduce the overall *energy consumption rate*[18] (Hypothesis 2.1). If no honest agents exist in a WMG system, the agents in the system are very likely to cheat – either because of their malicious preferences, or because their utility considerations indicate that without any punishment to fear, defection is the better option. As a result, all interaction attempts should result in not achieving cooperation. However, for the initial communication to find cooperation partners (that the agents can cheat on), energy is required (for sending the communication

---

[18]In the experiments we understand the term *energy consumption rate* as the quotient between the energy consumed in the simulation experiments (E_{sim}) and the battery power that would have been consumed if no WMG was in place and the agents would download everything themselves, without even considering cooperation (E_{noWMG}). An energy consumption rate of 1 therefore implies that in the simulation experiments the total energy consumption is equal to a scenario with no WMG, whereas a value higher / lower than 1 implies that the system is battery-wise disadvantageous / advantageous in comparison to a situation without a WMG.

messages). We hypothesize that this additional energy commitment will increase the overall energy consumption with all agents still having to download everything themselves. Therefore a completely unsuccessful WMG (in cooperation terms) – where every mobile phone user downloads everything himself without even trying to find cooperation partners – leads to higher energy consumption than in situations where no WMG is present in the first place (Hypothesis 2.2).

Hypothesis 2.1: The percentage of honest agents in the population is correlated to the energy consumption rate.

Hypothesis 2.2: In systems in which no honest agents are present, the overall energy consumption is worse than in scenarios without any WMG, i.e. when all agents download everything themselves.

Finally, we pointed out the importance of location in the earlier chapters of this dissertation. As a result of the bounded rationality assumption made, agents will not be able to perceive (and interact with) all other agents in the system, but only agents that are within their visibility radius at a given point of time. The more agents within the visibility radius, the more agents a particular agent can interact with. If no other agents are in the radius, the agent must download its required chunks on its own. We therefore hypothesize that $\rho_{neighbourhood}$ (i.e. the average number of other agents an agent can detect) has in impact on the cooperation attempts of the agents and therefore also on overall energy consumption.

Hypothesis 3: The neighbourhood density ($\rho_{neighbourhood}$) affects the overall data transfer and consequently the energy consumption.

For the experiments in the later chapters, it will be important how likely it is to interact with a given agent again. One parameter to determine this likelihood is $\rho_{neighbourhood}$. However, $\rho_{neighbourhood}$ on its own is not sufficient, as $|\mathcal{A}|$ can also influence the probability that agent pairs interact repeatedly. Thus, with a fixed $\rho_{neighbourhood}$, if $|\mathcal{A}|$ increases, the pool of agents the neighbours can come from increases as well. If agents move randomly, interacting with an agent again can be compared to randomly choosing $\rho_{neighbourhood}$ elements from $|\mathcal{A}|$. With an increasing $|\mathcal{A}|$ the likelihood for a particular element being chosen from $\rho_{neighbourhood}$ decreases. As we do not use any enforcement mechanism, which implies that the agents will not use information from past interactions with other agents, every interaction is like a new interaction to them. Therefore, we hypothesize that in the basic WMG scenario, the size of $|\mathcal{A}|$ will obviously affect the absolute number of interactions (and therefore also the number of defections), but does not affect relative measurements such as the cheating ratio or the energy consumption rate, etc.

Hypothesis 4: In the basic WMG simulation settings the absolute number of agents in the system ($|\mathcal{A}|$) does not affect the relative simulation results.

In order to test the hypothesis just presented, we execute factorial experiments using our simulation. Each of these experiments is structured similarly to the design-time model presented in Section 5.4. Each agent is repeatedly given the task of obtaining a file and then deciding how to go about doing this by using the decision process shown in Figure 7.1. In order to increase cooperation and make comparison between the results easier, we opted to have a large number of agents wanting to download the same file and to use only five different files, assigned at random to the agents. All these files have exactly the same file size, which is high enough to make cooperation interesting (with regard to the opportunity cost decision)[19]. With regard to the files in the simulations, we furthermore assume that after obtaining a file, the agents do not keep it permanently; hence, if downloading the same file again, they cannot rely on old chunks of the file being present on their mobile phone.

Besides this general setup, in order to test our hypotheses we must decide how to structure our experiments and which values to assign to the variables that define our simulation. Table 7.1 shows the setup of the Basic WMG Scenario experiments resulting from these considerations.

The first input parameter that we test corresponds to Hypothesis 4. It is the number of agents in the simulation ($|\mathcal{A}|$). The upper limit of $|\mathcal{A}|$ was chosen due to technical simulation restrictions (i.e. the computational resources available only allowed for executing stable simulations with a maximum of 800 agents), whereas the remaining two numbers were picked following the power distribution on page 144.

The number of interactions per agent was kept fixed at 50 throughout all experiments. It was chosen for statistical purposes, i.e. to have a large enough number of interactions to be able to apply means comparisons, which is required for ANOVA analysis (Field, 2009), and at the same time small enough to be able to execute a large set of different experiments.

The next three input parameters affect the composition of the agent population by determining which proportion of each agent behavioural type is present in the simulation. Given the full partitioning constraint, the value of the third percentage can always be determined by the first two. It is consequently not independent as required by ANOVA analysis. That is why in ANOVA analysis at most two agent behavioural type percentages are required to determine the remaining percentages being considered. Using the three percentages we have the input parameters required for Hypotheses 2–2.2. Finally, the input parameter $\rho_{neighbourhood}$ is relevant for Hypothesis 3. We have chosen the relatively high neighbourhood densities of 10 and 20 to ensure the likelihood of agents finding another agent within their vicinity that wants to obtain the same file, thus giving an incentive to cooperate and to better determine the effect of the parameter[20].

---

[19]This decision makes node 1 in Figure 7.1 always result in the agents checking their neighbourhood and not downloading the file themselves straight away because of high opportunity cost. One might argue that this decision step therefore could have been neglected in the simulation. We nevertheless kept it in the simulation for testing purposes as well, to leave options to extend the simulation later on with a larger variety of files.

[20]In the real world, limits to the number of maximum connection per handset can exist. These limits are not exceeded by the chosen $\rho_{neighbourhood} = 20$.

Table 7.1: Experimental Setup for Basic MAS Simulation of the WMG Scenario

Input Factors	Values
$\mid \mathcal{A} \mid$	200, 400, 800
Number of Interactions per Agent	50
Utility Agents as % of $\mid \mathcal{A} \mid$	0, 25, 50, 75, 100
Malicious Agents as % of $\mid \mathcal{A} \mid$	0, 25, 50, 75, 100
Honest Agents as % of $\mid \mathcal{A} \mid$	0, 25, 50, 75, 100
Enforcement Mechanism	None
Movement Pattern	Simple Boundless Random Waypoint
$\rho_{neighbourhood}$	10, 20
Measurements	**Type**
Number of Defection Actions	$\mathbb{N}$
Number of Cooperation Actions	$\mathbb{N}$
Communication-related Data Transferred per Agent per Interaction	$\mathbb{R}$
Download-related Data Transferred per Agent per Interaction	$\mathbb{R}$
Reception-related Data Transferred per Agent per Interaction	$\mathbb{R}$
Sending-related Data Transferred per Agent per Interaction	$\mathbb{R}$
Integrity Constraints	**Formula**
Full Partitioning	Utility Agents % + Honest Agents % + Malicious Agents % = 100

The $\rho_{neighbourhood}$ parameter was indirectly set in the experiments by adjusting the visibility radius of the agents depending on the number of agents.

Looking at the measurements, two sets of measurement parameters can be distinguished. The first set is comprised of numbers indicating the number of defects as well as cooperation actions in the simulation. This allows us to make statements about the relative and absolute defects and thus the cooperation success of the WMG. Therefore, it is important to note that the absolute figures depend on $|\mathcal{A}|$ and can only be compared for settings with equal numbers of agents.

The second set of measurements considers the amount of data transmitted for downloading, sending, receiving as well as communicating per agent. Using this data as well as the power value and data rate specifications given in Tables 3.1 and 3.2 on page 64, we are not only able to determine the overall energy consumption and energy per bit ratio for the overall system, but can also distinguish these figures according to the behavioural groups (or even single agents) and communication protocols. This enables us to verify Hypotheses 2–2.2 as well as 3 and 4. As a result of knowing the number of interactions each agent has to perform (i.e. the number of times it has to obtain a file) as well as the size of the files the agent has to obtain, we are able to calculate the energy consumption required if the agents do not try to participate in the WMG, but download everything themselves. This allows us to verify Hypothesis 1, i.e. the general underlying assumption of WMGs that higher cooperation leads to a better (i.e. lower) overall energy consumption rate.

7.3.2 Simulation Results

The simulation experiments consist of running 50 experiments for each parameter combination given in Table 7.1. This means that a total of 50 simulation runs for 90 parameter combinations (i.e. 4,500 simulation runs in total) are executed for the basic WMG simulation experiments for this chapter. The aggregated results for all runs per parameter combination are logged in txtfiles, which are used for the statistical analysis.

As previously mentioned, we used ANOVA to test the significance relationship between the independent variables (i.e. the parameters in the simulation) and the dependant variables (i.e. the number and ratio of defections as well as energy consumption rate) in our experiments and thus test Hypotheses 2–4. However, before performing the ANOVA analysis, we focus on Hypothesis 1; this predicts a correlation between the cooperation (or defection) level and the battery consumption. For the purpose of checking this correlation all cheating events as well as all interaction group sizes (both per simulation run and per parameter setting) are logged in the experiments. As pointed out in Section 5.4.1 on page 108, we set up the experiments in such a way that in each interaction group, each agent is assigned one chunk to download and share. As a result of this setup, the sum of the interaction group sizes gives us the number of chunks that are supposed to be sent in total by the agents. In combination with the sum of the cheating events per run this number is used to determine the defection rate per experiment, which is compared to the energy consumption in the respective run. The calculation of the correlation of the defection rates and the respective energy consumption results in the correlation value 0.9932, which specifies that the energy consumption and the defection rate are statistically strongly positively correlated, i.e. that high energy consumption rates result from high defection rates. As cooperation rates are the direct counterpart of defection rates, this also implies that an increase in cooperation results in lower energy consumption rates and affirms Hypothesis 1. With Hypothesis 1 being proven, we now focus the analysis on Hypotheses 2–4 in order to determine which factors influence the cooperation ratio and the energy consumption. For this purpose we perform a multi-way (n-way) ANOVA, which tests the effects of multiple factors (i.e. population composition, $\rho_{neighbourhood}$ and $|\mathcal{A}|$) on the mean of the battery consumption rates. Table 7.2 shows the results of the ANOVA for all three input factors for which we have formulated hypotheses. For each of these, the null hypothesis is that the factor does not have an effect on the energy consumption rate, whereas the alternative hypotheses state that an effect of the respective factor exists. ANOVA tests the null-hypothesis. When the p-value of the test is significant[21], then it is assumed that enough evidence is found to reject the null hypothesis and assume the alternative hypothesis.

Table 7.2: Analysis of Variance for Hypotheses 2–4

Source	Sum of Squares	Degrees of Freedom	Mean Squares	F	Prob > F (= p-value)
Population Composition	261.445	14	18.6747	25197.22	< 0.0001
$\rho_{neighbourhood}$	0.482	1	0.4817	649.88	< 0.0001
$|\mathcal{A}|$	0.056	2	0.0282	38	< 0.0001
Error	3.322	4482	0.0007		
Total	265.305	4499			

Table 7.2 shows that the p-value (given in the last column of the table) for all three input factors is approximately zero and is therefore significant. The n-way ANOVA generally tests the effects that the group of n given input factors (as a group) has on the dependent variable (i.e. the energy consumption rate in our simulation experiments). The n-way ANOVA focuses on results of the group of input factors. This implies that we can dismiss the null hypothesis that the group of input factors we have chosen does not have an effect on the energy consumption rate. However in order to determine the contribution of the different input factors, this effect needs further analysis in the form of a post hoc test. The test we choose to apply for the in-depth analysis is Tukey's test. If the sample sizes of the experiments are equally large and have similar group variances, and if the experiments consist of large sample sizes (which is the case

---

[21]The typical significance level chosen is 0.05, i.e. any p-value below or equal to that value is significant. We use this significance level throughout this dissertation.

in our simulation experiments), this test can be used to analyse differences in means in the input factors as well as for the variables chosen for the different input factors. Tables 7.3–7.5 show the results of this detailed post hoc test.

Table 7.3: Analysis of Variance in Comparison of the Population Composition – Post Hoc Test

Source	Sum of Squares	Degrees of Freedom	Mean Squares	F	Prob > F (= p-value)
Population Composition	261.445	14	18.6747	21699.7	< 0.0001
Error	3.86	4485	0.0009		
Total	265.305	4499			

Table 7.4: Analysis of Variance in Comparison of $|\mathcal{A}|$ – Post Hoc Test

Source	Sum of Squares	Degrees of Freedom	Mean Squares	F	Prob > F (= p-value)
$|\mathcal{A}|$	0.056	2	0.02816	0.48	0.6204
Error	265.249	4497	0.05898		
Total	265.305	4499			

Table 7.5: Analysis of Variance in $\rho_{neighbourhood}$ – Post Hoc Test

Source	Sum of Squares	Degrees of Freedom	Mean Squares	F	Prob > F (= p-value)
$\rho_{neighbourhood}$	0.482	1	0.48167	8.18	0.0043
Error	264.823	4498	0.05888		
Total	265.305	4499			

The results of the post hoc tests allow us to verify Hypotheses 2–4. The post hoc tests for Hypotheses 2 and 4 show a significant p-value (i.e. 0 and 0.0043 - see Tables 7.3 and 7.5). This indicates that we can reject the null hypotheses that these two values do not have an impact on the energy consumption rate, whereas the p-value for $|\mathcal{A}|$ is not significant (i.e. $0.6204 > 0.05$) in Table 7.4, indicating that for Hypothesis 3 we cannot reject the null hypothesis.

Analysing Tables 7.2–7.5 more closely, one further result of the ANOVA test is that the population composition is the factor that influences the energy consumption the most. This is indicated by its sum of squares value in Table 7.2 as well as the individual p-values of the three factors in the Tables 7.3–7.5, in which the population composition has the lowest and therefore most significant value, indicating that analysing the population composition in more detail is worthwhile. As a result of the simulation setup shown in Figure 7.1, 15 different compositions of the total population are checked in the simulation experiments. We use a Tukey's test to compare the means of the energy consumption rates of these 15 compositions. The results of the analysis are shown in Figure 7.5, which shows the means comparison of all population compositions in comparison to settings with 100% utility maximizing agents. In Figure 7.5 all means that are not significantly different from the means we compare with are coloured in black, whereas all means for population compositions with a significant difference are coloured in red. The results indicate that all population compositions with no honest agents have means that are not significantly different to each other and are higher than all means of compositions that include honest agents. The means are averaged around the value 1, i.e. at a point where $E_{sim} = E_{noWMG}$. With the increase of honest agents in the system the mean values decrease significantly, whereas the composition of the remainder of the population does not seem to be significantly relevant to the energy consumption, as all populations with the same percentage of honest agents are in the same means range (i.e. on one vertical line), independent of the percentages of utility or malicious agents. These findings support Hypothesis 2.1.

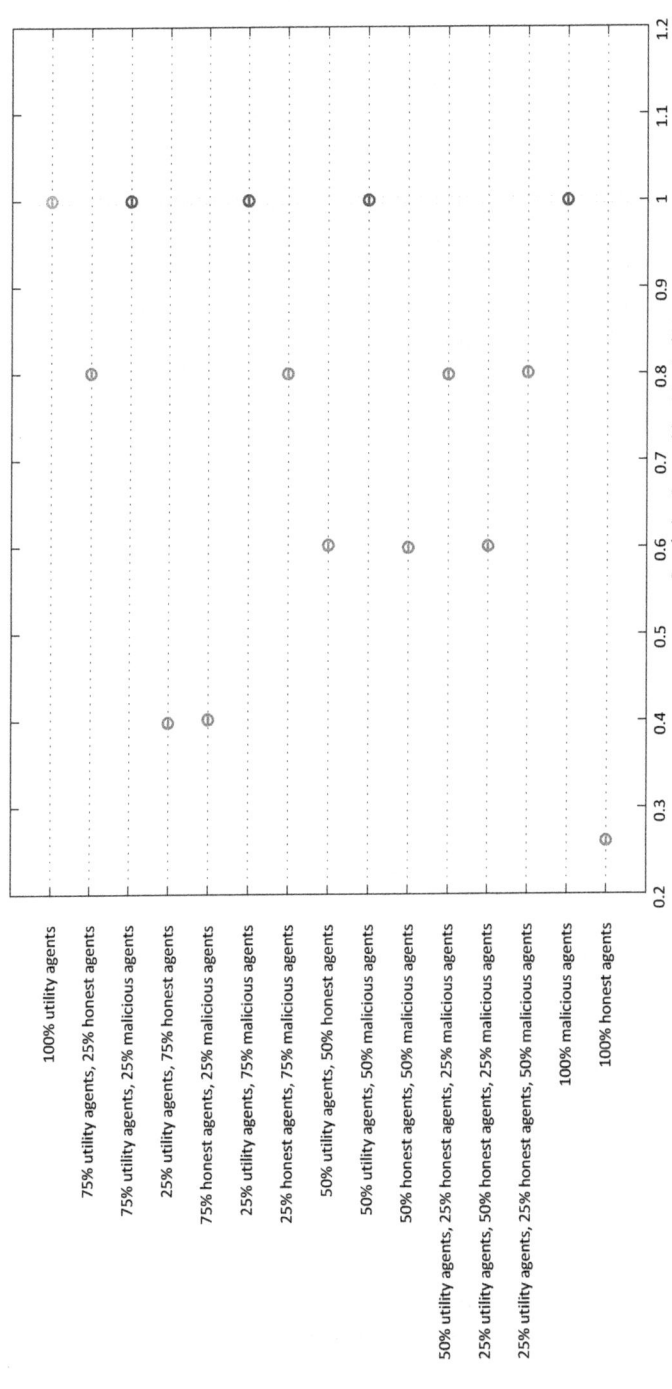

Figure 7.5: Multiple Comparison (Tukey's Test) Results of Population Composition Marginal Means – Post Hoc Test

To tests Hypothesis 2.2 we perform a one-sided t-test on the simulation runs with populations settings without any honest agents. In the t-test we compare two means: the means of the difference between the battery consumption with the WMG and without a WMG, and the means of the no-WMG scenario. The result of the t-test is a significant p-value, which allows us to conclude the correctness of Hypothesis 2.2.

Besides the population composition, as discussed above, $\rho_{neighbourhood}$ and $|\mathcal{A}|$ were candidates for factors having an impact on the energy consumption. Having run the simulations with different values for these two input factors, we now analyse to what extent the choice of values for the input factors affects the simulation results. For this purpose we use Tukey's test again to compare the means of the energy consumption rates for the different values of these two input factors. Figures 7.6 and 7.7 show the results of the test. In these figures the marginal means coloured in blue are the ones that the test used for the comparison. If no significant difference is found the marginal means are coloured in grey, otherwise they are red.

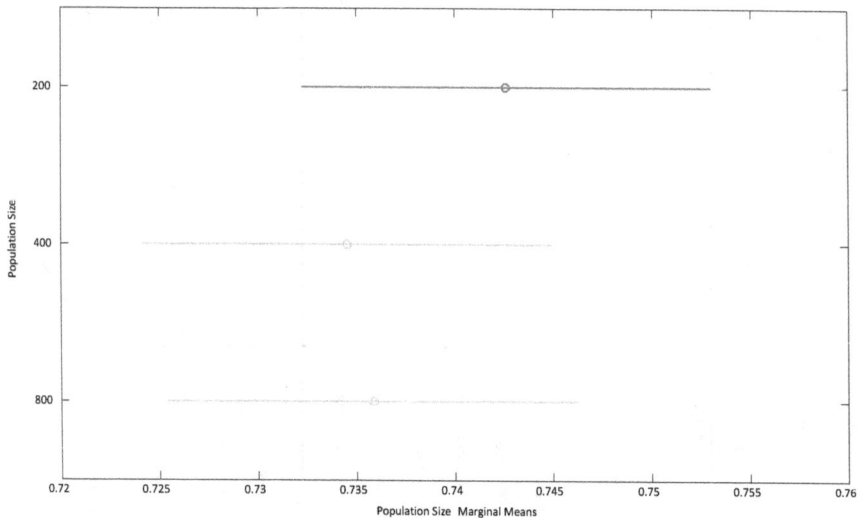

Figure 7.6: Multiple Comparison (Tukey's Test) Results of $|\mathcal{A}|$ Marginal Means – Post Hoc Test

From the test we see that the marginal means for the two different values of $\rho_{neighbourhood}$ are significantly different (the two bars do not overlap vertically), with the experiments with $\rho_{neighbourhood} = 20$ having lower marginal means than those with $\rho_{neighbourhood} = 10$. This indicates that a higher neighbourhood density results in lower energy consumption rates. Reasons for this are the potential larger number of cooperation partners an agent can choose from, and cooperate with in case of a higher $\rho_{neighbourhood}$. Thus if more agents can cooperate, the individual chunks each agent has to contribute is reduced, and if

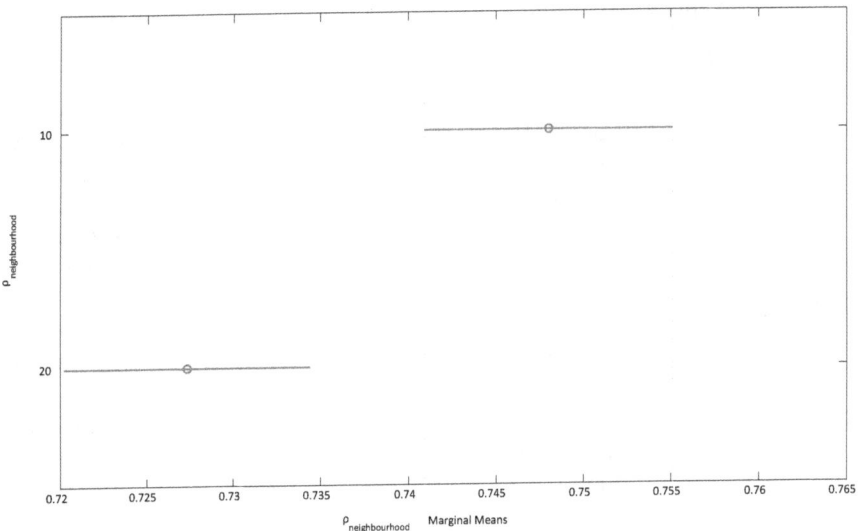

Figure 7.7: Multiple Comparison (Tukey's Test) Results of $\rho_{neighbourhood}$ Marginal Means – Post Hoc Test

the cooperation is successful the energy consumption rates decreases. In contrast to $\rho_{neighbourhood}$, for the different values of $|\mathcal{A}|$ no significant difference can be detected in the marginal means.

7.4 Summary

In this chapter we have introduced the basic WMG simulations without any enforcement mechanisms. We furthermore explained in detail how our simulation, as well as the experiments, are constructed. We have explained the general simulation implementation as well as the different simulation components, by using the underlying factors of the IAD framework (i.e. the biophysical characteristics, the attributes of the community and the rules-in-use) as presentation structure. We have conducted the first set of simulations with this basic WMG simulation. The results of these experiments show that the cooperation level in a system is strongly correlated with the population's energy consumption. Further results are summarized in Table 7.6.

The results show that in scenarios without any enforcement, the population composition (especially the percentage of honest agents) as well as the neighbourhood density are factors influencing the energy consumption, and thus all the hypotheses in this chapter can be assumed to be valid. For WMGs to be successful in the long run, the dependency on the percentage of honest agents in the system is a particular vulnerability. It calls for mechanisms to change this situation by encouraging agents to cooperate. In the next chapter we experiment

Table 7.6: Basic WMG Simulation Results Description Table

Input Factors	Measurement	α	Relationship
Population Composition	Energy Consumption Rate	> 0.99	The population composition influences the energy consumption in the WMG.
% Honest Agents	Energy Consumption Rate	> 0.99	A higher percentage of honest agents results in an overall reduction in energy consumption.
$\rho_{neighbourhood}$	Energy Consumption Rate	> 0.99	The neighbourhood density influences the energy consumption in the WMG.
$|\mathcal{A}|$	Energy Consumption Rate	0.38	The population size does not influence the energy consumption in the WMG significantly.

with enforcement mechanisms as one means to govern WMGs and thereby reduce the cooperation problems and dependencies on honest agents highlighted here. The results gained from the simulations without any enforcement mechanism conducted in this chapter thereby serve as a "worst case" reference point for the enforcement simulations.

Chapter 8

Enforcement Mechanisms for the Wireless Mobile Grid Simulation

The previous chapter showed that the WMG's energy consumption is particularly dependent on the number of honest agents in the system when no enforcement mechanisms are employed. Without any honest agents present, the WMG itself is at risk.

Enforcement offers an option to alter this situation, and in this chapter we explore the addition of enforcement mechanisms to the WMG simulation. Though WMGs are currently in an early prototyping stage, the results from these simulations provide a first impression of the effectiveness of the different enforcement mechanisms in reducing the cooperation dilemma in WMGs.

To begin our analysis of enforcement mechanisms and their effects on the WMG scenario, in Section 8.2 we present a taxonomy of enforcement mechanisms. We show how three of these mechanisms can be represented in the context of a WMG as well as how they may be implemented for our simulation purposes. Section 8.3 focuses on the simulation of the WMG, using the enforcement mechanisms. For this purpose – as in the previous chapter – we discuss the simulation hypotheses and the corresponding simulation setup first. This is followed by the presentation of the simulation results.

8.1 Retrospective: A Taxonomy for Fostering Norm Compliance

This section presents a taxonomy of different enforcement mechanisms based on which we select three mechanisms for implementation in the simulations.

We start our taxonomy by recalling Figure 2.1 on page 30 which shows the summary of the enforcement process described in Section 2.3.3.

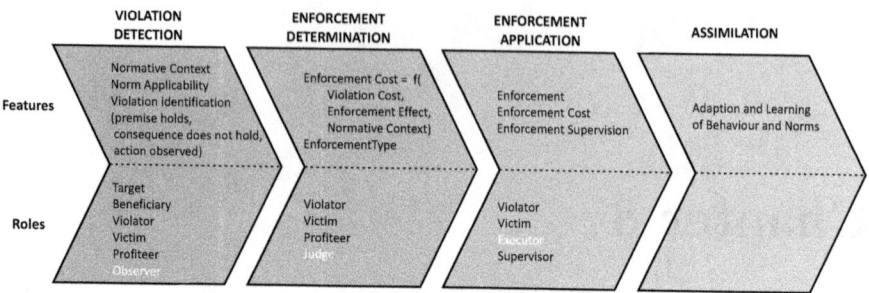

Figure 8.1: The Enforcement Process (Recalled)

In this enforcement process we distinguish four distinct phases, each associated with a set of roles for the participants in a system. The roles are not mutually exclusive, and system participants can play several roles simultaneously. The roles highlighted in white in Figure 8.1 are those that are responsible for the performance of the main enforcement-related actions in the phases. For the sake of simplicity, we assume the judge and the executor to be the same entity in an open distributed system without any central enforcement entity. This allows us to use the distinction of the different actors performing each highlighted role to arrive at a taxonomy for enforcement. This taxonomy is shown in Figure 8.2.

It is structured along two axes. Horizontally we have the phases of our enforcement process and vertically the distinction between regimentation as well as incentive-based enforcement (that – for simplicity – we refer to as enforcement) is shown. With respect to the enforcement process, we focus especially on the violation detection and the enforcement application stage. The assimilation stage is left out, because it is not a distinct part of the enforcement act, but the rather result of it. Due to the assumption that the judge that determines the enforcement sanction is also to the enforcing entity, we have no included the enforcement determination phase in detail in the taxonomy, and point only out examples in the fourth column.

The main taxonomy is derived from columns two and three, which specify actors for the roles highlighted in the violation detection phase and the enforcement application phase in Figure 8.1. The specified actors in the table are those presented in Section 2.3.3. By sensibly combining the different actors serving as observers and executors, we can distinguish six taxonomy elements (cells with a white background shown in the last column of Figure 8.2). We now describe these elements in more detail [1].

---

[1] The aim of this taxonomy is to present the basic elementary enforcement concepts. While any combination of them is conceivable, we present all taxonomy elements separately.

	observer	executor	sanction example	taxonomy (synthesis)
regimentation	infrastructure	infrastructure (mental states)	(impossibility of violation)	infrastructural control (white box)
		infrastructure (agent actions)		infrastructural control (black box)
incentive-based enforcement	normative framework empowered entities	normative framework empowered entities	infrastructural sanction	institutionalization normative framework empowered agents
	third-party observation (social forces)			normative framework assisted enforcement (third-party)
		social enforcement	vicarious retaliation / reciprocation	social control (third-party)
	second-party observation (agent acted upon)			social control (second-party)
		second-party enforcement	retaliation / reciprocation	promisee-enforced norms
		infrastructural entities	infrastructural sanction	normative framework assisted enforcement (second-party)
	first-party observation (actor)	first-party enforcement	self-sanction	self-control

Figure 8.2: A Taxonomy for Ensuring Normative Compliance

8.1.1   Utilization of Normative Empowered Entities

The utilization of normative empowered entities can be thought of in terms of the implementation of entities with normative power W (i.e. some kind of police) that participate in the system (in our example the WMG) and have permission from the system owner to punish negative behaviour (i.e. non-compliance) with sanctions, if detected. These normative empowered entities are given their special rights by the system designer who gives them the normative power W to perform specific sanction-related actions in the normative framework. However, in contrast to regimentation, the normative empowered entities do not control all actions but only act as enforcers if violations are detected. Detection of the violations is done by the normative entities themselves, who test the behaviour of entities and react to what they detect. Concerning the kind of sanctions that can be applied by the entities, several can be thought of, depending on the severity of the non-compliance, such as a complete exclusion from the WMG or penalty payments.

8.1.2   Normative Framework Assisted Enforcement (Second-Party / Third-Party)

The concepts of normative framework-assisted enforcement are very closely related to the idea of the utilization of normative framework entities. Normative empowered entities act as compliance enforcers, ensuring that entities can make use of sanctions. However, in contrast to the concept of the normative empowered entities, specifically empowered entities do not act as observers; this is done either by the entity that was acted upon, i.e. the one deceived by its transaction partner (second-party observer), or by a third-party, i.e. an entity that is not involved in the transaction but has observed the non-compliance of one actor. These observers then report their observations to the normative empowered entities to issue the sanction in order to assure compliance. Hence, in normative framework-assisted enforcement, additional communication is required. This gives rise to two problems. Firstly, the additional communication needed might result in a longer reaction time and, secondly, the normative entities empowered by the normative framework need to verify the testimonies made to them, as the reporting entities might purposefully lie in order to have sanctions issued to rival entities (and thereby profit themselves).

8.1.3   Social Control (Second-Party / Third-Party) and Promisee-Enforced Norms

Three other concepts that can be thought of where entities are directly involved in an interaction or third party entities act as observers are social control concepts (either with second and third party enforcement) and promisee-enforced norms. In these concepts, the observer entities act as executors, either by spreading negative reputation information about the violator in the system (in the case of social control) or by using the image information gained in new encounters with the same possible cooperation partner (in the case of promisee-enforced norms). In the case of promisee-enforced norms, if the cooperation partner does not perform as promised, that promisee can punish the non-compliance by for example not interacting with the agent once more, etc.

A contrasting case is that of social control, where third-party entities that observe a transaction form their own image of the transaction participants. Individual images of entities are shared between the system participants (either by the third party observers or the entities directly involved in the interaction) and hence are publicly criticised by society (e.g. with the help of gossip). Entities that did not comply with the norms of the normative framework therefore have to fear that every entity that receives the information about their non-compliance will not act with them in the future. Hence, in this example the whole society functions as enforcers. As was previously the case, one major shortcoming of these types of system is the spread of false information (a detailed description of the problems inherent in reputation systems was given in Section 2.4.3).

8.1.4 Self-Control

The last part of the taxonomy is self-control. In contrast to all other compliance-ensuring mechanisms presented so far, it does not include any additional party, but only the entity performing an action itself. This entity is assumed to have its own normative value system and to constantly check whether his actions are contradictory with the normative framework (i.e. the entity is its own observer). Based on the two normative value systems, the entity can then decide to punish itself. An example of such a self-control scenario in a WMG could be that a mobile phone user that did not contribute his assigned chunk, for example due to technical reasons, offers to resend the respective chunk to his transaction partner.

8.1.5 Selecting Enforcement Mechanisms

Having presented our taxonomy for mechanisms that help to ensure normative compliance, we must next determine which mechanisms will be used in this dissertation. Due to time limitations we cannot implement all mechanisms presented. When deciding which mechanisms to choose, we need to balance between covering as wide a range of different mechanisms as possible, as well as using mechanisms that are well established in the literature and therefore offer a solid basis for our simulations. As a result of these considerations we opted to implement the following three mechanisms: *normative empowered agents, promisee-enforced norms* (i.e. an image related concept) and *third-party social control* in the form of a reputation mechanism. We do intend to develop any completely new enforcement mechanisms, but as formulated earlier in the research questions: we want to show how enforcement mechanisms could be integrated in a WMG and what effects their integration might yield. This is why – especially for the reputation mechanisms implemented – we rely on existing work and include only existing and well-discussed mechanisms here.

8.2   Implementing the Enforcement Mechanisms

In the previous section we presented our taxonomy of mechanisms for ensuring compliance with the norms of a normative framework and selected three mechanisms to be used in our simulation. In this section we explain in detail how these three different mechanisms are integrated into the basic WMG simulation.

8.2.1   Normative Empowered Agents

When using normative empowered agents as enforcement mechanisms for the WMG simulation, in addition to the "normal" users with their mobile phones, agents that have been empowered to execute sanctions by the normative framework, and that police the WMG are added to the simulation. The idea of normative empowered agents is that some users participating in the wireless grids are helping to enforce cooperation in the system. They do so for some form of (financial) benefit. One typical example of normative empowered agents in the real world are police officers. They are paid by the state to help to ensure that the law is upheld. It is important to note is that they do not have superior physical abilities than "normal" citizens (i.e. no "superhero powers") but are given special rights and powers to carry out their job. Thinking of a WMG, enforcement agents might be mobile phone users that help to patrol the WMG system in exchange for free or reduced mobile phone service prices. As a result of the incentive to have cheaper mobile phone service, we assume the normative empowered agents always to act honestly in order not to lose this benefit.

The sanctions imposed could, for example, take the form of limitation of service or fines, or the sending of battery intensive messages to violators. The reception of such messages causes the battery consumption of the violators to rise and thus limits the gains from cheating. Here, we use the option of fining the cheating agents, where the amount of the fine is determined by the size of the chunks the cheating agents have not sent.

The implemented normative empowered agent in our simulation is in large parts similar to normal agents. One thing that differs however is that the normative empowered agent is not assigned a chunk (or perceives a chunk that it wants to download) at the beginning of a simulation and thus does not send cooperation proposals itself, but only becomes active whenever a cooperation proposal for any chunk is being sent. It will accept the cooperation proposal and join the group if is not already engaged in another group.

When assigned a chunk the normative empowered agent will always contribute its share. The energy consumption costs that the normative empowered agent incurs in doing so, are added to the total energy consumption in the WMG in order to account for the additional energy the enforcement mechanism consumes. After the deadline for cooperation has passed, it checks which other agents have cooperated by comparing the list of group members (i.e. all agents that were supposed to send a chunk) with the agents that it has received a chunk from. All the agents that have not sent any chunks to the normative empowered agent despite having been assigned one by the group are reported by the normative empowered agent and a sanction is applied to them. In our simulation experiments this sanction (for easier calculation) is measured in terms of energy consumption and was set to three times the relative gains the agents had from cheating, i.e. it was calculated in accordance with the size of the chunk the agent was supposed to sent. When determining whom to issue sanctions to, the normative empowered agent does not distinguish between the intentions of the agent, but only determine which agents have not sent a chunk to him. Consequently, agents that intended to send their assigned chunks but did not manage to do so before the deadline receive sanctions in the same way as agents that did intend not to send the chunk at all. It is important to note that the simulation ensures that agents cannot be punished twice for the same action, by checking whether an agent has already been punished for a particular act before the sanction is applied. If two normative empowered agents happen to observe the same interaction and detect the same cheaters, the cheaters only receive one sanction.

Besides the addition of the normative empowered agents, the remaining agents are the same as the ones used in the basic WMG simulation experiments. The only change that is made concerns the utility consideration of the utility agents. As before, the utility agents compare the costs of defection and cooperation in order to determine whether they will cooperate or not. However, the new potential punishment costs involved with cheating need to be accounted for by the utility agents. In order to do this, we have added a counter `cheating(X)` to the agents' code. The agents use this counter to determine the number of events `X` in which they have cheated so far. When calculating the possible costs associated with defection, the agents sum the fines they had to pay prior to the event and divide it by the number of times they have cheated. The result of this division is the average fine for cheating. If this number is higher than the possible gains of cheating, the utility agents will decide to cooperate, whereas otherwise they will decide to defect.

8.2.2 Image Information

In contrast to normative empowered agents, when image information is used as a means of enforcement, no additional entities are required in the system. Recalling Definition 7 (page 29), image information is information an agent has acquired about another agent through personal experiences with that agent (e.g. by cooperating with it). Agent using image information are very similar to the agents presented in Chapter 7, however when being offered collaboration by other agents, the agents of this class consider their past experiences.

When being sent a collaboration proposal by another agent, the agents will check the image information they have about the sender of the cooperation request. Using the information they have, they then decide on whether or not to collaborate with the other agent. If they have positive image information from past interactions, they will decide to cooperate with the agent, whereas in the opposite case, they will punish the agent by not interacting with it again.

The threshold up to which degree of cheating an agent will cooperate with other agents can theoretically be determined by each agent individually. For ease of analysis, we have however set this threshold uniformly for all agents in such a way that they will not interact with an agent again if they have previously been cheated by this agent. Agents that the agent has not interacted with before (i.e. has no image information about) are given the benefit of the doubt and considered to have good image information.

Resulting from this utilization of image information, the utility considerations of the utility agents with regard to the costs associated with defecting again need to be changed.

Again, we utilize a counter to keep track of information. This counter tracks the number of times the agent has been denied an interaction because of its bad past behaviour. To calculate the possible costs of defection the agent determines the number of times it was denied an interaction because of its image in respect to the total number of interactions it has performed so far. This quotient is then weighted together with the size of the chunk to send and the number of group members. The latter plays a significant role as a larger number of group members automatically results in more agents adding negative image information to their belief base, if the agent decides to cheat.

The advantage of image information is its reliability, which is why we have chosen to implement it in this dissertation. However, one problem that however is often associated with image mechanisms is that an agent first needs its own experience to acquire some form of image information, and consequently it can be cheated at least once by a cheater. Reputation mechanisms have been proposed as one way to avoid this problem.

8.2.3   Reputation

In reputation mechanisms the image information of the individual agents is circulated (see Definition 6 on page 28). In the literature a large number of reputation mechanisms have been proposed for different situations. In Balke et al. (2009), an overview of the mechanisms most commonly employed in MAS settings is given. In order to decide on a mechanism suitable for WMG, we start this section by examining the requirements for a reputation mechanism to be used in a WMG.

Despite this dissertation using a MAS simulation to represent the actors and interactions in a WMG, when being brought to market, it will obviously be humans that interact in it. To support human users, the mechanism employed needs to allow for subjective expressions of trustworthiness based on the different perceptions each individual has. Furthermore it needs to be suitable for human cognition, i.e. allowing for verbal (non numeric and non monotonic) expression of degrees of trustworthiness. Earlier on in this chapter we pointed out that one major problem for any reputation mechanism is to deal with the problem of incorrect information. The mechanism should therefore have the means to take into account the sources of information and to identify sources that deliberately provide false information.

One mechanism that meets all these criteria was proposed by Abdul-Rahman and Hailes (2000), which we use in this dissertation[2]. This mechanism is typically only referred to by the authors' names in the literature and we adhere to this convention. In their mechanism Abduhl-Rahman and Hailes propose that every agent carries a network of trust relationships in a database, which they can query for information themselves information or when being asked for feedback about another agent. Abduhl-Rahman and Hailes define a "trust-relationship" as a vectored connection between exactly two entities, which in some circumstances can be transitive. In this way they distinguish between direct trust relationships ("Alice trusts Bob.") and recommender trust relationships ("Alice trusts Bob's recommendations about the trustworthiness of other agents"), and thus allows entities to account for the source of reputation information as well as collecting and evaluating information about the reliability of recommenders. Another interesting contrast to the other formalizations lies in the fact that due to the qualitative nature of trust, Abduhl-Rahman and Hailes do not work with probability values or the $[-1, 1]$ interval, but use a multi-context recording model with abstract trust categories that are easier to understand by humans. These trust values relate to certain contextual information ("Alice trusts Bob, concerning "table"-transactions. However, she does not trust him when it comes to "chair"-transactions."). The trust categories used by Abdul-Rahman and Hailes can be seen in Table 8.1.

Table 8.1: Discrete trust value (Abdul-Rahman and Hailes, 1997, p. 53)

Value	Significance for direct trust relationship	Significance for recommender trust relationship
-1	Distrust - completely untrustworthy	Distrust - completely untrustworthy
0	Ignorance - cannot make trust-related judgement about entity	Ignorance - cannot make trust-related judgement about entity
1	Minimal - lowest possible trust	
2	Average - mean trustworthiness (most entities have this trust level)	
3	Good - more trustworthy than most entities	
4	Complete - completely trust this entity	

As a result of their context-dependant reputation considerations, Abduhl-Rahman and Hailes define reputation as a "troika" $(agent - ID, Trust -$

---

[2]The Java implementation of Abdul-Rahman and Hailes' reputation mechanism employed in this dissertation was developed as part of the EU-funded eRep project at the University of Bayreuth in collaboration with the Artificial Intelligence Research Institute (IIIA) in Barcelona, Spain.

Category, *Trust − Value*), with trust categories for example being "cooperation partner" or "recommender". Each agent stores such reputation information in its own database and uses it to articulate recommendations. For this database Abdul-Rahman and Hailes suggest two data structures, set Q and set R. In the former an agent should store all the agents it directly trusts for a specific context, whereas the latter contains experiences with recommenders. The core of Abduhl-Rahman and Hailes' papers propose a recommendation protocol that can be used to communicate recommendation requests and statements as well as updating inquiries within an MAS. In the protocol a recommendation request is passed on until one or more agents are found which can give information for the requested category and which is trusted by the penultimate agent in the chain. Based on this idea Abdul-Rahman and Hailes propose a mathematical algorithm for the rating phase in which the requesting agent can use the following equation to calculate the trustworthiness of a recommendation. For $tv(R_x)$ as the recommender trust value of the different recommendations of the involved agents and $rtv(T)$ as the trust value articulated by the last agent[3] the trustworthiness is given by the following equation:

$$tv_r(T) = \frac{tv(R_1)}{4} * \frac{tv(R_2)}{4} * \dots * \frac{tv(R_n)}{4} * rtv(T)$$

As we do not employ any routing considerations in our WMG and any routing of requests as envisioned by Abdul-Rahman and Hailes would result in large amounts of network traffic which would reduce the energy advantage of a WMG tremendously, we altered this component of the reputation mechanism to better suit the WMG. In our mechanism, the agent seeking a recommendation about a target sends out one broadcast message to its neighbours and if no answers are given in response to this request, the agent will not wait for further information, but as in the case of image information will cooperate with the target.

In their decision process the agents use the reputation mechanisms as follows: whenever being approached to cooperate, the agents check their own image information about the potential cooperation partner. If this query does not yield any results, i.e. if they have not interacted with the agent before, the agents will send a request for recommendations to their neighbouring agents and wait for replies. If no replies arrive, they will decide to interact with the other agent. Otherwise, they will check the information they have about the information sources. If the information about the recommender is negative, the agents will not rely on the information, whereas otherwise they will take it into consideration. If the recommendations about the target agent are negative (and the recommendation is being considered), the agent will decide against cooperation, otherwise it will cooperate.

This general procedure is the same for all agents, nevertheless in respect to answering recommendation requests, differences between the agent types are made. Whereas the utility maximizing agent prefers not to send any information,

---

[3]In case an agent receives more than one recommendation about another agent, the values are averaged.

as energy consumption is involved in answering a message, honest as well as malicious agents answer on average one request per interaction event. This limitation to one message per interaction is made in order to reduce the energy consumption involved in answering message requests. Despite this similarity, the malicious and the honest agents differ significantly in the content of their answers to the request. If not being asked about itself, the malicious agent will always give negative feedback for the target in order to appear better in comparison to the other agents. In contrast, an honest agent will report truthfully if it has any information about the target, trying to help to improve information in the system.

8.3 Simulation Experiments

In the last section we described how the three different enforcement mechanisms chosen are implemented in our simulation. In this section we hypothesize how these different mechanisms might effect the WMG and how the results will compare to our basic scenario presented in the previous chapter.

8.3.1 Simulation Hypotheses and Setup

Beginning with the simulation hypotheses, the primary reason for testing enforcement mechanisms is to improve cooperation and thereby energy consumption in the WMG. Our first hypothesis covers this aspect by assuming that a difference in the energy consumption rate with and without enforcement exists, and that this difference is such that the introduction of enforcement mechanisms results in lower energy consumption rates.

Hypothesis 5: The introduction of enforcement mechanisms reduces the average energy consumption rate in comparison to the basic WMG simulation experiments without enforcement mechanisms.

Although we assume enforcement mechanisms to have an effect on the energy consumption rate, one question remains. That is, to what extent the significant effects that we observed relating to the values of the simulation input factors in the basic WMG simulation experiments remain the same with the introduction of enforcement mechanisms. As – with $\rho_{neighbourhood}$ constant – the size of the population determines the likelihood by which an agent cooperates again, we form the hypothesis that, in contrast to the experiments conducted earlier, $|\mathcal{A}|$ has an impact on the energy consumption rate. For the remaining two input parameters, we can see no reason for any change in their significance with respect to the energy consumption rate. Given that, we formulate the following three hypotheses:

Hypothesis 5.1: The success (in terms of the average energy consumption rate) of a WMG employing normative empowered agents as enforcement mechanism is dependant not only on the population composition and the neighbourhood density $\rho_{neighbourhood}$, but also on the population size.

Hypothesis 5.2: The success (in terms of the average energy consumption rate) of a WMG using image-based enforcement is dependant not only on the population composition and the neighbourhood density $\rho_{neighbourhood}$, but also on the population size.

Hypothesis 5.3: The success (in terms of the average energy consumption rate) of a WMG using reputation-based enforcement is dependant not only on the population composition and the neighbourhood density $\rho_{neighbourhood}$, but also on the population size.

The final hypothesis we test in the simulation experiments is concerned with normative empowered agents. So far in the simulations experiments the one input factor for which we alter values is $|\mathcal{A}|$, i.e. the size of the population. As well as being able to adjust the population size, in the simulation experiments we can also test whether the number of normative empowered agents influences the energy consumption rate, i.e. whether more normative empowered agents automatically result in lower energy consumption rates. Assuming this correlation to be true, that is, more normative empowered agents are able to observe more interactions and therefore can detect and punish more defecting agents, we formulate Hypothesis 5.4.

Hypothesis 5.4: An increase in the percentage of normative empowered agents will result in a reduction of the average energy consumption ratio.

In order to test the hypothesis just formulated, we use the experimental setup shown in Table 8.2. In this table, all the changes to the experimental setup from the previous chapter (i.e. Table 7.1) are highlighted in bold. In addition to the obvious difference that enforcement mechanisms are now in place, three more changes have been made. The first of these is directly linked to the enforcement mechanisms. In order to test Hypothesis 5.4 we introduce two different percentages of normative empowered agents (the percentages are given in relation to the total number of agents in the system), i.e. 1% and 5%. Another change is the exclusion of settings with 100% honest or 100% malicious agents from the experiments[4], as these two agents groups have fixed strategies that are not influenced by changes in the normative framework. Therefore results with all agents coming from either of the two groups would not be meaningful[5]. The final change we make is the extension of the dependant variables being tested by three new values. The value "Number of Detected Defection Actions" specifies how many defection action the normative empowered agents have detected, whereas the "Number of Punishment Actions" refers to the punishments that were issued. While it might seem at a first glance that they point to the same information, the two numbers can have different values. If two normative

---

[4]This change is not directly highlighted in bold as both numbers are excluded from the table.

[5]As a result of this change – in order to avoid any distortion of results – we excluded all experimental results from settings with 100% honest or malicious agents from Chapter 7 in the comparative analyses in this chapter.

Table 8.2: Experimental Setup for Enforcement Experiments

Input Factors	Values
$\mid \mathcal{A} \mid$	200, 400, 800
Number of Interactions per Agent	50
Utility Agents as % of $\mid \mathcal{A} \mid$	0, 25, 50, 75, 100
Malicious Agents as % of $\mid \mathcal{A} \mid$	0, 25, 50, 75
Honest Agents as % of $\mid \mathcal{A} \mid$	0, 25, 50, 75
Enforcement Mechanism	**Normative Empowered Agents, Image Information, Reputation Information**
Percentage of Normative Empowered Agents as % of $\mid \mathcal{A} \mid$	**1, 5**
Movement Pattern	Simple Boundless Random Waypoint
$\rho_{neighbourhood}$	10, 20
Measurements	**Type**
Number of Defection Actions	$\mathbb{N}$
Number of Cooperation Actions	$\mathbb{N}$
Number of Detected Defection Actions	$\mathbb{N}$
Number of Punishment Actions	$\mathbb{N}$
Number of Correctly Punished Malicious Actions	$\mathbb{N}$
Communication-related Data Transferred per Agent per Interaction	$\mathbb{R}$
Download-related Data Transferred per Agent per Interaction	$\mathbb{R}$
Reception-related Data Transferred per Agent per Interaction	$\mathbb{R}$
Sending-related Data Transferred per Agent per Interaction	$\mathbb{R}$
Integrity Constraints	**Formula**
Full Partitioning	Utility Agents % + Honest Agents % + Malicious Agents % = 100

empowered agents detect the same defection for example, two detected defection actions are added. However, as a result of the previously stated rule that each defection should receive only one sanction, the second sanction is not counted. Thus, the difference between the two numbers tells us the number of extra work done by the normative empowered agents. The last number specifies the number of correctly punished malicious actions, i.e. the number of times the normative empowered agents issued a sanction to an agent that did not send a chunk deliberately. Using this number in combination with the total number of cheating events and the number of sanction events we are able to compute the false positive and the false negative rates for the experiments with normative empowered agents involved. These two rates will not be part of the analysis in this chapter as they are not covered by any hypothesis, however they are discussed with regard to the multi-stakeholder impact in Chapter 10.

8.3.2   Simulation Results

After presenting our hypotheses in the previous section we turn to the analysis of the simulation results in order to test the hypotheses made[6].

We first look at differences in the mean energy consumption rates in simulation experiments with and without enforcement and compare these with the help of ANOVA, by testing whether the null hypothesis that the enforcement mechanisms account for no difference can be rejected or not. Table 8.3 shows the results of this comparison.

Table 8.3: Analysis of Variance Comparing Simulation Experiments with and without Enforcement

Source	Sum of Squares	Degrees of Freedom	Mean Squares	F	Prob > F (= p-value)
Enforcement	224.759	4	56.1898	1539.94	< 0.0001
Error	709.842	19454	0.0365		
Total	934.601	19458			

As the significant p-value suggests, we can reject the null hypothesis and conclude that the utilization of enforcement results in a difference in the average energy consumption rates. However, looking at the table more closely, a high error rate can be detected, indicating that a difference exists between the three enforcement mechanisms that are grouped in the ANOVA. In order to examine this effect more closely, as well as to determine the extent to which each enforcement mechanism contributes to this difference, we perform Tukey's test as post hoc analysis. Figure 8.3[7] shows the results of this analysis and Table 8.4 gives on overview of the respective statistical values per enforcement mechanism.

---

[6]For testing the hypotheses formulated in this chapter, we use the same statistical tools and file setup as in the previous chapter.

[7]Normative empowered agents are denoted as "Police" in the figure.

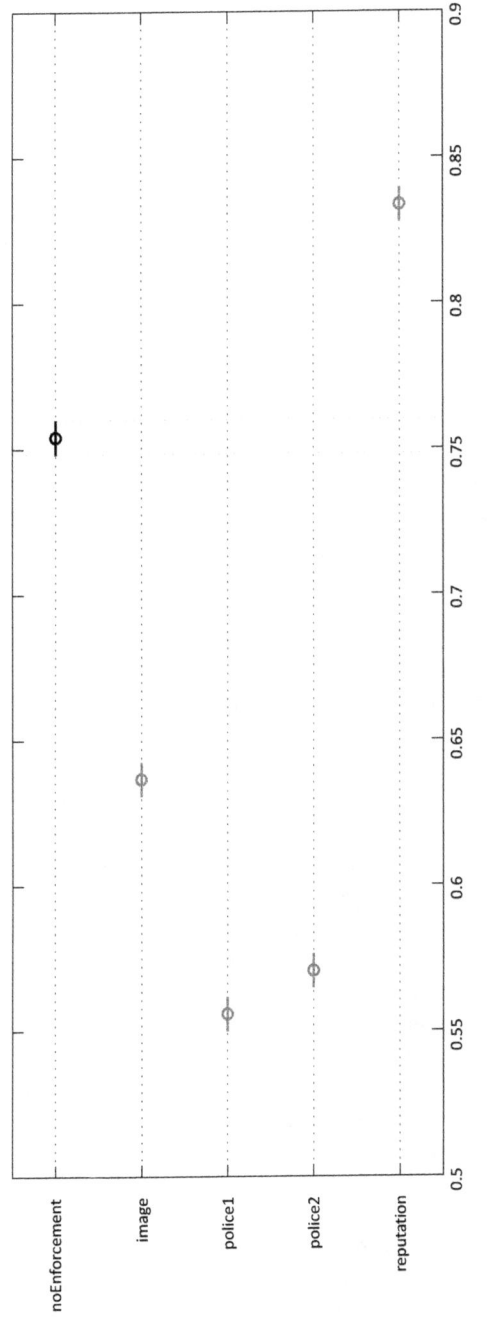

Figure 8.3: Multiple Comparison (Tukey's Test) Results of Marginal Means for Comparing Simulation Experiments with and without Enforcement – Post Hoc Test

Table 8.4: Analysis of Variance in Comparison of the Enforcement Mechanisms –
Post Hoc Test

Source	Sum of Squares	Degrees of Freedom	Mean Squares	F	Prob > F (= p-value)
Image-Information	26.969	1	26.969	684.73	< 0.0001
Normative Empowered Agents (1%)	76.276	1	76.6272	1871.18	< 0.0001
Normative Empowered Agents (5%)	65.642	1	65.6416	1623.26	< 0.0001
Reputation	12.308	1	12.3078	332.01	< 0.0001

As the p-values in Table 8.4 show, all mechanisms have an average energy consumption rate significantly different to the experiments with no enforcement, however looking at the Tukey's test results it is clear that the results for the reputation mechanism stand out. Thus, whereas the results indicate that we can confirm the Hypothesis 5 for image-related information and normative empowered agents, the Tukey's test for the reputation mechanism show that the experiments using this mechanism, have an on average higher mean than when no enforcement is used. This implies that the utilization of reputation information *increased* the average energy consumption rate. We examine this initially rather surprising effect later when discussing Hypothesis 5.3 in more detail, and summarize the results for Hypothesis 5 as follows: enforcement mechanisms with image information as well as with normative empowered agents help to decrease the average energy consumption ratio, but Hypothesis 5 is incorrect for the reputation-information based enforcement mechanism implemented.

A second interesting effect we can observe in Figure 8.3 is that our Hypothesis 5.4 (that an increase in the percentage of normative empowered agents will result in a reduction of the average energy consumption ratio) seems incorrect. In the figure, the average mean energy consumption rate for experiments with 1% normative empowered agents is lower than the one with 5%. A one-tailed paired-sample t-test[8] performed based on this observation, in order to test for significant evidence against the null hypothesis that the means of the results with 1% normative empowered agents are not smaller than the results with 5% normative empowered agents, results in p-value of 1.7889×10^{-18}. This implies that we have to reject the null hypothesis, which in turn means that the lower average energy consumption values for results with 1% normative empowered agents are not a result of chance. Looking for reasons for this effect, comparing the number of cheat events in the simulation experiments

---

[8]We opt for the one-tailed version of the t-test as we have a hypothesis which of the two means is larger. As a result of our assumptions that the means of the cases with 5% normative empowered agents are higher, we choose the left-tailed option in the one-tailed test.

with normative empowered agents by performing a t-test, only a slight, and not significant, advantage for experiments with 5% normative empowered agents can be found (the p-value of the one-tailed t-test is 0.3980), i.e. that in these settings normative empowered agents only added slightly to the total energy consumption. This implies that the improved detection of violations resulting from the larger number of normative empowered agents, is outweighed by the additional energy they consume. In economic terms this means that the lower percentage of normative empowered agents perform better with regard to satisficing cooperation under consideration of the energy consumption rate. In economics, *satisficing* refers to a decision-making strategy that attempts to meet criteria for adequacy, rather than to identify an optimal solution (Simon, 1956). Thus, although not optimal with regard to the detection of violations (1% normative empowered agents will detect less than 5%) the costs associated with them (i.e. the energy they consume for performing their observation and punishing actions) are significantly lower, making them more advantageous in terms of the overall energy saving.

Analysing the results to check whether Hypothesis 5.1 is correct, the ANOVA gives the values shown in Tables 8.5 and 8.6:

Table 8.5: Analysis of Variance for Hypothesis 5.1, 1% Normative Empowered Agents

Source	Sum of Squares	Degrees of Freedom	Mean Squares	F	Prob > F (= p-value)
Population Composition	122.391	12	10.1993	1667.62	< 0.0001
$\rho_{neighbourhood}$	2.957	1	2.95707	80.49	< 0.0001
$|\mathcal{A}|$	0.0313	2	0.15625	4.17	0.0154

Table 8.6: Analysis of Variance for Hypothesis 5.1, 5% Normative Empowered Agents

Source	Sum of Squares	Degrees of Freedom	Mean Squares	F	Prob > F (= p-value)
Population Composition	109.739	12	9.14495	1096.28	< 0.0001
$\rho_{neighbourhood}$	2.828	1	2.82762	79.1	< 0.0001
$|\mathcal{A}|$	0.205	2	0.10231	2.81	0.0604

As the p-values in both tables indicate, as hypothesized, similarly to Chapter 7, for both $\rho_{neighbourhood}$ as well as the population composition we can reject the null hypothesis and therefore assume the input parameters to be relevant for the energy consumption rate. Concerning $|\mathcal{A}|$ Tables 8.5 and 8.6 suggest a biased picture. Whereas $|\mathcal{A}|$ is significant enough to reject the null hypothesis in case of 1% normative empowered agents, this significance is marginally missed (at an level of $p < 0.05$) in case of 5% normative empowered agents. Thus, a clear tendency towards a significance can be seen, but it is not reached. Looking at reasons for the marginally non-significance value in case of 5% normative empowered agents, the extra energy consumption by the greater number of normative empowered agents seems to have reduced the general effect. In summary, we have established that Hypothesis 5.1 is correct for $\rho_{neighbourhood}$ and the population composition, but only can be partially proven for $\mid A \mid$, as the increased energy consumption caused by the normative empowered agents weakens the effect.

Looking at the ANOVA results for image and reputation information based scenarios, the values in Tables 8.7 and 8.8 can be determined from the experimental results.

Table 8.7: Analysis of Variance for Hypothesis 5.2, Image Information

Source	Sum of Squares	Degrees of Freedom	Mean Squares	F	Prob > F (= p-value)
Population Composition	100.244	12	8.35369	963.03	< 0.0001
$\rho_{neighbourhood}$	0.57	1	0.5704	16.67	< 0.0005
$\mid A \mid$	1.545	2	0.77266	22.74	< 0.0001

Table 8.8: Analysis of Variance for Hypothesis 5.3, Reputation Information

Source	Sum of Squares	Degrees of Freedom	Mean Squares	F	Prob > F (= p-value)
Population Composition	77.117	12	6.42641	663.29	< 0.0001
$\rho_{neighbourhood}$	0.031	1	0.03118	1.05	0.3052
$\mid A \mid$	10.166	2	5.08315	188.08	< 0.0001

For simulations in which image information is being used, as hypothesized, we can reject the null hypothesis for all input factors (i.e. have enough evidence to assume that Hypothesis 5.2 is correct). For the simulations with reputation information this is not the case. Although – as expected – for both the population composition and $\mid A \mid$ we can reject the null hypothesis that they have no impact on the energy consumption rate, the null hypothesis for $\rho_{neighbourhood}$ can clearly not be rejected with a p-value of 0.3052. We performed post hoc tests (as before we use Tukey's test) for $\rho_{neighbourhood}$ and the population composition. The latter is analysed in more detail due to it having the highest impact on the energy consumption rate of all three input factors. Figures 8.4 and 8.5 show the results of the post hoc tests.

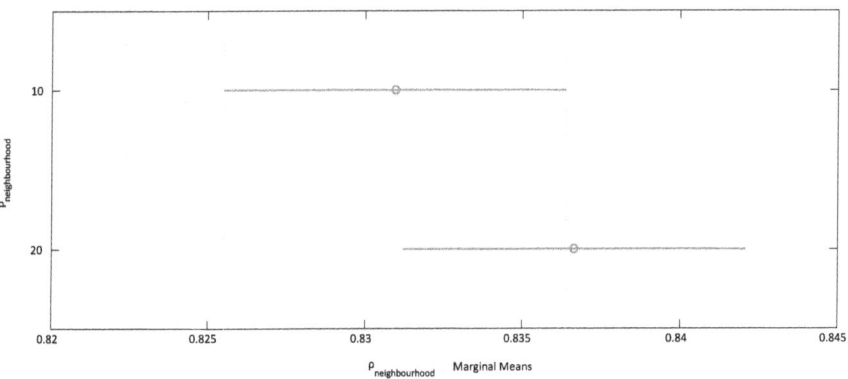

Figure 8.4: Multiple Comparison (Tukey's Test) Results of $\rho_{neighbourhood}$ Marginal Means in Settings with Reputation Information – Post Hoc Test

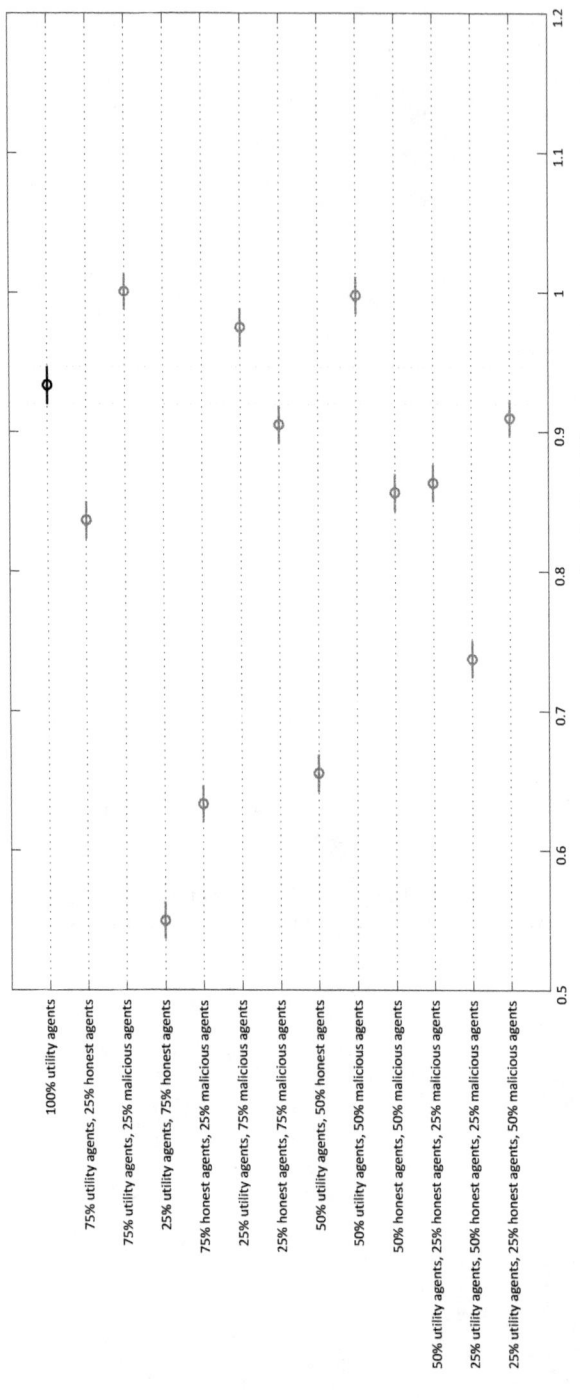

Figure 8.5: Multiple Comparison (Tukey's Test) Results of Marginal Means for the Population Composition in Settings with Reputation Information – Post Hoc Test

Looking at Figure 8.5 first and comparing it to Figure 7.5 (page 175), several interesting differences can be observed, which also explain the unexpected mean difference between reputation information and no enforcement discovered earlier (see Figure 8.4).

The first difference to notice is that the means in the settings with reputation information have a wider distribution around each mean value (indicated by the horizontal bars), which implies that the results in the simulation runs with identical settings resulted in a larger variance in energy consumption rates. The second aspect to notice is that minor (but not significant) improvements of the energy consumption rates for the settings which performed worse in the basic WMG experiments can be detected. Thus, at the upper end of the energy consumption spectrum, consumption decreases as a result of the enforcement can be detected. The problem in case of reputation, however, is the cases with lower energy consumption rates, which overshadow this positive result by an increase in the consumption rate.

Digging deeper into the differences in the energy consumption rate the reason for this effect becomes apparent. As a result of the large amount of additional messages with reputation information being transmitted, the communication costs are disproportionately high and thereby worsen the energy consumption. Thus, especially in cases with large numbers of honest agents which would prefer cooperation even without enforcement mechanisms, the reputation requests and answers do not improve enforcement but rather produce additional energy consumption.

This effect of the additional communication costs can also be seen when looking at the Tukey's test result for $\rho_{neighbourhood}$ (i.e Figure 8.4), which shows that – in contrast to Chapter 7, Figure 7.7 – a higher $\rho_{neighbourhood}$ tends to result in a worse energy consumption rate (the difference is not significant as the two bars in Figure 8.4 overlap), which can again be attributed to the raised number of cooperation messages in these settings (with a higher $\rho_{neighbourhood}$ more agents get and answer messages). A second effect that had negative consequences for the reputation mechanism was the negative information induced by the malicious agents. Our simulation experiments were set up in such a way that if in doubt (i.e not verifiable), reputation information was considered to be correct. As a result of this, especially in the beginning of each simulation experiment, by giving negative reputation information, malicious agents were able to discourage agents from cooperating with potential rivals, which also had the side-effect that agents could gather less image information on their own. As a consequence, they were more likely to have to trust uncertain reputation information in the following interactions causing problems for the overall reputation mechanism. Summarizing, for Hypothesis 5.3 we can only verify the hypothesized effect for $|\mathcal{A}|$ and the population composition, whereas for the neighbourhood density the null hypothesis cannot be rejected.

8.4   Summary

Starting with the results from the simulation experiments in Chapter 7 as a basis for comparison in this chapter we introduced enforcement mechanisms to the WMG. Table 8.9 summarizes the main findings.

Table 8.9: Simulation Results Description Table for Enforcement Mechanisms

Input Factors	Measurement	α	Relationship
Image Information	Energy Consumption Rate	> 0.99	The utilization of image information helps to improve the energy consumption rate compared to settings without enforcement.
Normative Empowered Agents	Energy Consumption Rate	> 0.99	The utilization of normative empowered agents helps to improve the energy consumption rate compared to settings without enforcement.
Reputation Information	Energy Consumption Rate	> 0.99	The utilization of reputation information does not help to improve the energy consumption rate compared to settings without enforcement.
% of Normative Empowered Agents	Energy Consumption Rate	< 0.0001	A decrease from 5% to 1% decreases the energy consumption rate.
Population Composition	Energy Consumption Rate	> 0.99	The population composition has a significant impact on the energy consumption rate. This result is independent of the enforcement mechanism chosen.

Of the enforcement mechanisms introduced, the normative empowered agents especially help to improve the energy consumption rate. Although the results we present are strongly linked to the specific mechanisms chosen, nevertheless some general findings can be made.

The first of the findings is that, similar to the basic WMG experiments, the population composition again accounts for the majority of the impact of the input factors on the energy consumption rate. Secondly, the costs associated with the enforcement can outweigh the benefits. In the case of normative empowered agents this resulted in the situation that fewer agents produced a better absolute result in terms of the energy consumption rate, while "only" satisficing the detection of violation actions. Similar effects could be seen in experiments with reputation information, where the message overhead produced by the reputation request and answers outweigh the benefits of the mechanism. One further aspect that influenced the performance of our reputation information-based enforcement mechanism negatively is false information. This is particularly important as we implemented an adaptation of a mechanism that tried to account for this problem. However it did not have sufficient interaction numbers to have any significant effect, as the mechanism by Abdul-Rahman and Hailes – like any reputation mechanism – works better the higher the number of repeated interactions. Unfortunately this seems unlikely in a WMG, suggesting that reputation mechanisms in general might be unsuitable for such settings.

Chapter 9

On the Effects of Movement in the Wireless Mobile Grid Simulation

In the predeceasing chapters we emphasised the importance of location and vicinity. As a result of the limited range of WLAN transmission and the bounded rationality assumption made in this dissertation, the vicinity of the agents in the simulation determines their possible interaction partners. It therefore has a huge impact on the WMG. To represent the movement behaviour of mobile phone users as accurately as possible in the simulation experiments, ideally real movement traces would be required. However, as a result of a lack of tools as well as legal problems for monitoring time-resolved movement of individuals (González et al., 2008), no real movement traces are available for this dissertation. This necessitates the use of abstract models for determining the movement of the mobile phone users. Up to this point, the simulations have used a simple one-dimensional boundless random waypoint movement model that is characterized by agents choosing random locations in the interval [0;1] and stopping after each move to decide on some new action (which could either be to move again or to obtain a file).

Despite random waypoint or random walk[1] models often being used in (wireless) telecommunication systems simulations (see Bettstetter (2001b) for a list of works using the random waypoint or random movement model), the model itself was often criticized, mainly for its simplifying assumptions (Yoon et al., 2003). Following Camp et al. (2002) and Bettstetter (2001a), who amongst others show that the choice of movement model can have significant effects on

---

[1]Random walk models differ from random waypoint models in the respect that they do not explicitly assume the moving entities to stop for a certain period of time after a walk, but that the moving entities directly move into a new direction with a step. A random waypoint model can be transformed into a random walk model by setting the pauses between two movement steps to zero.

simulation results, we now move away from the random waypoint movement model used in the simulation experiments conducted so far. Instead we employ three different movement models in the simulation experiments and test to what extent the movement models effect the results found earlier.

We first give a brief overview of the movement models currently employed in simulations of user movement in mobile telecommunication networks. Afterwards we explain the three models we have chosen to implement for the simulation experiments in more detail (both from a theoretical as well as an implementation perspective). As in the two previous chapters, we will then briefly outline the simulation experiment hypotheses and setup. We conclude this chapter by presenting and discussing the results of the simulation experiments.

9.1   Extending the Basic Movement Pattern

9.1.1   Movement Patterns for Wireless Mobile Grids

Looking at surveys dealing with movement models for open distributed systems in general or focusing on human mobile phone user movement in particular, a broad distinction between two groups of movement models can be found in these surveys:

- entity movement models, and

- group movement models (Camp et al., 2002).

Entity movement models view the movement of the individual mobile phone users as independent of each other, whereas group movement models assume some form of dependency (i.e. some form of social structure) between the individuals. For both groups of movement models, a large number of variants with different specifications exist. Camp et al. (2002) for example discuss seven different synthetic entity movement models[2]:

Random Walk Movement Models (and their many derivatives): In models belonging to this category, new locations and possibly the speed by which the location is reached are determined at random.

Random Waypoint Movement Models: As the name indicated, in these models, waypoints are chosen at random. Furthermore, this model extends the random walk movement model by introducing stops between two movement step.

Random Direction Movement Models: This mode assumes the area in which the mobile phone users move to be bounded (typically having the form of a rectangle). The individuals in these models have to travel to one edge of the area first, before they can change directions (and speed) again.

---

[2]Camp et al. (2002) uses the term "mobility" rather "movement" when referring to the locomotion of individuals. For reasons of consistency with the terminology used in this dissertation we use movement instead of mobility when describing the research presented by Champ et al.

Boundless Simulation Area Movement Models: In these models the 2D rectangular space is converted to a boundless area by connection of the opposite edges of the rectangle. This shape is referred to as a torus. Although described as separate model by Camp et al. (2002), it can be implemented as variant to any of the model described here.

Gauss-Markov Movement Models: The Gauss-Markov movement models are based on the random models. However they use a so-called tuning parameter to enable the utilization of different degrees of randomness in the mobility patterns (Tolety, 1999).

Probabilistic Versions of the Random Walk Movement Models: These models utilize a set of additional probability values to determine the next positions the mobile phone users move to.

City Section Movement Models: In the city section movement models the simulation area in which the mobile phone users can move is a street network that represent a city in which the mobile network exists. With the help of speed limits, etc. for certain streets, different city structures can be represented.

In addition to these seven movement models classified by Camp et al. another movement model that is characterised by its assumption that mobile phone users exhibit sedentary movement behaviour can be found in literature(González et al., 2008). Sedentary behaviour implies that mobile phone users typically have a couple of locations (e.g. home or work) that they regularly return to, whereas random movement is assumed otherwise. Due to a lack of a common term in literature, we refer to the models exhibiting these characteristics as *sedentary movement models*.

With regard to the group movement models, Camp et al. (2002) distinguish five different sub-models:

Exponential Correlated Random Movement Models: In these kind of group movement models a motion function which is applicable for the whole group is used to determine new locations for the members of the group.

Column Movement Models: These models are characterized by the members of the group being always located uniformly along a line. This movement is uniformly forward, keeping the uniform distribution between the individuals.

Nomadic Community Mobility Models: In contrast to other group movement models, in these models the whole group does not always move together to a new location, but only a subset of the group. Individuals can still move around freely but have a certain probability to arrange their movement to return to the group's location.

Pursue Movement Models: As the name implies, in this kind of group movement models, the mobile phone users follow a single target.

Reference Point Group Movement Models: In reference point group movement models, a local center of a group is determined. When members of the group move, the local center changes and the remaining group members will try to adjust their location around the new logical center.

Looking at all the briefly described movement models, there are to many to integrate them all into the WMG simulation. Instead we selected a few models. The selection criteria that we used for our decision, are two fold: (i) we want to represent a wide range of different models, both accounting and not accounting for social structures between the individuals, (ii) the chosen models should represent real movement patterns as well as possible, without unnecessarily increasing the complexity of the simulation.

Based on these two criteria we chose to implement three different movement models. These movement models are, a probabilistic version of the basic movement model, a sedentary movement pattern as well as a movement model combining the advantages of the pursue movement model and the nomadic group approach. The models were chosen based on comparative study by Camp et al. (2002) who analyse movement models with regard to their realism. They conclude that for entity movement models especially boundless simulation areas – which we use in our simulations – increase realism and furthermore highlight probabilistic models being good in their portrayal of realism despite being very simple. In addition we decided to follow the ideas presented by González et al. (2008) who were the first research team that was able to obtain real mobile phone user movement data from a telecommunications provider. Based on this data, González et al. concluded that especially nomadic properties in combination with daytime considerations are important to model human movement. Thus, humans are likely to be at home during nighttime and travel to work in the morning, etc. As we use an event-based, rather than a continuous time definition in this dissertation, we do not include the time aspect González et al. describe but focused on a nomadic movement model.

With respect to the group movement models Camp et al. (2002) point out the realism of both nomadic group and pursue movement models, however they criticize the lack nomadic character in the latter. One problem with nomadic group models is that few specifications are given on which part of the group to focus on when determining a location. This makes them difficult to implement. We therefore combine the nomadic group and the pursue movement model in this dissertation and thereby try to benefit from the advantages of both models.

Having decided to extend our simulations with these three models, we provide a more detailed description of them in the next section and furthermore we explain how we implement them in the simulation. Afterwards we will focus on the simulation experiments. We provide the simulation hypothesis for the simulation experiments before discussing the resulting simulation setup.

9.1.2 The Extended Random Waypoint Pattern

We start with on extension of the original movement pattern to include probabilistic values that distinguish the movement of the individuals. The probabilistic value is a *mobility index* that determines the size of the steps a single individual can move by when changing its location. This probabilistic value was first discussed in the Gauss-Markov movement models (Bettstetter, 2001b) and expresses the idea that humans have different probabilities to travel large distances, either for example due to physical restraints or as a result of social constraints. The idea of the mobility index is that each individual has a certain likelihood for travelling large distances. Businessmen for example might be more likely to travel large distances on a regular basis (e.g. by plane) then elderly people whose mobility could be restricted due to physical problems. The mobility index (or maximum step size) accounts for this by giving a mobility level to every individual. It determines how far it will move at maximum within one movement step. Making no assumptions about which population groups are more likely to move short or far distances, in the simulation experiments we assign mobility values to each individual by using random numbers. When using the extended random waypoint movement model, at the beginning of each simulation experiment, each agent picks a random number $r_m \in [0,1]$ (with the number in the interval being uniformly distributed) that is used as the agent's mobility index throughout one simulation run (i.e. for the complete 50 interactions). In the movement process when determining a new location and obtaining r_d, the agent multiplies $r_d \times r_m$ to determine the final distance it will move. As a result of this multiplication, agents with a low r_m are more likely to choose new locations that are close to the old location than agents with a high r_m. The multiplication also has the effect that the distances the agents move, instead of being uniformly distributed between -1 and 1, will get closer to a normal distribution with mean 0.

9.1.3 The Nomadic Pursue Group Pattern

The second additional movement pattern we include in the simulation experiments is a nomadic pursue group pattern. This pattern represents members of a group moving together with some members of the group performing individual steps in between (Dalu et al., 2008). An example of such a movement in the real world is a group of friends that move closely together very often when they do something together. While the group tends to move together, the friends will not always do everything together but at certain times some of them might be in another location and do something else instead, and join the group later again.

To simulate this movement pattern, at the beginning of the simulation experiments, the agents are distributed into groups. For the size of these groups we choose sizes that are uniformly distributed in the interval [8,12] and have a mean that is the same as the lower $\rho_{neighbourhood}$ value used in the simulations so far. In nomadic group movement models a probability value r_α is used to determine whether the agents follow the group or not. At each step, the agents pick a random number and if that number is lower or equal to r_α they follow the group, otherwise they choose a random location independent of the group. In our simulation experiments r_α had a value of 0.5, to allow for a strong group relationship, but at the same time give agents a chance to perform individual movements[3]. The problem in nomadic group movement models is how to define the group location at a given point in time. Averaging the location of all group members is not a sufficient solution as the result will be distorted by the group members that have decided to choose a location away from the group. In order to solve this problem, in the simulation experiments we combined the nomadic group movement model with another movement model that was identified as having a high realism by Camp et al. (2002): the pursue movement model. In this model the group pursues a single predetermined target at every step of the simulation, i.e. the group follows the target to whatever location it moves to. The targets are chosen at random at the beginning of the experiment and remain constant. When the agents are sorted into groups, we assign the first agent sorted to a specific group to be the target for that group. Hence when choosing a new location, the agents check for this "group leader" (i.e. the target) and follow it to a location within its visibility radius. The target chooses its location using the simple random waypoint movement pattern we employed in the experiments so far.

9.1.4   The Sedentary Movement Pattern

The third additional movement model we test in this chapter is a sedentary movement model. For reasons of simplicity and because the representation of time is based on discrete events, we did not associate any temporal parameters with the sedentary locations as suggested by Nguyen et al. (2011). Instead, in the simulation experiments, we attribute two random sedentary locations to each agent and define a sedentary level r_{sed}[4]. When the sedentary movement model is being used, if an agents wants to determine its new location it picks a random number ($\in [0, 100]$). If the number is higher than r_{sed}, the agent picks a random location according to the random waypoint model, otherwise it will chose one of the two sedentary location[5].

9.2   Simulation Experiments

9.2.1   Simulation Hypothesis and Setup

In the introduction to this chapter we pointed out that works by Camp et al. (2002) as well as Bettstetter (2001a) suggest that the choice of movement model

---

[3]We use $r_\alpha = 50\%$ in all simulation experiments that employ a nomadic pursue group movement pattern. The simulation is however written in such as way that r_α can be set to any value in $[0, 100]$ and different values for r_α can be tested.

[4]Similar to r_α, in the simulation experiments we set r_{sed} to 50%, but developed the simulation in such as way, that it can be started with any $r_{sed} \in [0, 100]$.

[5]Both sedentary locations have approximately equal chance of being picked by the agent.

does have an effect on the results in a simulation. We take their statements as hypothesis for our experiments.

Hypothesis 6: A change in the movement model used in the simulation experiments will effect the enforcement mechanism results.

As a result of this hypothesis, the experimental setup shown in Figure 9.1 is the same as in Chapter 8, except for being extended with respect to the movement model employed (highlighted in bold print). In the movement experiments conducted, we repeated the experiments conducted in Chapter 8 three times, using a different movement pattern each time.

9.2.2 Simulation Results

After presenting our hypothesis, we now focus on the analysis of the simulation experiments. Due to a lack of computational resources, we were not able to determine the results of the n-way ANOVA for the three enforcement mechanisms (in comparison to the situation without any enforcement). So for the analysis we compared the results for each single enforcement mechanism separately and will draw our conclusions form these results. Figures 9.1–9.4 show the results of the ANOVA for each movement model (analysed by enforcement mechanism) and the respective post hoc Tukey's test results. In all four figures, the bar representing the means of the initial setting (with the random waypoint movement) is coloured in cyan. If no significant evidence against the null hypothesis can be found (i.e. that no difference between the respective movement model and the random waypoint model exists), the respective bars are coloured grey, otherwise they are coloured red.

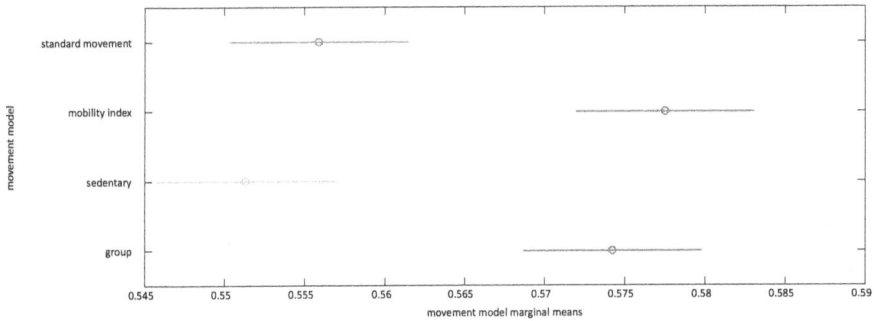

Figure 9.1: Multiple Comparison (Tukey's Test) Results of Movement Model Marginal Means for Experiments with 1% Normative Framwork Empowered Agents – Post Hoc Test

Looking at the figures, no coherent picture for the movement patterns can be determined, but their performance varies a lot and we have to drop Hypothesis 6. Of the twelve simulation experiments with different movement patterns, in

Table 9.1: Experimental Setup for Movement Pattern Experiments

Input Factors	Values
$|\mathcal{A}|$	200, 400, 800
Number of Interactions per Agent	50
Utility Agents as % of $|\mathcal{A}|$	0, 25, 50, 75, 100
Malicious Agents as % of $|\mathcal{A}|$	0, 25, 50, 75
Honest Agents as % of $|\mathcal{A}|$	0, 25, 50, 75
Enforcement Mechanism	Normative Empowered Agents, Image Information, Reputation Information
Percentage of Normative Empowered Agents as % of $|\mathcal{A}|$	1, 5
Movement Pattern	**Simple Boundless Random Waypoint, Sedentary Movement, Reference Point (Nomadic) Group Movement**
r_{α}, $r_{sedentary}$	**50%**
$\rho_{neighbourhood}$	10, 20
Measurements	**Type**
Number of Defection Actions	$\mathbb{N}$
Number of Cooperation Actions	$\mathbb{N}$
Number of Detected Defection Actions	$\mathbb{N}$
Number of Punished Actions	$\mathbb{N}$
Number of Correctly Punished Malicious Actions	$\mathbb{N}$
Communication-related Data Transferred per Agent per Interaction	$\mathbb{R}$
Download-related Data Transferred per Agent per Interaction	$\mathbb{R}$
Reception-related Data Transferred per Agent per Interaction	$\mathbb{R}$
Sending-related Data Transferred per Agent per Interaction	$\mathbb{R}$
Integrity Constraints	**Formula**
Full Partitioning	Utility Agents % + Honest Agents % + Malicious Agents % = 100

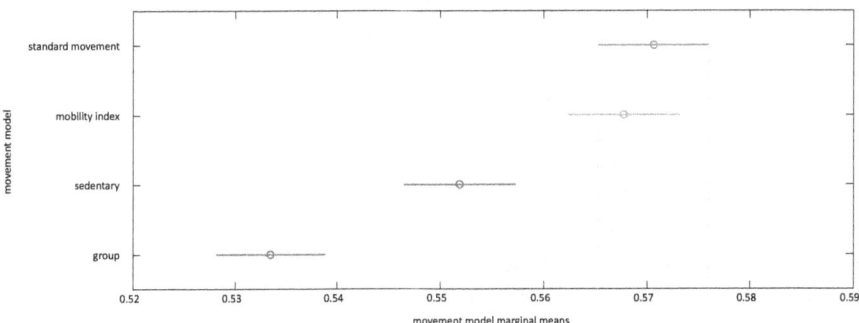

Figure 9.2: Multiple Comparison (Tukey's Test) Results of Movement Model Marginal Means for Experiments with 5% Normative Framwork Empowered Agents – Post Hoc Test

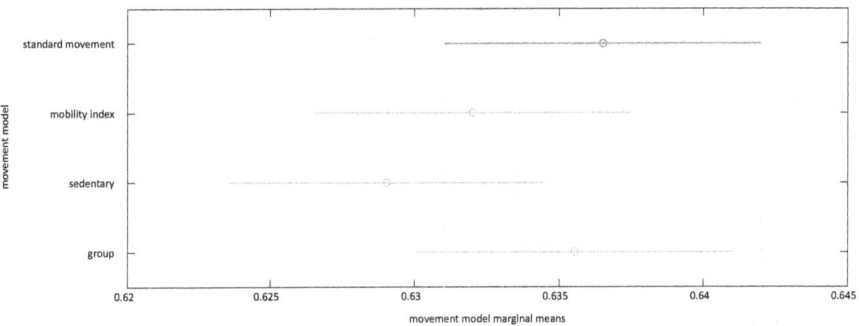

Figure 9.3: Multiple Comparison (Tukey's Test) Results of Movement Model Marginal Means for Experiments with Image Information – Post Hoc Test

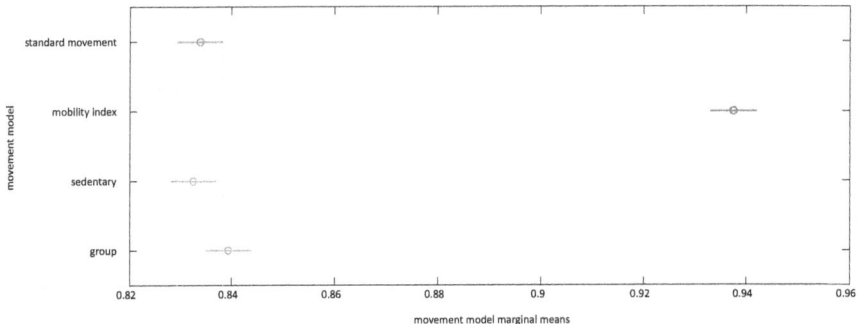

Figure 9.4: Multiple Comparison (Tukey's Test) Results of Movement Model Marginal Means for Experiments with Reputation Information – Post Hoc Test

seven cases (i.e. more than half) no significant difference could be found, two times the average consumption rate decreased and three times it increased. In detail, the introduction of the mobility index twice resulted in a worse energy consumption rate, while on two occasions no significant difference was found; the sedentary movement produced a better result once and three times no significant difference; and the nomadic group movement pattern resulted in one worse, one better and twice a not significantly different energy consumption rates.

Looking at the individual enforcement mechanisms, the mechanism using image information is the most stable mechanism in all movement model conditions. A similar statement is true for the reputation mechanism which has no significant difference, except when the mobility index is employed. In contrast, when normative empowered agents are used as enforcement mechanism, the result is very diverse between the two settings with different percentages of these agents tested in the simulation experiments. Whereas in the case of 5% normative empowered agents, the energy consumption rate improves for the different movement patterns, in case of 1% a different picture appears. Except for one mechanism with no significant difference, the performance of the enforcement mechanism with respect to the energy consumption rate decreases. One reason for this poor performance of the mechanism with 1% normative empowered agents might be that it already produced very good results, and further improvement was difficult, but without further experiments and analysis no general conclusions can be drawn. Despite the seemingly negative impact of movement models on the normative empowered agents, in absolute terms, these still performed the best of all four enforcement mechanisms. The reputation information based mechanism showed the worst performance. The only difference now is that systems with 1% normative empowered agents performed similarly to systems with 5%, which implies that the advantage (in terms of energy consumption rate) that the 1% normative empowered agents had in the previous experiments is lost.

9.3　Summary

To make the simulation experiments more realistic, in this chapter we introduced movement patterns to the simulation. We implemented three different movement patterns, including a pattern which focused on social (i.e. group) movement and one that incorporated sedentary information. Analysing the results of the simulation experiments with enforcement, normative empowered agents still help to reduce the energy consumption rate the most effective. Concerning the impact of the movement in general, unfortunately no clear pattern could be detected. Overall the image-information based enforcement mechanism showed the most stable performance (with respect to the energy consumption rate) whereas for the remaining three enforcement mechanisms at least one movement pattern resulted in a significant difference. Due to the large number of experiments, it appears unlikely that these significant differences are a product of chance, but to be able to draw more specific conclusions, further tests with more fine-grained settings are required. These however are beyond the scope of this dissertation.

Part IV

Evaluation

Chapter 10

Further Simulation Analysis

The previous chapter concluded the description of the simulation experiments conducted in this dissertation. The analysis of the simulation results has so far focused in particular on the energy consumption rates, i.e. on the energy efficiency of the scenarios. In this chapter we examine the results in a broader context by analysing them using the evaluation criteria defined in Section 6.3. First, the analysis focuses on the WMG business case study described in Chapter 3. In Section 10.2 we broaden the view and discuss the implications of the findings for open distributed systems in general.

10.1 Implications of the Simulation Results for the Business Case Study

Summing up the results from the simulation experiments so far, we have compared three different enforcement mechanisms (with one mechanism being tested with different input parameters), of which the normative empowered agents performed best as regards to the energy consumption rate. In contrast the reputation-information based mechanism that we implemented suffered from false information as well as severe message-based energy consumption overhead, making the energy consumption rate higher than in cases where no enforcement is used.

In their IAD framework, Ostrom et al. (1994) argue that patterns of interaction resulting from interactions in the action arena (which are the outcomes of the simulation experiments in this dissertation) always have to be seen in the light of the evaluation criteria relevant for the system or governance decision being evaluated. In Section 6.3 we highlight that for real world business initiatives, the evaluation criteria need to account for the various interests of the multiple stakeholders in the system, and incorporate inter-dependencies between them. Resulting from this challenge we established six major evaluation criteria covering technical as well as economic and social aspects, namely:

- scalability,

- robustness,

- infrastructural cost savings,

- enforcement costs and their distribution across stakeholder groups,

- cost and benefit distribution between users, and

- false positive and false negative rates.

In this section, we will in turn discuss the performance of the four enforcement mechanisms presented with respect to these criteria. When the presentation of the results of all experimental settings would distract from the finding of our analysis (e.g. due to the mere number of results), we do not analyse all four enforcement mechanisms, but concentrate on normative empowered agents because of their good performance in previous experiments. Starting with scalability as the first criterion and recalling the p-values for the population size affecting the enforcement mechanisms in Tables 8.5–8.8, of the mechanisms presented, normative empowered agents were the one least affected by population size $| \mathcal{A} |$ and the chosen reputation mechanism was affected the most. This implies that in terms of scalability, normative empowered agents are likely to perform equally well in small as well as in large settings if their percentage is kept constant. Although being advantageous in that respect, in real world settings, keeping the percentage of normative empowered agents constant and scaling with the number of system participants will not always be easily possible. As expressed beforehand in Section 8.2.1, in a WMG setting, normative empowered agents are likely to be normal mobile phone users that help to control the WMG for some form of financial benefit such as a a free mobile phone contract. As these incentives typically have a long term nature it may be difficult to increase (and especially decrease) the number of mobile phone users required to keep the percentage of normative framework empowered agents constant in a system quickly. The burden of determining the number of normative empowered agents that will be required in the WMG lies with the iPr. Any over or underestimation can have negative consequences: an underestimation can result in the loss of the control of the system, while an overestimation can result in unnecessarily high enforcement costs, both financial and in terms of the energy consumption rate. In contrast, reputation and image based mechanisms scale automatically with the number of users in the WMG. These mechanisms however can have practical implementation problems relating to scalability. Being decentralized enforcement mechanisms, both rely on information stored with the agents. In a WMG, the image/reputation information the agents acquire will therefore need to be stored on the mobile phones of the agents. This storing of information is therefore restricted by the amount of physical storage capacity on the phones that users are prepared to allocated, which sets an upper bound to the amount of information that can be dealt with.

Looking at the robustness criterion, all enforcement mechanisms are affected by malicious agents and perform significantly worse if more malicious agents join the system. Of the enforcement mechanisms tested the normative framework empowered agents functioned best in punishing the malicious agents, but were not able to stop the malicious behaviour. This is not surprising, because the malicious agents in the simulations preferred always to cheat, no matter what the associated costs. In the simulation experiments we tested with up to 75% malicious agents to determine the robustness of the system. In a real world setting, it seems very unlikely that such a high percentage of agents would adopt the malicious strategy. The mechanism that was worst affected, was the reputation mechanism. As outlined in Section 8.3, one major reason for this is the spreading of false image information by malicious agents, which the mechanism could not tolerate. As discussed in the previous chapter, this problem might be dealt with by choosing another reputation mechanism.

Considering the economic / financial criteria, the first criterion is especially interesting for the iPr: the infrastructural costs saving resulting from less data being requested from the base stations. As we pointed out in Section 6.3, according to Buck (2010) approximately 31% of the costs per user for an iPr are related to infrastructural costs and of these costs approximately 50% are attributable to the base stations Gruber (2005). As a consequence, any saving with respect to base stations is very advantageous for the iPr if it can be achieved without having any negative effects on the mobile phone users. Figures 10.1 and 10.2 show the average reduction in the data transmission in the experiments without any enforcement (Figure 10.1) and in the experiments with the highest reduction in the energy consumption rate, i.e. experiments with 1% normative empowered agents (random waypoint movement model) (Figure 10.2)[1]. The values in the figures are given in percentages.

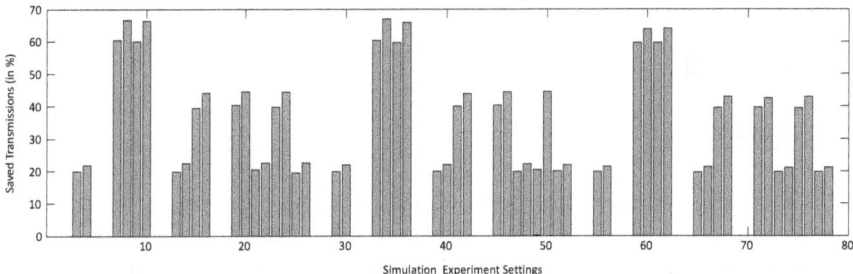

Figure 10.1: Base Station Transmission Savings (in %) – No Enforcement

Looking at Figure 10.1, the first aspect to notice is that in some runs no saving were made and all the data that would have been downloaded in the absence of WMGs. These runs did not include any honest agents in their configuration and therefore no cooperation took place. This is evidence that cooperation is not only advantageous for energy consumption, but also helps the

---

[1]In order to be able to compare the two different data sets, in the analysis of the data of the experiments without any enforcement we excluded all settings with only honest and only malicious agents.

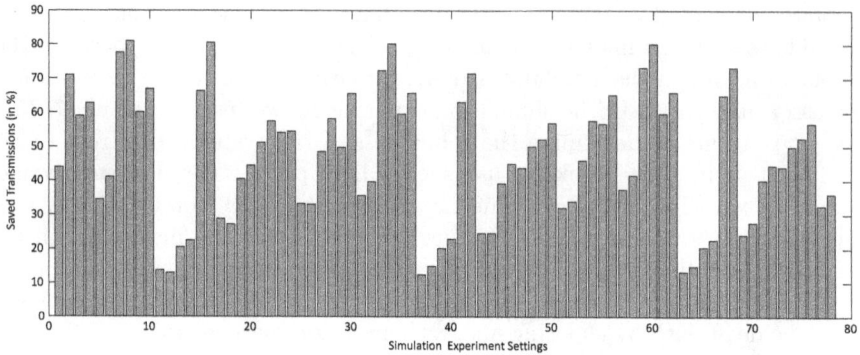

Figure 10.2: Base Station Transmission Savings (in %) – 1% Normative empowered
Agents

iPr by reducing the base station load. Looking at the base station savings when
1% normative empowered agents were introduced to the simulation experiments
(i.e. comparing Figures 10.1 and 10.2) a significant increase in the base station
transmission savings can be observed. In more detail, the minimum, maximum
and mean base station transmission savings[2] specified in Table 10.1 can be
calculated for the two simulation experiments.

Table 10.1: Comparison of Base Station Transmission Savings

	No Enforcement	**1% Normative empowered Agents**
Minimum Value	0%	12.4%
Mean Value	25.0%	46.4%
Maximum Value	66.9%	80.9%

Besides confirming the already visually obvious increase in transmission
savings with normative empowered agents, the numbers also give a first insight
about the scale of the savings possible with a WMG, which in case of the
utilization of normative empowered agents increase by approximately 20%. For
the largest mobile telecommunications provider in Germany – T-Mobile – which
in the mobile telecommunications sector had a total revenue of € 8,349 million,
an EBIT margin of 19.6% and approximately 34.7 million customers (T-Mobile,
2010, pp. 77-79), a 20% increase in base station infrastructural costs accounts
for several million euro. These values always have to be discussed keeping
in mind that they are results of simulation experiments and as such have a
validity strongly linked to the simulations and its assumptions. Thus, in the
experiments we for example assumed very highly multicastable traffic. As

---

[2]For the individual figures, the mean energy savings have little informative value, as the
simulation settings within each simulation experiment are not comparable. We only specify
the mean values to show the large differences between the two figures.

Scellato et al. (2011) suggest examining geographical social cascades for twitter video streams, only a small fraction of traffic has these properties. Furthermore, base station transmission savings cannot directly be translated to savings in infrastructural costs, as we cannot account for fixed infrastructural costs with the saving calculation for example. Nevertheless, the numbers are indicative that a functioning WMG could reduce infrastructural costs and thereby increase EBIT margins.

One set of costs that can counter these cost savings are the additional costs required to maintain the WMG, namely those for enforcing cooperation. With respect to these costs, the three different enforcement mechanisms we have presented have very different characteristics: whereas for image and reputation information based enforcement mechanisms the costs are mainly energy, which are carried by users and offset yb more energy-efficient downloads, for normative empowered agents the costs have to be paid by the iPr. Furthermore, besides the additional energy costs (accounted for in the simulation experiments) for normative empowered agents additional costs could accumulate. When first describing the idea of normative empowered agents in Section 8.2.1, we explained that one incentive for them to help control the system could be financial benefits such as free mobile phone contracts. Currently, mobile telecommunication providers have an average monthly revenue per user of approximately € 15 (ie Market Research, 2010), which is mainly generated from mobile phone contracts. Assuming that the provider employs 1% normative empowered agents, for an iPR like T-Mobile with 34,694,000 customers, this results in costs of approximately € 5.2 million, which need to be weighed against benefits of the WMG (e.g. energy savings or possibly higher subscriptions numbers if users embrace the WMG idea). As a result of the scale of these costs, it is very unlikely that enforcement in WMGs is economically viable under the assumptions made in this dissertation. Instead of offering normative empowered agents completely free service contracts, it might be of benefit to the providers to give less generous incentives to the normative empowered agents, i.e. to only offer them the marginal costs of their contract to reduce the costs for this kind of enforcement. Furthermore, it is likely that if iPr employ normative empowered agents, they will only employ a smaller percentage and combine the normative empowered agents with other less expensive enforcement mechanisms (e.g. image information). As we did not analyse any combined mechanisms in this dissertation, this analysis is subject to future work.

The final set of evaluation criteria we discussed in Section 6.3 focuses on social aspects of the chosen enforcement mechanisms. For these criteria, the Lorenz curve with the respective Gini Coefficients and the false-positive / false-negative rates are of interest. For the limitations mentioned on page 218, we cannot present the social data of all simulation experiments with all input parameter settings. Instead we focus on one setting only which will be analysed as an example. The criteria for this one setting being used are (i) that all behavioural groups of agents are represented in the example and (ii) that the majority of agents are utility maximizing, as this is the type of agent that is most likely to be encouraged to change its behaviour. Furthermore we opt for a small population of only 200 agents, because computational restrictions limit the handling of anything larger. Summing up, the setting we used has the following basic features: (i) $| \mathcal{A} |= 200$, (ii) utility agents $= 50\%$, honest agents $= 25\%$, malicious agent $= 25\%$, (iii) $\rho_{neighbourhood} = 10$, and (iv) random waypoint movement model. Figure 10.3 shows the Lorenz Curves and Gini Coefficients for energy consumption in the agent population in the basic WMG simulation experiments plus those with enforcement mechanisms. Besides showing the basic shape of the Lorenz curves, each behavioural type is encoded with one colour (i.e. blue for honest agents, red for utility agents, yellow for malicious agents and black for normative empowered agents[3]).

---

[3]Resulting from the inclusion of normative empowered agents in the respective experiments, the bottom two sub-figures show more than $| \mathcal{A} |= 200$ agents.

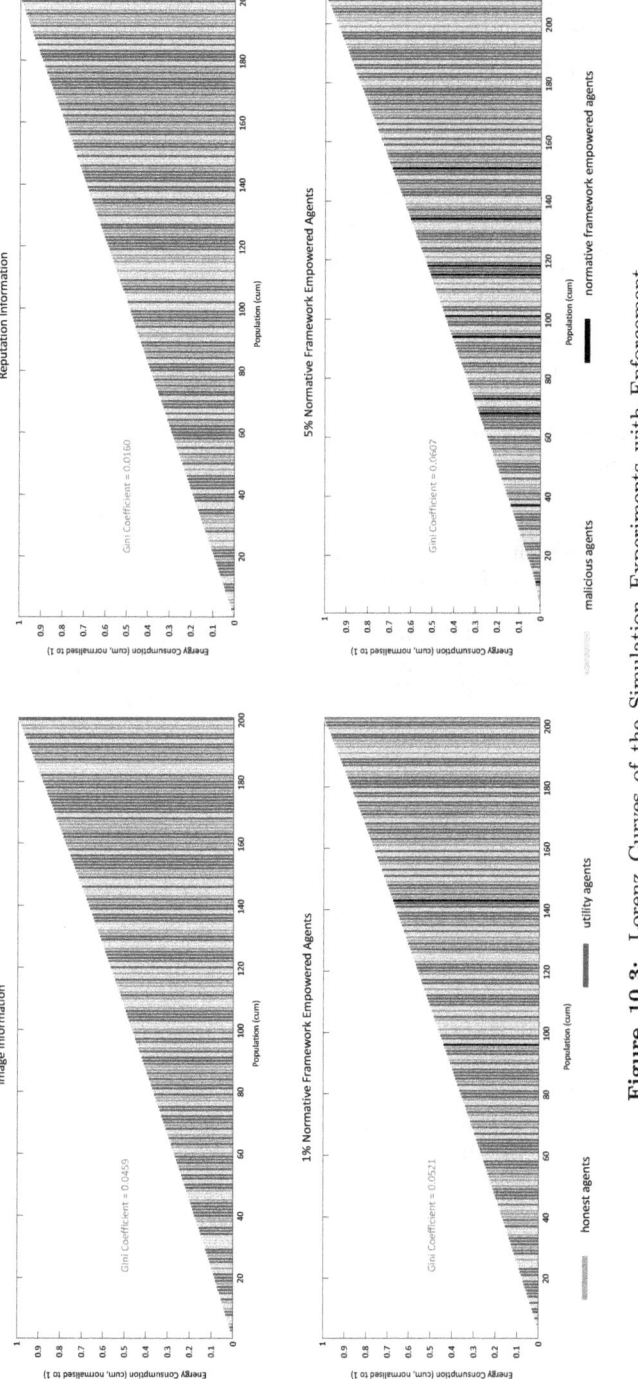

Figure 10.3: Lorenz Curves of the Simulation Experiments with Enforcement (Random Waypoint Movement Model)

As Figure 10.3 shows and the low Gini Coefficients corroborate, with the above settings, all Lorenz curves have progressions that are close to the line of perfect equality (i.e. the 45-degree-line), with the normative empowered agents having the highest Gini Coefficient values. Looking at the distribution of agent types in the Lorenz curves, there is no group of agents that is particularly advantaged or disadvantaged. This is an indicator of equally distributed energy consumption. Although from an enforcement point of view it is desirable that malicious agents are disadvantaged and honest agents advantaged in the system, the almost equally distributed energy consumption is positive for the users (and therefore indirectly positive also for the iPr) to the extent that cheaters are not favoured.

The final evaluation criterion we examine is the false negative / false positive rate. This rates indicate the percentage of cheating events not detected by the normative empowered agents, and to what extent agents have been wrongly punished for cheating when they did not cheat. Table 10.2 shows the results of these experiments.

Table 10.2: False Positive and False Negative Values for Simulation Settings with 1% and 5% Normative Framework Empowered Agents

	1% Normative empowered Agents	5% Normative empowered Agents
Cheat Events	3,935	4,500
Punishment Events	533	1,074
Correctly Punished	533	1,074
False-Positive	0	0
False-Negative	3,402	3,426

The values in Tables 10.2 indicate that the false positive problem we were initially concerned about does not occur at all. This is advantageous for both the iPr as well as the mobile phone users in the WMG. In contrast, we have false negative effects in both experimental settings, i.e. with 1% and 5% normative empowered agents. Whereas in the former case only 13.6% of all defection actions have been detected, in the latter the case the deployment of five times as many normative empowered agents resulted in a detection rate of 23.9% (i.e. approximately a doubling of the detection rate). These numbers support the conclusion drawn in Section 8.3.2, that although in absolute terms 5% normative empowered agents are able to detect more violations, a 1% proportion of agents has a better marginal detection rate, i.e. on average more detections per normative framework agent. Concerning the absolute percentage numbers, despite the numbers appearing to be relatively low, we cannot draw any conclusions to what extent detection rates of approximately 13.6% or 23.9% are sufficient for the general (and not only energy savings-related) success of a WMG and further analyses are required.

10.2   Generalisation of the Results for Open Distributed Systems

In Chapter 1, we set out the two theses for this dissertation:

1. Governance decisions such as the introduction of enforcement mechanisms can help to reduce the cooperation and collaboration problems in open distributed systems.

2. Normative frameworks and multi-agent system simulations can be used to model and reason about the interaction of users with one another as well as with the norms in an open distributed system.

In order to test these theses, we have used a WMG case study which served as an example throughout this dissertation. In this section we focus on the generalisation of the results from the specific case study. Before doing so we first briefly recall the properties of open distributed systems and explain how WMGs fit into them.

Definition 5 on page 20, characterises open distributed systems with the help of the following features:

- They contain autonomous entities which are free to join and leave the system and which perform actions, based on their own goals as well as their bounded rational perception of the system.

- The actions of these entities contribute to a global system result, which in contrast to the actions of the individual entities in the system, can be perceived externally.

Examining WMGs against this definition, human mobile phone users interact in a WMG. They are autonomous, can freely decide to participate in the WMG and base their decisions on their limited perception of the network as well as on what they perceive to suit their needs and goals best. The individual actions of the users (in particular their decisions to cooperate or defect) result in the global system result which is the failure or the success of the WMG. In contrast to the decision of the actors in the system, this success/failure result can be observed from outside the system (e.g. through reports of the results). The WMG therefore fulfills all the criteria of an open distributed system and we can continue to analyse to what extent the results that we have found for this particular open distributed system can be transferred to other systems. It needs to be noted that we used WMG specific values for the simulation experiments, such as the specific energy consumption rates of the Nokia N95 for certain transmission protocols. As a consequence, when generalising we cannot draw any conclusions about absolute values, but rather focus on relative results.

Starting with the most high level result, we can observe that successful enforcement increases cooperation and improves sharing of resources. Other generalisations concern the input factors of our simulations. We were able to show that $\rho_{neighbourhood}$ had a significant impact on the success of enforcement as it determined the likelihood that two participants would interact again. Another parameter important for the success of enforcement is population composition. We consider the importance of these two input parameters to be generally true for all open distributed systems.

In our experiments, we were able to demonstrate that the costs associated with enforcement need to be considered, as they can significantly alter the cost/benefit ratio of a mechanism. We believe this finding also to be generally true.

In case of the WMG we focused on a scenario very similar to P2P networking file sharing systems in which files are collaboratively obtained. As presented in Table 2.1 on page 22, file sharing is not the only collaborative idea in open distributed systems and shared bandwidth/transmissions are not the only shared resources, but a variety of open distributed systems specifications exist. Despite their differences, with respect to cooperation they all have the same problem: the contribution of shared resources needs to be encouraged whereas free-riding (i.e. obtaining resources without contributing any) needs to be discouraged. The type of resources that are being shared has very little influence on actual results. The only problem we see with respect to the different shared resources of open distributed systems applications concerns virtual teams. All shared resources of the other presented systems are physical resources that can be divided to some degree making their sharing, distribution and allocation relatively easy. In contrast in virtual teams the resources can be special character properties and skills, whose division is considerably more difficult, which in turn demands other collaboration models.

A second difficulty we see is that in our simulations we focused on direct reciprocation only, in which collaboration always focuses on immediate quid pro quo (i.e. for sending its share an agent expected the receiving of the shares of its partners in the same transaction). This neglects the idea that the contribution of an agent to the system and its benefit from the system might have a time discrepancy. In routing scenarios for example, an agent might act as a relay node passing on messages on one occasion, hoping that in future when it needs to send information other nodes will do the same. The agents are not in a direct debitor-creditor relationship, but instead indirect reciprocity is required. The mechanisms presented here all relied on direct observations and interactions. Removing direct interaction makes enforcement harder to apply, which is why we see difficulties with the generalisation of our results for all application with indirect reciprocity.

These two constraints apart, we believe the general conclusions drawn from our simulation experiments to be valid for open distributed systems in general. This is supported by our simulation results confirming some of the known enforcement problems in open distributed systems, such as the slow spreading of image information and the significant negative effect of badmouthing on the performance of reputation mechanisms.

10.3   Summary

In this section, we expanded our analysis of the results of our simulation experiments from a pure focus on energy consumption rates (Chapters 7–9) to include criteria that we view as particularly important in a multi-stakeholder context. These criteria covered technical, economic / financial as well as social aspects. Summing up the results of this analysis, it becomes apparent that these criteria have an effect and can even change the preference order for enforcement mechanisms for the iPr as well as mobile phone users. This becomes especially apparent when looking at the normative empowered agents, i.e. the enforcement mechanism that has previously shown the best energy efficiency values. Despite performing well in respect to scalability as well as robustness, infrastructural costs savings and costs/benefit distribution between the users, under the given model assumption, the costs associated with this mechanism may make it unattractive for an implementation by the iPr. Instead from an iPr's perspective it might be favourable to lower the incentives for the normative empowered agents (e.g. by only only offering them the marginal costs of their contract), or to combine this mechanisms with an image-based mechanism in order to reduce the percentage of normative empowered agents required in the system and thereby reducing the costs of enforcement.

Furthermore, we discussed the extentto which our WMG specific simulation results can be generalised for open distributed systems. The negligence of indirect reciprocity as well as the assumption of easy divisibility of the shared resources limits the generalisation partially. Nevertheless, we demonstrated that as long as no absolute values need to be transferred, a large degree of the general simulation findings (e.g. which input parameters are important for the success of enforcement) are applicable in all open distributed systems.

Chapter 11

Summary and Conclusions

Having presented all the supporting research, we now reflect on the results and establish whether the theses set out in the introduction of this dissertation have been successfully satisfied.

11.1 Contributions

The research reported is driven by the two theses that we formulated in Chapter 1:

1. Governance decisions realized through the introduction of enforcement mechanisms can help to reduce the cooperation and collaboration problems in open distributed systems.

2. Normative frameworks and multi-agent system simulations can be used to model and reason about the interaction of users with one another as well as with the norms in an open distributed system.

In the pursuit of the proof of these theses, we have presented and analysed three different enforcement mechanisms in order to test, in an early prototyping stage of WMGs, to what extent these mechanisms could be used to address the cooperation problem in WMGs. The results of this analysis successfully show that enforcement mechanisms can reduce the cooperation problem in open distributed systems and can reduce the load on shared resources. This is true even if the system participants act selfishly and have private utility functions that are not known to the system designer (Chapter 8). This proves the first thesis of this dissertation.

In order to prove the second thesis, we presented normative frameworks and demonstrated how they can effectively be applied to model the cooperation problem in WMGs in the form of a design-time model (Chapter 5). Our model shows that one can develop complete (though simplified) realistic scenarios. We show that the development of a design-time model is an easily comprehensible way of developing powerful first normative models to analyse the impact of governance decisions. These models can be used to evaluate whether it makes sense to pursue the governance decision at all and whether it is worth investigating its impact further.

The design-time model provides a useful first approach to explore governance mechanisms, but cannot properly account for user autonomy. Consequently, thanks to the positive results from the design-time model, we moved into a second phase, deriving a run-time model, and in doing so a novel methodology for the design to run-time model derivation process. We furthermore showed how the run-time model can efficiently be combined with MAS simulations that generate the exogenous events for the it (Chapter 7). To the best of our knowledge, this run-time model is the first of its kind. We successfully demonstrated its functionality by implementing it for the WMG case study. With this implementation and the design and run-time methodology, we provide practical, powerful and easily accessible tools to assess the impact of governance decisions at an early prototyping stage. This successfully confirms our second thesis.

Using the run-time model of the WMG we demonstrated the effectiveness of our approach by performing a comparative analysis of three different enforcement mechanisms: image information, reputation information and normative empowered agents. The results of this comparison provided us with good insights into WMGs as well as the role of different input parameters on enforcement (Chapters 8 and 9).

Governance decisions such as the introduction of enforcement mechanisms always have impact on several stakeholder groups in a system. Therefore we identified potential criteria of importance for the stakeholders of a WMG by analysing them with respect to the simulation results, creating a better understanding of the significance and impact of multi-stakeholder interests (Chapter 10).

Summing up, we conclude that we have fully established the two theses on which this dissertation rests. Furthermore we have made the following important contributions:

- We gave in-depth insight into wireless mobile grids and the cooperation problem in them.

- We successfully demonstrated how to use normative frameworks to develop a design-time model of a normative system.

- We provided a detailed methodology on how to generate a run-time model from the design-time model and how to combine it effectively with a multi-agent system simulation.

- We significantly increased the understanding of how different enforcement mechanisms affect cooperation in open distributed systems and how different stakeholders' perspectives need to be incorporated in governance decisions.

11.2   Future Work

Our design and run-time models have shown themselves to be powerful, easy to build and particularly useful in the early prototyping stage, and we have demonstrated effectiveness by performing our comparative analysis of enforcement mechanisms. However, we could only some of the many aspects of modelling and analysing governance decisions (in particular the introducing of enforcement mechanisms) in open distributed systems.

11.2.1   Relaxing Model Assumption

As in every model representing real world phenomena (even though the technology we discuss is still in a prototyping stage), we had to make assumptions with respect to all three components of the underlying factors of the IAD framework to keep the model simple and significant. Lifting these assumptions is direction for future work.

With respect to the biophysical characteristics, we relied on the flat earth model and made a full coverage assumption for the base stations, i.e. assuming that every spot in the simulation area is covered at least by one mobile phone base station and that consequently the simulated mobile phone users have access to the 3G mobile phone network at any point in time (given that the base station is not busy otherwise). By removing this assumption we can more appropriately simulate rural or urban areas (with low or high coverage) and analyse the effects on these specific areas. Furthermore, we currently assume that after having obtained a file in a simulation run, the agents delete it. Hence, in an interaction the agents cannot rely on existing files in the system but need to start obtaining files from scratch. By keeping track of which files an agent already possesses, we could for example allow for indirect reciprocal situations in which an agent that already possesses the file another agent wants, sends this file to the other agent without directly benefiting from the interaction. The agent might nevertheless contribute his file, hoping that other agents behave the same way and it will benefit in another interaction. Thus would allow us to check to which extent this reduces the amount of data the base stations have to send as well as how fast data spreads in the system.

Concerning the attributes of the community, in future work the simulation experiments could be extended by employing more sophisticated agents. In our simulation experiments we have employed three kinds of agents that pursue very different strategies in order to test how sensitive the simulation is to very one-sided behaviour (e.g. always defect or always cooperate). In an actual deployment of a WMG such a one-side behaviour might not be very realistic. Therefore in future work agents with more sophisticated reasoning processes that exhibit more diverse responses to the successes or failures of cooperation situations are needed. One extension could for example be to allow malicious agents to cooperate occasionally in order to make them harder to detect for other agents, or to allow for variations in the reactions to sanctions by the utility maximizing agents.

Finally with regard to the rules-in-use, i.e. the third component of the underlying factors of the IAD framework, extensions of the enforcement mechanisms employed are a good direction for future work. In this dissertation we have implemented three different enforcement mechanisms and have given our rationale for selecting them. We assumed that only one of them can be employed at any given time and that the choice of mechanism is constant throughout a simulation experiment. We plan to extend the work by relaxing this assumption and combining different enforcement concepts in the simulation experiments. One combination to test could be to employ both image information and normative empowered agents at the same time, in order to support the good results from the latter with the easy to obtain image information from the different agents.

11.2.2   Extending the Simulation Experiments

Besides merely relaxing assumptions made in the course of the modelling process, we might also the simulation experiments, both with respect to the parameters and set-up as well as their scale (i.e. the number of agents simulated).

The enforcement mechanism that produced the best results (in terms of reducing the energy consumption rate) are normative empowered agents. For this mechanism, parameters that are good candidates for further exploration are the percentage of normative empowered agents in the system, and the level of fines. Furthermore, the introduction of coordination between the normative empowered agents seems sensible. Our simulation experiments suggested that – with respect to the energy consumption rate – 1% normative empowered agents are more efficient in the simulated WMG than 5% of these agents, because the energy costs produced by the normative empowered agents in the latter case outweighed their detection benefits. It would be interesting to investigate how the system performs for any percentages between these two numbers (e.g. 3%) or below 1% in order to determine in more detail at which point the detection advantages and the energy costs cancel each other out. As a second extension, when calculating the fines for the defecting agents, the normative empowered agents could take into account the number of times an agent already has been detected cheating, and increase fines for repeat offenders. Currently the individual normative empowered agents all act independently without coordinating with one another. Despite this lack of coordination, resulting from the relatively small percentages of these agents chosen in the experiments, we only had very few occasions on which two normative empowered agents observed the same interaction. Nonetheless, for more effective enforcement, a limited level of coordination between these agents could be useful, such as pre-assigning specific areas to specific normative empowered agents.

On a technical level, directions for future work include the scaling of experiments to include more agents and thus allow for simulations of more realistically sized WMG scenarios as well as advancing the single normative framework model to one allowing for several (possibly interacting) normative frameworks. Concerning the former extension, the Jason simulation environment allows for running the simulation distributed on several computers in order to scale up simulation experiment size. One platform that might be suitable candidate for this purpose could be AgentScape[1], which is an open distributed agent middleware developed mainly at the TU Delft, as AgentScape offers specific support for Jason agents. In real world situations, we observe that often more than one normative framework is active within a system. Extending our run-time model to account for multiple normative frameworks is therefore another direction future work can take. In Cliffe et al. (2007b), the authors present the concept of multi-normative frameworks where events in one normative framework cause events in another or change another normative framework's state. Extension of the Governor to accommodate reasoning about several possibly interlinked normative frameworks is an important part of future work, along with the issue of using conventional distributed systems techniques, such as replication, as a means to avoid the Governor becoming a bottleneck and a single point of failure.

One further direction is the inclusion of more real world data in the simulation, once this is available. As already outlined in Chapter 9, due to a lack of mobile phone user movement traces, we had to abstract from real human movement behaviour with the help of generic movement models. Obtaining more realistic movement traces – and to implement them in the simulation experiments – therefore is a goal for future work. One further area where we hope to be able to obtain real data concerns the demands of mobile phone users in respect to WMGs. In this area currently a study is being developed by C. Buck based on the findings in (Buck, 2010). The results of this study could help to improve the choice of simulation parameters as well as the analysis of the simulation results with respect to the multiple stakeholders. One final aspect we would like to consider in future work is the inclusion of larger number of iPr. In this dissertation, we assumed that only one iPr exists and that all mobile phone users have their contracts with this particular iPr. By allowing for different iPr which provide different networks, it would be interesting to analyse effects between both for the mobile phones users as well as the iPr.

11.2.3 Adaptive Learning of Norms

One final direction for future work is the adaptive learning of norms. In our simulation experiments the designer of the simulation had to determine the governance changes in advance of the simulation. The changes had to be implemented and tested in order to get an initial idea whether the proposed governance decision have the desired influence on the actors' decisions and whether they will contribute to the system goals in the way expected. Even if an effect can be asserted for the tested governance decisions, it is far from clear whether the changes introduced by the governance decisions are already the ideal setting for inducing the desired system behaviour.

The brute-force mechanism of testing all possible options for a governance decision is one solution to this problem, but this approach would be very time

---

[1]For more information on AgentScape see http://www.agentscape.org/.

and resource consuming and might even be infeasible because of the large number of different setting combinations for governance decisions to be tested. In order to tackle this problem, future work could be directed at developing an adaptive governance mechanism for these systems that constantly tries to induce behaviour of the actors in the systems in a way that they behave in accordance with the systems objectives by updating the norms at run-time. In detail we envision this process of adapting norms at run-time to consist of four stages:

1. First of all the overall desired macro system results need to be identified. If this is not feasible at least desired actions by the actors need to be determined and a first set of norms been drafted.

2. Once the norms are incorporated in the system and the system is running, the results of the running system compared to the desired results.

3. Based on the comparison of the actual and desired system results, conclusions about the effectiveness of the norms as well as deductions on the preferences and utilities of the actors can be drawn.

4. Having acquired these two components, the adaption – if required – needs to take place. The process for this adaption is a learning process, including the learning of the actors' preferences as well as their respective reactions to the norms in the system.

Using this four-stage process we might therefore be able to eliminate the strong interdependence between the initial governance decisions and the system results, and to determine good governance decisions more effectively at run-time and thereby arrive at better enforcement mechanisms.

Summing up, we believe that developing better tools for designing and testing governance decisions for open distributed system is a valuable and highly important undertaking, however for its successful accomplishment, much work is still required.

11.3   Concluding Remarks

In this dissertation we have focused on the modelling and analysis of governance decisions in open distributed systems. By its nature, governing open distributed systems in which actors with a multitude of different interests interact is complex and difficult. We have contributed to addressing this by proposing a practical, powerful and easily accessible methodology that can be used to analyse norms and their impact on the interactions in these systems. We proved the effectiveness of this methodology by successfully implementing a prototype which we used for analysing the effects of introducing different enforcement mechanisms to an open distributed system.

Bibliography

Abdul-Rahman, A. and S. Hailes (1997). Using recommendations for managing trust in distributed systems. In *IEEE Malaysia International Conference on Communication*. xxi, 187

Abdul-Rahman, A. and S. Hailes (2000). Supporting trust in virtual communities. In *HICSS*. 187

Abea, H., A. Suzuki, M. Etoh, S. Sibagaki, and S. Koike (2008). Towards systematic innovation methods: Innovation support technology that integrates business modeling, roadmapping and innovation architecture. In *Portland International Conference on Management of Engineering Technology, 2008. PICMET 2008.*, pp. 2141–2149. 44

Adar, E. and B. A. Huberman (2000). Free riding on gnutella. *First Monday 5*. 42

Alchourrón, C. E. and E. Bulygin (1971). *Normative Systems*. Number 5 in Library of Exact Philosophy. Springer. 75

Aldewereld, H. M. and V. Dignum (2010). Operetta: Organization-oriented development environment. In *Proceedings of the International Workshop on LAnguages, methodologies and Development tools for multi-agent Systems (LADS 2010)*. 91

Allen, J. (1991). Time and time again: The many ways to represent time. *International Journal of Intelligent Systems 6*(4), 341–356. 84

Allen, S. M., G. Colombo, and R. M. Whitaker (2010, February). Cooperation through self-similar social networks. *ACM Transactions on Autonomous and Adaptive Systems 5*(1), 4:1–4:29. 35

Andersson, C., D. Freeman, I. James, A. Johnston, and S. Ljung (2006, January). *Mobile Media and Applications, From Concept to Cash : Successful Service Creation and Launch*. John Wiley & Sons. 55, 57

André, F. J., M. A. Cardenete, and C. Romero (2007). Using compromise programming for macroeconomic policy making in a general equilibrium framework: Theory and application to the spanish economy. *Journal of the Operational Research Society 59*, 875–883. 140

Andrighetto, G., R. Conte, P. Turrini, and M. Paolucci (2007). Emergence in the loop: Simulating the two way dynamics of norm innovation. In G. Boella, L. van der Torre, and H. Verhagen (Eds.), *Normative Multi-agent Systems*, Number 07122 in Dagstuhl Seminar Proceedings. 15

Andrighetto, G., D. Villatoro, R. Conte, and J. Sabater Mir (2010). Simulating the relative effects of punishment and sanction in the achievement of cooperation. In *Proceedings of 8th European Workshop on Multi-Agent Systems*. 39

Arrow, K. J. (1994). Methodological individualism and social knowledge. *American Economic Review 84*(2), 1–9. 21

Artikis, A. (2003, November). *Executable Specification of Open Norm-Governed Computational Systems*. Ph. D. thesis, Department of Electrical & Electronic Engineering, Imperial College London. 92

Artikis, A., M. Sergot, and J. Pitt (2007, April). Specifying norm-governed computational societies. *ACM Transactions on Computational Logic*. 92

Axelrod, R. (1981). The emergence of cooperation among egoists. *The American Political Science Review 75*(2), 306–318. 69

Axelrod, R. (1984). *The Evolution of Cooperation*. New York: Basic Books. 28, 35

Axelrod, R. (1986, December). An evolutionary approach to norms. *The American Political Science Review 80*(4), 1095–1111. 18

Axelrod, R. (1997). Advancing the art of simulation in the social sciences. In R. Conte, R. Hegselmann, and P. Terna (Eds.), *Simulating Social Phenomena*, pp. 21–40. Springer. 126

Axelrod, R. and D. Dion (1988, December). The further evolution of cooperation. *Science 242*(4884), 1385–1390. 36

Balci, O. (1994). Validation, verification, and testing techniques throughout the life cycle of a simulation study. In *WSC '94: Proceedings of the 26th conference on Winter simulation*, San Diego, CA, USA, pp. 215–220. Society for Computer Simulation International. 134

Balke, T. (2009). A taxonomy for ensuring institutional compliance in utility computing. In G. Boella, P. Noriega, G. Pigozzi, and H. Verhagen (Eds.), *Normative Multi-Agent Systems*, Number 09121 in Dagstuhl Seminar Proceedings. Schloss Dagstuhl - Leibniz-Zentrum für Informatik, Germany. 7, 25, 31, 32

Balke, T., M. De Vos, and J. A. Padget (2011, October). Analysing energy-incentivized cooperation in next generation mobile networks using normative frameworks and an agent-based simulation. *Future Generation Computer Systems Journal 27*(8), 1092–1102. 8

Balke, T., M. De Vos, J. A. Padget, and F. Fitzek (2011). Using a normative framework to explore the prototyping of wireless grids. In M. De Vos, F. Nicoletta, J. V. Pitt, and G. Vouros (Eds.), *Coordination, Organizations, Institutions, and Norms in Agent Systems VI*, Volume 6541 of *Lecture Notes on Computer Science*, pp. 95–113. Springer. 8

Balke, T., M. De Vos, J. A. Padget, and D. Traskas (2011a). Normative run-time reasoning for institutionally-situated bdi agents. In *Proceedings of the 13th International Workshop on Coordination, Organization, Institutions and Norms in Agent Systems (COIN) @ WI-IAT2011*. 9

Balke, T., M. De Vos, J. A. Padget, and D. Traskas (2011b). On-line reasoning for institutionally-situated bdi agents. In K. Tumer, P. Yolum, L. Sonenberg, and P. Stone (Eds.), *Proceedings of 10th International Conference on Autonomous Agents and Multiagent Systems (AAMAS 2011)*, pp. 1109–1110. 8, 9

Balke, T. and T. Eymann (2009, February). Using institutions to bridge the trust-gap in utility computing markets - an extended "trust game". In H. R. Hansen, D. Karagiannis, and H.-G. Fill (Eds.), *Proceedings of the 9. Internationale Tagung Wirtschaftsinformatik*, Volume 2 of *books@ocg.at*, Vienna, pp. 213–222. Österreichische Computer Gesellschaft. 30

Balke, T. and T. Eymann (2010). Challenges for social control in wireless mobile grids. In T. Doulamis (Ed.), *GridNets2009*, Volume 25 of *Lecture Notes of ICST*, pp. 147–154. Springer. 8

Balke, T., S. König, and T. Eymann (2009). A survey on reputation systems for artificial societies. Bayreuther Arbeitspapiere zur Wirtschaftsinformatik 46, University of Bayreuth. 7, 28, 40, 186

Balke, T. and D. Villatoro (2011). Operationalization of the sanctioning process in hedonic artificial societies. In *Workshop on Coordination, Organization, Institutions and Norms in Multiagent Systems @ AAMAS 2011, Taiwan*. 7, 18, 27, 28, 30

Bansal, S. and M. Baker (2003). Observation-based cooperation enforcement in ad hoc networks. In *CoRR*. 41

Baral, C. (2003). *Knowledge Representation, Reasoning, and Declarative Problem Solving*. New York, NY, USA: Cambridge University Press. 92, 93

Bardossy, A., I. Bogardi, and L. Duckstein (1985). Composite programming as an extension of compromise programming. In P. Serafini (Ed.), *Mathematics of Multiple Objective Optimization*, pp. 375–408. Springer Wien. 141

Becker, G. S. (1968). Crime and punishment: An economic approach. *The Journal of Political Economy 76*(2), 169–217. 34

Bertsch, V. and J. Geldermann (2008). Preference elicitation and sensitivity analysis in multi-criteria group decision support for industrial risk and emergency management. *International Journal of Emergency Management 5*(1/2), 7–24. 140

Bettstetter, C. (2001a, July). Mobility modeling in wireless networks: Categorization, smooth movement, and border effects. *ACM Mobile Computing and Communications Review 5*(3), 55–67. 203, 208

Bettstetter, C. (2001b). Smooth is better than sharp: a random mobility model for simulation of wireless networks. In *Proceedings of the 4th ACM International Workshop on Modeling, Analysis and Simulation of Wireless and Mobile Systems*, New York, NY, USA. ACM. 203, 207

Bingham, M. (2010, March, 28). Smartphone overload – the explosion in mobile internet access means the networks are fast running out of capacity. The Sunday Times. 62

Boissier, O. and B. Gâteau (2007). Normative multi-agent organizations: Modeling, support and control, draft version. In G. Boella, L. van der Torre, and H. Verhagen (Eds.), *Normative Multi-agent Systems*, Number 07122 in Dagstuhl Seminar Proceedings, pp. Internationales Begegnungs– und Forschungszentrum fuer Informatik (IBFI), Schloss Dagstuhl, Germany. 91

Bolton, G. E. (1991). A comparative model of bargaining: Theory and evidence. *American Economic Review 81*(5), 1096–1136. 37

Bolton, G. E., J. Brandts, and A. Ockenfels (1998, December). Measuring motivations for the reciprocal responses observed in a simple dilemma game. *Experimental Economics 1*(3), 207–219. 37

Bolton, G. E. and A. Ockenfels (2000, March). Erc: A theory of equity, reciprocity, and competition. *American Economic Review 90*(1), 166–19. 37

Bonabeau, E. (2002, May). Agent-based modeling: Methods and techniques for simulating human systems. *Proceedings of the National Academy of Sciences of the United States of America 99*(10), 7280–7287. 130

Bordini, R. H., J. F. Hübner, and R. Vieira (2005). Jason and the golden fleece of agent-oriented programming. In R. H. Bordini, M. Dastani, J. Dix, and A. E. F. Seghrouchni (Eds.), *Multi-Agent Programming*, Volume 15 of *Multiagent Systems, Artificial Societies, and Simulated Organizations*, Chapter 1, pp. 3–37. Springer US. 162

Bordini, R. H., J. F. Hübner, and M. Wooldridge (2007). *Programming Multi-Agent Systems in AgentSpeak using Jason*. Wiley Series in Agent Technology. John Wiley & Sons. 133, 142, 161

Bradshaw, J. M., S. Dutfield, B. Carpenter, R. Jeffers, and T. Robinson (1995). Kaos: A generic agent architecture for aerospace applications. In *Proceedings of the CIKM '95 Workshop on Intelligent Information Agents*. 26

Broersen, J., F. Dignum, V. Dignum, and J.-J. C. Meyer (2004). Designing a deontic logic of deadlines. In A. Lomuscio and D. Nute (Eds.), *Deontic Logic in Computer Science – Proceedings of the 7th International Workshop on Deontic Logic in Computer Science (DEON 2004)*, Volume 3065 of *LNCS*, pp. 43–56. Springer. 17

Buchanan, J. T. and H. G. Daellenbach (1987). A comparative evaluation of interactive solution methods for multiple objective decision models. *European Journal of Operational Research 29*(3), 353–359. 140

Buchegger, S. and J.-Y. L. Boudec (2002). Performance analysis of the confidant protocol (cooperation of nodes - fairness in dynamic ad-hoc networks). In *Proceedings of MobiHoc 2002*. 41

Buchegger, S. and J. Chuang (2007). Encouraging cooperative interaction among network entities – incentives and challenges. In F. H. P. Fitzek and M. D. Katz (Eds.), *Cognitive Wireless Networks – Concepts, Methodologies and Visions Inspiring the Age of Enlightenment of Wireless Communications*, pp. 87–108. Springer Netherlands. 21

Buck, C. (2010). Zu den auswirkungen von wireless mobile grids auf die kostenstruktur von mobilfunknetzen – eine kostenrechnerische betrachtung von infrastruktur- und netzwerkkosten. Master's thesis, Universität Bayreuth. 138, 219, 233

Bundesnetzagentur (2010). Jahresbericht 2009. 54

Buttyán, L. and J.-P. Hubaux (2000). Enforcing service availability in mobile ad-hoc wans. In *MobiHoc '00: Proceedings of the 1st ACM international symposium on Mobile ad hoc networking & computing*, Piscataway, NJ, USA, pp. 87–96. IEEE Press. ISBN 0-7803-6534-8. 41

Camp, T., J. Boleng, and V. Davies (2002). A survey of mobility models for ad hoc network research. *Wireless Communications and Mobile Computing 2*, 483–502. 159, 203, 204, 205, 206, 208

Capgemini Consulting (2005, September). The future of 3g: Assessing the threat of alternative wireless technologies. Technical Report 10, Capgemini TME Strategy Lab. 55

Centeno, R., H. Billhardt, R. Hermoso, and S. Ossowski (2009). Organising mas: a formal model based on organisational mechanisms. In *SAC '09: Proceedings of the 2009 ACM symposium on Applied Computing*, New York, NY, USA, pp. 740–746. ACM. 150

Chamberlin, E. H. (1948). An experimental imperfect market. *The Journal of Political Economy 56*(2), 95–108. 37

Checkland, P. (1999). *Soft Systems Methodology in Action*. Wiley & Sons. 44

Cisco Systems (2010). Cisco visual networking index 2009–2014: Forecast and methodology. 54

Cliffe, O. (2007). *Specifying and Analysing Institutions in Multi-agent Systems Using Answer Set Programming*. Ph. D. thesis, University of Bath. 17, 75, 79, 82, 84, 86, 92, 96, 105

Cliffe, O., M. De Vos, and J. A. Padget (2006). Specifying and analysing agent-based social institutions using answer set programming. In O. Boissier, J. A. Padget, V. Dignum, G. Lindemann, E. T. Matson, S. Ossowski, J. S. Sichman, and J. Vázquez-Salceda (Eds.), *Coordination, Organizations, Institutions, and Norms in Multi-Agent Systems – AAMAS 2005 International Workshops on Agents, Norms and Institutions for Regulated Multi-Agent Systems, ANIREM 2005, and Organizations in Multi-Agent Systems, OOOP 2005, Utrecht, The Netherlands, July 25-26, 2005, Revised Selected Papers*, Volume 3913 of *Lecture Notes in Computer Science*, pp. 99–113. Springer Berlin / Heidelberg. 92, 101

Cliffe, O., M. De Vos, and J. A. Padget (2007a). Answer set programming for representing and reasoning about virtual institutions. In K. Inoue, K. Satoh, and F. Toni (Eds.), *Computational Logic in Multi-Agent Systems*, Volume 4371 of *Lecture Notes in Computer Science*, pp. 60–79. Springer Berlin / Heidelberg. 92, 96, 97

Cliffe, O., M. De Vos, and J. A. Padget (2007b). Specifying and reasoning about multiple institutions. In *Coordination, Organizations, Institutions, and Norms in Agent Systems II*, Volume 4386 of *LNAI*, pp. 67–85. Springer Berlin / Heidelberg. 93, 233

Cliffe, O., M. De Vos, and J. A. Padget (2009). Modelling normative frameworks using answer set programing. In *LPNMR*, pp. 548–553. 92, 100

Cohen, B. (2003). Incentives build robustness in bittorrent. In *Proceedings of the 1st Workshop on Economics of Peer-to-Peer Systems*. 43

Coleman, J. S. (1998, August). *Foundations of Social Theory*. Belknap Press. 35

Conte, R. and M. Paolucci (2002, October). *Reputation in Artificial Societies: Social Beliefs for Social Order*. Springer. 29

Courcoubetis, C. and R. Weber (2006). Incentive for large peer-to-peer systems. *IEEE Journal of Selected Areas in Communications 24*, 1034–1050. 43

Dalu, S. S., M. K. Naskar, and C. K. Sarkar (2008). Hardware implementation of a topology control algorithm for manets using nomadic community mobility model. In *International Conference on Wireless Networks & Embedded Systems (WECON-2008)*. 207

Davidsson, P. (2001). Categories of artificial societies. In *Proceedings of the Second International Workshop on Engineering Societies in the Agents World II*, Volume 2203 of *Lecture Notes in Computer Science*, London, UK, pp. 1–9. Springer Verlag. 19, 127

Dignum, F., J. Broersen, V. Dignum, and J.-J. C. Meyer (2004). Meeting the deadline: Why, when and how. In *Formal Approaches to Agent-Based Systems*, pp. 30–40. 17

Dignum, V. (2003). *A Model for Organizational Interaction: Based on Agents, Founded in Logic*. Ph. D. thesis, Utrecht University. 90

Dignum, V., J.-J. Meyer, F. Dignum, and H. Weigand (2003). Formal specification of interaction in agent societies. In M. Hinchey, J. Rash, W. Truszkowski, C. Rouff, and Gordon-Spears (Eds.), *Formal Approaches to Agent-Based Systems (FAABS)*, Volume 2699 of *Lecture Notes in Artificial Intelligence*. Springer. 90

Dooley, K. (2002). Simulation research method. In J. Baum (Ed.), *Companion to Organizations*, pp. 829–848. London: Blackwell. 126, 127

Dreber, A., D. Rand, D. Fudenberg, and M. Nowak (2008). Winners don't punish. *Nature 452*, 348–351. 35

EITO, E. I. T. O. (2010, August). More than five billion mobile phone users. Press Release. 54

Ellickson, R. C. (2005, June). *Order without Law: How Neighbors Settle Disputes*. Harvard University Press. 27

Esteva, M., D. de la Cruz, and C. Sierra (2002). Islander: an electronic institutions editor. In *AAMAS '02: Proceedings of the first international joint conference on Autonomous agents and multiagent systems*, New York, NY, USA, pp. 1045–1052. ACM. 26, 89

Esteva, M., J. A. Padget, and C. Sierra (2002). Formalizing a language for institutions and norms. In *ATAL '01: Revised Papers from the 8th International Workshop on Intelligent Agents VIII*, London, UK, pp. 348–366. Springer-Verlag. 89

Esteva, M., J. A. Rodríguez-Aguilar, J. L. Arcos, C. Sierra, and P. García (2000). Formalizing agent mediated electronic institutions. In *Proceedings of the Congrès Català d'Intel.ligència Artificial*, pp. 329–338. 89

Esteva, M., J. A. Rodríguez-Aguilar, C. Sierra, P. Garcia, and J. L. Arcos (2001). On the formal specifications of electronic institutions. In *Agent Mediated Electronic Commerce, The European AgentLink Perspective*, pp. 126–147. Springer. 89

Esteva, M., B. Rosell, J. A. Rodríguez-Aguilar, and J. L. Arcos (2004). Ameli: An agent-based middleware for electronic institutions. In N. e. a. Jennings (Ed.), *Proceedings of the Third International Joint Conference on Autonomous Agents and Multiagent Systems (AAMAS 2004)*, Volume 1, Washington, DC, USA, pp. 236–243. IEEE Computer Society. 90

Eymann, T. (2000). *AVALANCHE - Ein agentenbasierter dezentraler Koordinationsmechanismus für elektronische Märkte*. Ph. D. thesis, Institut für Informatik und Gesellschaft, Albert-Ludwigs-Universität Freiburg im Breichsgau. 40

Fasli, M. (2001). On commitments, roles and obligations. In *CEEMAS '01: Revised Papers from the Second International Workshop of Central and Eastern Europe on Multi-Agent Systems*, London, UK, pp. 93–102. Springer. 82

Fehr, E. and S. Gächter (2000). Cooperation and punishment in public goods experiments. *The American Economic Review 90*(4), 980–994. 35

Fehr, E., S. Gächter, and G. Kirchsteiger (1997). Reciprocity as a contract enforcement device, experimental evidence. *Econometrica 65*(4), 883–860. 37

Ferber, J. (1999). *Multi-Agent Systems – An Introduction to Distributed Artificial Intelligence* (1st ed.). Boston, MA, USA: Addison-Wesley Longman Publishing Co., Inc. 129

Field, A. (2009). *Discovering Statistics using SPSS*. Sage Publications. 142, 169, 284

Fisher, R. A. (1918). The correlation between relatives on the supposition of mendelian inheritance. *Philosophical Transactions of the Royal Society of Edinburgh 52*, 399–433. 143

Fitzek, F. H., M. V. Pedersen, J. Heide, and M. Médard (2010). Network coding: applications and implementations on mobile devices. In *Proceedings of the 5th ACM workshop on Performance monitoring and measurement of heterogeneous wireless and wired networks*, New York, NY, USA, pp. 83–87. ACM. 63

Fitzek, F. H. P. (2007). External energy consumption measurements on mobile phones. In F. H. P. Fitzek and F. Reichert (Eds.), *Mobile Phone Programming*, pp. 441–447. Springer Netherlands. 57, 58

Fitzek, F. H. P. and M. D. Katz (2007a). Cellular controlled peer to peer communications: Overview and potentials. In F. H. P. Fitzek and M. D. Katz (Eds.), *Cognitive Wireless Networks*, pp. 31–59. Springer. 59, 60, 65

Fitzek, F. H. P. and M. D. Katz (Eds.) (2007b). *Cognitive Wireless Networks – Concepts, Methodologies and Visions Inspiring the Age of Enlightenment of Wireless Communications*. Springer Netherlands. 58

Fitzek, F. H. P. and F. Reichert (2007). Introduction to mobile phone programming. In F. H. P. Fitzek and F. Reichert (Eds.), *Mobile Phone Programming*, pp. 3–20. Springer Netherlands. 60

Freedman, D., R. Pisani, and R. Purves (2007). *Statistics* (4th ed.). W.W. Norton & Company. 142

Friedman, J. W. (1971). A non-cooperative equilibrium for supergames. *Review of Economic Studies 38*(1), 1–12. 36

García-Camino, A. (2009). *Normative regulation of open multi-agent systems.* Ph. D. thesis, Artificial Intelligence Research Institute (IIIA). 39

Gâteau, B., O. Boissier, D. Khadraoui, and E. Dubois (2005). $moise^{Inst}$: An organizational model for specifying rights and duties of autonomous agents. In *1st International Workshop on Coordination and Organisation (CoOrg 2005) affiliated with the 7th International Conference on Coordination Models and Languages.* 91

Gebser, M., R. Kaminski, B. Kaufmann, M. Ostrowski, T. Schaub, and S. Thiele (2008). Engineering an incremental asp solver. In M. Garcia de la Banda and E. Pontelli (Eds.), *Logic Programming*, Volume 5366 of *Lecture Notes in Computer Science*, pp. 190–205. Springer Berlin / Heidelberg. 104

Gebser, M., B. Kaufmann, A. Neumann, and T. Schaub (2007). Conflict-driven answer set solving. In *Proceedings of the 20th International Joint Conference on Artifical Intelligence*, San Francisco, CA, USA, pp. 386–392. Morgan Kaufmann Publishers Inc. 93

Gelfond, M. (2008). Answer sets. In F. van Harmelen, V. Lifschitz, and B. Porter (Eds.), *Handbook of Knowledge Representation*, pp. 285–316. Elsevier. 94

Gelfond, M. and V. Lifschitz (1988). The stable model semantics for logic programming. In R. Kowalski, Bowen, and Kenneth (Eds.), *Proceedings of International Logic Programming Conference and Symposium*, pp. 1070–1080. MIT Press. 92, 93

Gelfond, M. and V. Lifschitz (1991). Classical negation in logic programs and disjunctive databases. *New Generation Computing 9*(3–4), 365–386. 96

Gibbs, J. P. (1965, March). Norms: The problem of definition and classification. *The American Journal of Sociology 70*(5), 586–594. 13

Gibson, C. C. (2005). In pursuit of better policy outcomes. *Journal of Economic Behavior & Organization 57*(2), 227 – 230. Polycentric Political Economy. 45

Gilbert, N. (1995). Emergence in social simulation. In *Artificial Societies: The Computer Simulation of Social Life*, pp. 144–156. UCL Press. 15

Gilbert, N. (2008). *Agent-Based Models.* Number 153 in Quantitative Applications in the Social Science. SAGE Publications. 133

Gilbert, N. and K. G. Troitzsch (2005). *Simulation for the Social Scientist* (2nd ed.). Open University Press. 131

Gini, C. (1912). Variabilità e mutabilità. In *Studi Economico-Giuridici dell'Università di Cagliari*, Volume 3, pp. 1–158. Reprinted in Memorie di metodologica statistica (Ed. Pizetti E, Salvemini, T). Rome: Libreria Eredi Virgilio Veschi. 138

Giunchiglia, E., J. Lee, V. Lifschitz, N. McCain, and H. Turner (2004, March). Nonmonotonic causal theories. *Artificial Intelligence Journal – Special Issue on Logical Formalizations and Common Sense Reasoning 153*(1–2), 49–104. 92, 93

Goldman, A. I. (1976). *A Theory of Human Action*. Princeton University Press. 78

González, M. C., C. A. Hidalgo, and A.-L. Barabási (2008, June). Understanding individual human mobility patterns. *Nature 453*(7196), 779–782. 159, 203, 205, 206

Gordillo, G. and K. Andersson (2004). From policy lessons to policy actions: Motivation to take evaluation seriously. *Public Administration and Development 24*, 305–320. 45

Grossi, D. (2007). *Designing invisible handcuffs. Formal investigations in institutions and organizations for multi-agent systems*. Ph. D. thesis, Utrecht University. 75

Grossi, D., H. Aldewereld, and F. Dignum (2007). Ubi lex, ibi poena: Designing norm enforcement in e-institutions. In *Coordination, Organizations, Institutions, and Norms in Agent Systems II*, Volume 4386, pp. 101–114. Springer Berlin / Heidelberg. 25, 39

Gruber, H. (2005). *The Economics of Mobile Telecommunications*. Cambridge University Press. 138, 219

Gruber, H. and F. Verboven (2001). The diffusion of mobile telecommunications services in the european union. *European Economic Review 45*, 577–588. 53

Gulyás, L. (2005). *Understanding Emergent Social Phenomena: Methods, Tools and Applications for Agent-Based Modeling*. Ph. D. thesis, Computer and Automation Research Institute, Hungarian Academy of Sciences, Budapest, Hungary. 127

Gurerk, O., B. Irlenbusch, and B. Rockenbach (2006, April). The competitive advantage of sanctioning institutions. *Science 312*(5770), 108–111. 35

Güth, W. (1995). An evolutionary approach to explaining cooperative behavior by reciprocal incentives. *International Journal of Game Theory 24*(4), 323–344. 37

Hämäläinen, R. P. (2004). Reversing the perspective on the applications of decision analysis. *Decision Analysis 1*(1), 26–31. 140

Hannoun, M., O. Boissier, J. S. Sichman, and C. Sayettat (2000). Moise: An organizational model for multi-agent systems. In M. Monard and J. Sichman (Eds.), *Advances in Artificial Intelligence*, Volume 1952 of *Lecture Notes in Computer Science*, pp. 156–165. Springer Berlin / Heidelberg. 91

Hardin, G. (1968). The tragedy of the commons. *Science 162*, 1243–1248. 23

Hastenteufel, H. G., A. Daembke, and M. Tybus (2009). $1 + 1 = 1$: Network sharing – redefining telecom with a structured approach. Technical report, A.T. Kearney. 54

Hayek, F. A. (1996). *Individualism and Economic Order*. University Of Chicago Press. originally published in 1948. 20, 48

Hess, C. and E. Ostrom (2007). *Understanding Knowledge as a Commons – From Theory to Practice*. The MIT Press. 45

Hill, W. C., J. D. Hollan, D. Wroblewski, and T. McCandless (1992). Edit wear and read wear. In *Proceedings of the SIGCHI conference on Human factors in computing systems*, CHI '92. 2

Ho, T., M. Médard, R. Koetter, D. Karger, M. Effros, J. Shi, and B. Leong (2006). A random linear network coding approach to multicast. *IEEE Transactions on Information Theory 52*(10), 4413–4430. 63

Hohfeld, W. N. (1913). Some fundamental legal conceptions as applied in judicial reasoning. *The Yale Law Journal 23*(1), 16–59. 16

Holland, J. H. (1992). *Adaptation in natural and artificial systems*. Cambridge, MA, USA: MIT Press. 129

Hollander, C. D. and A. S. Wu (2011). The current state of normative agent-based systems. *Journal of Artificial Societies and Social Simulation 14*(2). 14, 18

Horne, C. (2001). Sociological perspectives on the emergence of norms. In M. Hechter and K.-D. Opp (Eds.), *Social Norms*, pp. 3–34. New York, New York, USA: Russell Sage Foundation. 13

Hsueh, K.-S. and A. Mosleh (1996). The development and application of the accident dynamic simulator for dynamic probabilistic risk assessment of nuclear power plants. *Reliability Engineering & System Safety 52*(3), 297–314. 131

Hu, J. (2005). Cooperation in mobile ad hoc networks. Technical report, Computer Science Department, Florida State University. 41

Hu, J. and M. Burmester (2006). Lars: a locally aware reputation system for mobile ad hoc networks. In *ACM-SE 44: Proceedings of the 44th annual Southeast regional conference*, New York, NY, USA, pp. 119–123. ACM. ISBN 1-59593-315-8. 41

Hübner, J. F., R. H. Bordini, G. P. Gouveia, R. H. Pereira, G. Picard, M. Piunti, and J. S. Sichman (2009). Using jason, moise, and cartago to develop a team of cowboys. In J. Dix, M. Fisher, and P. Novak (Eds.), *Proceedings of 10th International Workshop on Computational Logic in Multi-Agent Systems (CLIMA 2009), Agent Contes*, pp. 203–207. 91

Hübner, J. F., J. S. Sichman, and O. Boissier (2007, December). Developing organised multiagent systems using the *moise*⁺ model: programming issues at the system and agent levels. *International Journal of Agent-Oriented Software Engineering 1*(3/4), 370–395. 91

ie Market Research (2010). 3q10 germany mobile operator forecast, 2010 – 2014. 221

Johansson, M. and H. Verhagen (2010). Modelling the social fabric for normative npcs in mmogs. In *Proceedings of the Combined International Symposium on Social Network Analysis and Norms for MAS @ AISB AGM 2010*, pp. 79–84. 24

Johnson, D. B., D. A. Maltz, and J. Broch (2001). Dsr: The dynamic source routing protocol for multi-hop wireless ad hoc networks. In *Ad hoc networking*, pp. 139–172. Addison-Wesley Longman Publishing Co., Inc. 41

Jones, A. J. I. and M. J. Sergot (1996). A formal characterisation of institutionalised power. *Logic Journal of the IGPL 4*(3), 427–443. 16, 80, 92

Jones, N. R. (2011). A framework for institutional models in multi-agent systems. Technical report, University of Bath, Department of Computer Science. 104

Katz, M. D. and F. H. P. Fitzek (2006). Cooperation in 4g networks - cooperation in a heterogenous wireless world. In F. H. P. Fitzek and M. D. Katz (Eds.), *Cooperation in Wireless Networks: Principles and Applications*, pp. 463–496. Springer. v, 55, 56

Keeney, R. L. and H. Raiffa (1976). *Decisions with Multiple Objectives – Preferences and Value Trade-Offs* (1st ed.). John Wiley & Sons. Reprinted in 1993 by Cambridge University Press. 140

Kelsen, H. (1960). What is the pure theory of law? *Tulane Law Review 34*, 269–276. 17

Khan, A. H., M. A. Qadeer, J. A. Ansari, and S. Waheed (2009). 4g as a next generation wireless network. *Future Computer and Communication, International Conference on*, 334–338. 55

Kim, Y. K. and R. Prasad (2006). *4G Roadmap and Emerging Communication Technologies*. Artech House Publishers. 55

Kiser, L. and E. Ostrom (1982). The three worlds of action: A metatheoretical synthesis of institutional approaches. In E. Ostrom (Ed.), *Strategies of Political Inquiry*, pp. 179–222. Sage Publications. 45, 46, 246

Kiser, L. and E. Ostrom (2000). The three worlds of action: A metatheoretical synthesis of institutional approaches. In M. D. McGinnis (Ed.), *Polycentric Games and Institutions: Readings from the Workshop in Political Theory and Policy Analysis*, Chapter 2. University of Michigan Press. originally published in Kiser and Ostrom (1982). 46

Klügl, F. (2008). A validation methodology for agent-based simulations. In *SAC '08: Proceedings of the 2008 ACM symposium on Applied computing*, New York, NY, USA, pp. 39–43. ACM. 134

Kotz, D., C. Newport, R. S. Gray, J. Liu, Y. Yuan, and C. Elliott (2004). Experimental evaluation of wireless simulation assumptions. In *Proceedings of the 7th ACM international symposium on Modeling, analysis and simulation of wireless and mobile systems*, New York, NY, USA, pp. 78–82. ACM. 151

Kowalski, R. and M. Sergot (1986, January). A logic-based calculus of events. *New Generation Computing 4*(1), 67–95. 92

Kowalski, R. A. and F. Sadri (1997, April–June). Reconciling the event calculus with the situation calculus. *Journal of Logic Programming 31*(1–3), 39–58. 93

Leavitt, N. (2009). Is cloud computing really ready for prime time? *Computer 42*(1), 15–20. 23

Leibowitz, N., M. Ripeanu, and A. Wierzbicki (2003). Deconstructing the kazaa network. In *Proceedings of the Third IEEE Workshop on Internet Applications*. IEEE Computer Society. 43

Levene, H. (1960). Robust tests for the equality of variances. In *Contributions to Probability and Statistics*, pp. 278–292. Stanford University Press. 143

Lindman, H. W. (1974). *Analysis of Variance in Complex Experimental Design*. W. H. Freeman & Co. Ltd. 143

López y López, F. (2003, June). *Social Power and Norms: Impact on Agent Behaviour*. Ph. D. thesis, University of Southampton. 13, 19

López y López, F. and M. Luck (2003). Modelling norms for autonomous agents. In E. Chavez, J. Favela, M. Mejia, and A. Oliart (Eds.), *Fourth Mexican International Conference on Computer Science*, pp. 238–245. IEEE Computer Society. 39

Lukes, S. (1968, June). Methodological individualism reconsidered. *British Journal of Sociology 19*(2), 119–129. 20

Machado, R. and R. H. Bordini (2001). Running agentspeak(l) agents on sim_agent. In J.-J. C. Meyer and M. Tambe (Eds.), *Intelligent Agents VIII, 8th International Workshop, ATAL 2001 Seattle, WA, USA, August 1–3, 2001, Revised Papers*, Volume 2333 of *Lecture Notes in Computer Science*, pp. 158–174. Springer. 133, 142

Makinson, D. (1986). On the formal representation of rights relations: Remarks on the work of stig kanger and lars lindahl. *Journal of Philosophical Logic 15*(4), 403–425. 16

Malczewski, I., R. Moreno-Sanchez, L. A. Bojórquez-Tapia, and E. Ongay-Delhumeau (1997). Multicriteria group decision-making model for environmental conflict analysis in the cape region, mexico. *Journal of Environmental Planning and Management 40*(3), 349–374. 140

Marti, S., T. J. Giuli, K. Lai, and M. Baker (2000). Mitigating routing misbehavior in mobile ad hoc networks. In *Proceedings of the 6th annual international conference on Mobile computing and networking*, New York, NY, USA, pp. 255–265. ACM. 41

Mason, B. and S. Thomas (2008). A million penguins research report. Technical report, Institute of Creative Technologies, De Montfort UNiversity, Leicester. 1, 2

McKnight, L., W. Lehr, and J. Howison (2007). Coordinating user and device behaviour in wireless grids. In F. H. Fitzek and M. D. Katz (Eds.), *Cognitive Wireless Networks – Concepts, Methodologies and Visions Inspiring the Age of Enlightenment of Wireless Communication*, pp. 679–697. Springer Netherlands. 137

Merino, G. G., D. D. Jones, D. L. Clements, and D. Miller (2003). Fuzzy compromise programming with precedence order in the criteria. *Applied Mathematics and Computation 134*(1), 185–205. 140

Merton, R. K. (1968). *Social Theory and Social Structure*. Free Press. 14

Meyer, J.-J. C. (1988). A different approach to deontic logic: deontic logic viewed as a variant of dynamic logic. *Notre Dame Journal of Formal Logic 29*(1), 109–136. 17

Miceli, M. and C. Castelfranchi (2000). The role of evaluation in cognition and social interaction. In K. Dautenhahn (Ed.), *Human cognition and social agent technology*. Amsterdam: Benjamins. 29

Michiardi, P. and R. Molva (2002). Core: A *Co*llaborative *Re*putation mechanism to enforce node cooperation in mobile ad hoc networks. In *Communications and Multimedia Security*, pp. 107–121. 41

Michiardi, P. and R. Molva (2003). A game theoretical approach to evaluate cooperation enforcement mechanisms in mobile ad hoc networks. In *In Modeling and Optimization in Mobile, Ad Hoc and Wireless Networks*, pp. 3–5. 41

Milojicic, D. S., V. Kalogeraki, R. Lukose, K. Nagaraja, J. Pruyne, B. Richard, S. Rollins, and Z. Xu (2002). Peer-to-peer computing. Technical Report HPL-2002-57R1, Hewlett-Packard Company. 23

Minsky, N. H. (1991a, February). The imposition of protocols over open distributed systems. *IEEE Transactions on Software Engineering 17*(2), 183–195. IEEE Press. 26

Minsky, N. H. (1991b, September). Law-governed systems. *Software Engineering Journal - Special issue on software process and its support 6*(5), 285–302. 18, 25

Morris, R. T. (1956, October). A typology of norms. *American Sociological Review 21*(5), 610–613. 13

Nash, J. F. (1950). Equilibrium points in n-person games. *Proceedings of the National Academy of Sciences 36*(1), 48–49. 35

Neumann, M. (2008). Homo socionicus: a case study of simulation models of norms. *Journal of Artificial Societies and Social Simulation 11*(4), 6. 14

Nguyen, A. D., P. Sénac, V. Ramiro, and M. Diaz (2011). Steps - an approach for human mobility modeling. In J. Domingo-Pascual, P. Manzoni, S. Palazzo, A. Pont, and C. M. Scoglio (Eds.), *NETWORKING 2011 - 10th International IFIP TC 6 Networking Conference, Valencia, Spain, May 9-13, 2011, Proceedings, Part I*, Volume 6640, pp. 254–265. Springer. 208

Nielsen, J. (2008). Participation inequality: Lurkers vs. contributors in internet communities. 2

Niemelä, I. and P. Simons (1997, July). Smodels - an implementation of the stable model and well-founded semantics for normal lp. In J. Dix, U. Furbach, and A. Nerode (Eds.), *Proceedings of the 4th International Conference on Logic Programing and Nonmonotonic Reasoning*, Volume 1265 of *LNAI*, Berlin, pp. 420–429. Springer. 93

Noriega, P. (1997). *Agent mediated auctions: The Fishmarket Metaphor*. Ph. D. thesis, Universitat Autònoma de Barcelona. 164

North, D. C. (1990, October). *Institutions, Institutional Change and Economic Performance (Political Economy of Institutions and Decisions)*. Cambridge University Press. 15

North, D. C. (1993, September). Institutions, transaction costs and productivity in the long run. Economic History 9309004, EconWPA. 15

Oakerson, R. J. (1990). Analyzing the commons: A framework. In *Designing Sustainability on the Commons, the First Biennial Conference of the International Association for the Study of Common Property*. 45

Oram, A. (Ed.) (2001). *Peer to Peer: Harnessing the Power of Disruptive Technologies*. O'Reilly Media. 22

Ormerod, P. and B. Rosewell (2009). Validation and verification of agent-based models in the social sciences. In *Epistemological Aspects of Computer Simulation in the Social Sciences*, Volume 5466, pp. 130–140. Springer Berlin / Heidelberg. 134

Ostrom, E. (1990). *Governing the Commons: the Evolution of Institutions for Collective Action*. Cambridge University Press. 18th printing (2006). 23, 38

Ostrom, E. (1999, June). Coping with tragedies of the commons. *Annual Review of Political Science 2*, 493–535. Workshop in Political Theory and Policy Analysis; Center for the Study of Institutions, Population, and Environmental Change, Indiana University, Bloomington, Indiana 47408-3895; e-mail: ostrom@indiana.edu. 69

Ostrom, E. (2000). Collective action and the evolution of social norms. *The Journal of Economic Perspectives 14*(3), 137–158. 45

Ostrom, E. (2005). *Understanding Institutional Diversity*. Princeton University Press. 38, 45, 46

Ostrom, E., R. Gardner, and J. Walker (1994). *Rules, Games, and Common-Pool Resources*. University of Michigan Press. 45, 46, 48, 50, 217

Ostrom, E. and J. Walker (Eds.) (2003). *Trust and Reciprocity: Interdisciplinary Lessons from Experimental Research*. New York: Russell Sage Foundation. ISBN: 0-87154-647-7. 45

Ostrom, V. and E. Ostrom (1977). Public goods and public choices. In E. S. Savas (Ed.), *Alternatives for Delivering Public Services: Towards Improved Performance*, pp. 7–49. Westview Press. 38

Paolucci, M., T. Balke, R. Conte, T. Eymann, and S. Marmo (2006). Review of internet user-oriented reputation applications and application layer networks. Technical report, Social Science Research network (SSRN). 40

Paolucci, M., T. Eymann, W. Jager, J. Sabater-Mir, R. Conte, S. Marmo, S. Picascia, W. Quattrociocchi, T. Balke, S. König, T. Broekhuizen, D. Trampe, M. Tuk, I. Brito, I. Pinyol, and D. Villatoro (2009). Social knowledge for e-governance: Theory and technology of reputation. ISTC-CNR, Rome. 40

Perkins, C. E. and E. M. Royer (1999, February). Ad-hoc on-demand distance vector routing. In *Proceedings of the 2nd IEEE Workshop on Mobile Computing Systems and Applications*, pp. 90–100. 41

Perreau de Pinninck Bas, A. (2010). *Techniques for Peer Enforcement in Multiagent Networks*. Phd thesis, Universitat Autónoma de Barcelona. 25, 26

Perreau de Pinninck Bas, A., C. Sierra, and M. Schorlemmer (2010). A multiagent network for peer norm enforcement. *Autonomous Agents and Multi-Agent Systems 21*, 397–424. 10.1007/s10458-009-9107-8. 39

Perrucci, G. P. (2009). *Energy Saving Strategies on Mobile Devices*. Ph. D. thesis, Aalborg University. 59

Perrucci, G. P., F. H. Fitzek, and M. V. Petersen (2009). Energy saving aspects for mobile device exploiting heterogeneous wireless networks. In *Heterogeneous Wireless Access Networks*. Springer US. v, 57, 61, 63, 152

Pidd, M. (2008). Why modelling matters. In S. J. Mason, R. R. Hill, L. Mönch, O. Rose, T. Jefferson, and J. W. Fowler (Eds.), *Winter Simulation Conference*, pp. 10. WSC. 131

Poundstone, W. (1992). *Prisoner's Dilemma*. Doubleday. 35

Prahalad, C. K. and V. Ramaswamy (2004). *The Future of Competition: Co-Creating Unique Value With Customers*. Harvard Business Press. 1, 2

Railsback, S. F., S. L. Lytinen, and S. K. Jackson (2006, September). Agent-based simulation platforms: Review and development recommendations. *Simulation 82*(9), 609–623. Society for Computer Simulation International. 133

Rao, A. S. (1996). Agentspeak(l): Bdi agents speak out in a logical computable language. In *MAAMAW '96: Proceedings of the 7th European workshop on Modelling autonomous agents in a multi-agent world: agents breaking away*, Secaucus, NJ, USA, pp. 42–55. Springer-Verlag New York, Inc. 133, 142, 161

Rao, A. S. and M. P. Georgeff (1995). Bdi-agents: from theory to practice. In *Proceedings of the First International Conference on Multiagent Systems*. 133, 161

Rawls, J. (1955, January). Two concepts of rules. *The Philosophical Review 64*(1), 3–32. 75

Reardon, M. (2009, December). At&t considers incentives to curb heavy data usage. 54

Regan, K. and R. Cohen (2005). Indirect reputation assessment for adaptive buying agents in electronic markets. In *Proceedings of the Business Agents and Semantic Web (BASeWEB05)*, Victoria, Canada. 29

Riggs, J. L. (1982). *Engineering economics* (2nd ed.). McGraw-Hill. 2

Royer, E. M. (1999, April). A review of current routing protocols for ad hoc mobile wireless networks. *IEEE Personal Communications*, 46–55. 41

Rubinstein, A. (1998). *Modeling Bounded Rationality*. MIT Press. 31

Russell, S. and P. Norvig (2002). *Artificial Intelligence - A Modern Approach*. Prentice Hall. 25

Saaty, T. L. (1977). A scaling method for priorities in hierarchical structures. *Journal of Mathematical Psychology 15*(3), 234–281. 140

Sabater-Mir, J., M. Paolucci, and R. Conte (2006). Repage: Reputation and image among limited autonomous partners. *Journal of Artificial Societies and Social Simulation 2*(2). 29, 40

Sabater-Mir, J. and C. Sierra (2001). Regret: A reputation model for gregarious societies. In *AGENTS '01: Proceedings of the fifth international conference on Autonomous agents*, New York, NY, USA, pp. 194–195. ACM. 40

Sabater-Mir, J. and C. Sierra (2002). Social regret, a reputation model based on social relations. *SIGecom Exch. 3*(1), 44–56. 29

Scariano, S. M. and J. M. Davenport (1987, May). The effects of violations of independence assumptions in the one-way anova. *The American Statistician 41*(2), 123–129. 143

Scellato, S., C. Mascolo, M. Musolesi, and J. Crowcroft (2011). Track globally, deliver locally: improving content delivery networks by tracking geographic social cascades. In *Proceedings of the 20th International Conference on World wide web*, New York, NY, USA, pp. 457–466. ACM. 221

Schaaf, J. (1989). *Governing a Monopoly Market under Siege: Using Instittional Analysis to Understand Competitive Entry into Tellecommunications Markets.* Ph. D. thesis, Indiana University. 45

Schimanoff, S. B. (1980). *Communication rules: Theory and Research.* Sage Publications. 18

Schneck, A., F. Haakh, and U. Lang (2004). Multikriterielle optimierung der grundwasserbewirtschaftung - dargestellt am beispiel des wassergewinnungs-gebiets donauried. *Wasserwirtschaft 94*(12), 32–39. 140, 141

Schumpeter, J. A. (1908). *Das Wesen und der Hauptinhalt der theoretischen Nationaloökonomie.* Duncker & Humblot. 20

Searle, J. R. (1997). *The Construction of Social Reality.* Free Press. 76, 78

Sergot, M. J. (2001, October). A computational theory of normative positions. *ACM Transactions on Computational Logic 2*(4), 581–622. 16, 92

Shannon, R. E. (1998). Introduction to the art and science of simulation. In D. J. Medeiros, E. F. Watson, J. S. Carson, and M. Manivannan (Eds.), *Proceedings of the 30th Conference on Winter Simulation*, Los Alamitos, CA, USA, pp. 7–14. IEEE Computer Society Press. 126

Shapiro, S. S. and M. B. Wilk (1965, December). An analysis of variance test for normality (complete samples). *Biometrika 52*(3/4), 591–611. 143

Siebers, P.-O. and U. Aickelin (2007). Introduction to multi-agent simulation. In F. Adam and P. Humphreys (Eds.), *Encyclopedia of Decision Making and Decision Support Technologies*, pp. 554–564. Pennsylvania, US: IDEAS Group. 127

Siebers, P.-O., C. M. Macal, J. Garnett, D. Buxton, and M. Pidd (2010). Discrete-event simulation is dead, long live agent-based simulation! *Journal of Simulation 4*(3), 204–210. 131

Simon, H. A. (1956). Rational choice and the structure of the environment. *Psychological Review 63*(2), 129–138. 195

Smith, V. L. (1990). *Experimental Economics. Schools of Thought in Economics.* Hants, Vermont, USA: Edward Elgar Publishing. 37

Smyth, D. S. and P. B. Checkland (1976). Using a systems approach: The structure of root definitions. *Journal of Applied Systems Analysis 5*, 75–83. 44

Snidal, D. (1991, September). Relative gains and the pattern of international cooperation. *The American Political Science Review 85*(3), 701–726. 35

Stanford Encyclopedia of Philosophy (2007, October). Prisoner's dilemma. 35

Stanford Encyclopedia of Philosophy (2010, April). Non-monotonic logic. 93

Sun, L., L. Jiao, Y. Wang, S. Cheng, and W. Wang (2005). An adaptive group-based reputation system in peer-to-peer networks. In X. Deng and Y. Ye (Eds.), *Internet and Network Economics*, Volume 3828 of *Lecture Notes in Computer Science*, pp. 651–659. Springer Berlin / Heidelberg. 40

T-Mobile (2010). Geschäftsbericht (annual report). 220

Tanenbaum, A. S. and M. van Steen (2007). *Distributed systems – principles and paradigms* (2nd ed.). Pearson Education. 20

Taylor, M. (1976). *Anarchy and Cooperation.* John Wiley & Sons Ltd. 49

The Institute of the Future of the Book (2007, July). A million penguins: A wiki-novelty. 1

The Nobel Foundation (2009, October). The sveriges riksbank prize in economic sciences in memory of alfred nobel 2009. 37

Therborn, G. (2002). Back to norms! on the scope and dynamics of norms and normative action. *Current Sociology 50*, 863–880. 15

TNS (2004, September). Two-day batter life tops wish list for future all-in-one phone device. Technical report, Taylor Nelson Sofres. 56

Tolety, V. (1999). Load reduction in ad hoc networks using mobile servers. Master's thesis, Colorado School of Mines. 205

Tuomela, R. (1995). *The Importance of Us: A Philosophical Study of Basic Social Notions.* Stanford University Press. 19

Tuomela, R. and M. Bonnevier-Tuomela (1995). Norms and agreements. *European Journal of Law, Philosophy and Computer Science 5*, 41–46. 15, 19, 49

Vasconcelos, W. W., M. Esteva, C. Sierra, and J. A. Rodríguez-Aguilar (2004). Verifying norm consistency in electronic institutions. In *AOTP: The AAAI-04 Workshop on Agent Organizations: Theory and Practice*, pp. 8–14. AAAI: AAAI. 17

Vázquez-Salceda, J., H. Aldewereld, and F. Dignum (2005). Norms in multiagent systems: from theory to practice. *International Journal of Computer Systems Science & Engineering 20*(4), 225–236. 28, 83

VDI (2010). Frequenzauktion: Ebay für fortgeschrittene. *VDI Nachrichten 15*, 4. 54

Viganò, F. and M. Colombetti (2007). Model checking norms and sanctions in institutions. In J. S. Sichman, J. A. Padget, S. Ossowski, and P. Noriega (Eds.), *Coordination, Organizations, Institutions, and Norms in Agent Systems III*, Volume 4870 of *Lecture Notes in Computer Science*, pp. 316–329. Springer. 17

Vingelmann, P., F. H. P. Fitzek, and D. E. Lucani (2010, May). Application-level data dissemination in multi-hop wireless networks. In *2010 IEEE International Conference on Communications Workshops (ICC)*, pp. 1–6. 66

Vishnumurthy, V., S. Chandrakumar, and E. G. Sirer (2003). Karma : A secure economic framework for peer-to-peer resource sharing. In *Workshop on the Economics of Peer-to-Peer Systems*, pp. 2003. 43

von Wright, G. H. (1951, January). Deontic logic. *Mind 60*(237), 1–15. 17

Wallenius, J., J. S. Dyer, P. C. Fishburn, R. E. Steuer, S. Zionts, and K. Deb (2008). Multiple criteria decision making, multiattribute utility theory: Recent accomplishments and what lies ahead. *Management Science 54*(7), 1336–1349. 140

Wang, Y., V. C. Gurika, and M. Singhal (2004). A fair distributed solution for selfish nodes problem in wireless ad hoc networks. In I. Nikolaidis, M. Barbeau, and E. Kranakis (Eds.), *Proceedings of the Third International Conference on Ad-Hoc, Mobile, and Wireless Networks (ADHOC-NOW)*, Volume 3158 of *Lecture Notes in Computer Science*, pp. 211–224. 42

Weber, J. and U. Schäffer (2005). *Einführung in das Controlling*. Schäffer Poeschel. 137

Websters Electronic Dictionary (2011, April). Definition of norm. 14

Weiss, G. (Ed.) (1999). *Multiagent Systems: A Modern Approach to Distributed Artificial Intelligence*. MIT Press. 25

White, J. L. (1991). Knowledge and deductive closure. *Synthese 86*(3), 409–423. 95

Wooldridge, M. J. and N. R. Jennings (1995). Intelligent agents: Theory and practice. *The Knowledge Engineering Review 10*(2), 115–152. 25, 128

Wrona, K. (2005). *Cooperative Communication Systems*. Ph. D. thesis, RWTH Aachen. 22, 35

Wrona, K. and P. Mähönen (2004, October). Analytical model of cooperation in ad hoc networks. *Telecommunication Systems 27*(2–4), 347–369. 56

Wrona, K. and P. Mähönen (2006). Stability and security in wireless cooperative networks. In *Cooperation in Wireless Networks: Principles and Applications – Real Egoistic Behavior is to Cooperate!*, pp. 313–363. Springer Netherlands. 69

WWRF (2011). Wireless world research forum. 56

Yarbrough, B. V. and R. M. Yarbrough (1999). Governance structures, insider status, and boundary maintenance. *Journal of Bioeconomics 1*, 289–310. 28

Yoon, J., M. Liu, and B. Noble (2003). Random waypoint considered harmful. In *IEEE Societies INFOCOM. Twenty-Second Annual Joint Conference of the IEEE Computer and Communications*, Volume 2. 203

Yu, B. and M. P. Singh (2002). Distributed reputation management for electronic commerce. *Computational Intelligence 18*(4), 535–549. 29, 40

Zacharia, G., A. Moukas, and P. Maes (1999). Collaborative reputation mechanisms in electronic marketplaces. In *Proceedings of the 32nd Hawaii International Conference on System Sciences, Wailea Maui*, Washington, DC, USA. IEEE Computer Society. 29, 40

Zhang, Q., J. Heide, M. V. Pedersen, and F. H. Fitzek (2010, February). Network coding and user cooperation for streaming and download services in lte networks. 67

Zhong, S., J. Chen, and R. Yang (2003). Sprite: A simple, cheat-proof, credit-based system for mobile ad-hoc networks. In *Proceedings of IEEE INFOCOM*, pp. 1987–1997. 41

Part V

Appendix

Appendix A

The Jason Simulation Code

This Appendix presents the Jason agent code used for the simulation experiments.

Listing 1: sim_agent.asl class

```
1   /**
2    * Agent sim_agent in project Thesis_simulation
3    * This agent controls the simulation execution and termination.
4    *
5    * @author Tina Balke
6    * @version 12-07-2011
7    *
8    */
9
10  /* initial goals */
11  !start.
12
13  /* initial rules */
14  transactions_ended(M):- num_agents(X) & M < X.
15  not_all_agents_kill(K):- all_agents(A) & K < A.
16
17  /* plans */
18  +!start: true
19     <- /* creation of all other agents */
20        create_agent;
21        set_start;
22        /* additional end event generator if an agent dies and therefore not
             all
23        * not all agents will send their end message */
24        .at("now + 30 minutes", "+!forced_stop").
25
26  /* if receiving end message by agent, check whether all agents are finished */
27  +end[source(Agent)]: true
28     <- .count(end[source(_)],M);
29        !check_stop(M).
30
31  /* if all agents have finished their interactions, stop the MAS */
32  +!check_stop(M): not transactions_ended(M)
33     <- /* the killing of all agents before stopping the MAS is required to
34        * reduce the risk of threat locking porblems when logging the
35        * simulation data */
36        !killAllAgents.
37
38  /* otherwise do nothing */
39  +!check_stop(M): transactions_ended(M)
40     <- true.
41
42  /* perceive all agents to kill (i.e. all agents in the MAs besides from the
43  * environment */
44  +!killAllAgents: true
45     <- getAllAgents.
46
47  /* if percept to kill an agent (from the environment), kill agent and check
48  * whether all agents have been killed */
49  +agentToKill(Agent): true
50     <- .kill_agent(Agent);
51        .count(agentToKill[source(_)],A);
52        !check_killed_agents(K).
53
54  /* stop the MAS, once all agents are killed */
55  +!check_killed_agents(K): not not_all_agents_killed(K)
56     <- .stopMAS.
57  +!check_killed_agents(K): not_all_agents_killed(K)
58     <- true.
59
60  /* if time for MAS execution expires, stop the MAS */
61  +!forced_stop: true
62     <- .stopMAS.
63
```

Listing 2: base_station.asl class

```
1   // Agent base_station in project Thesis_simulation
2
3   /* Initial beliefs and rules */
4
5   // Encoding: c1 = channel 1, c2 = channel 2, ...
6   free(c1).
7   free(c2).
8   free(c3).
9   free(c4).
10  free(c5).
11  free(c6).
12  free(c7).
13  free(c8).
14  free(c9).
15  free(c10).
16  free(c11).
17  free(c12).
18  free(c13).
19  free(c14).
20  free(c15).
21  free(c16).
22  free(c17).
23  free(c18).
24  free(c19).
25  free(c20).
26  free(c21).
27  free(c22).
28  free(c23).
29  free(c24).
30  free(c25).
31
32
33  /* Initial goals */
34
35
36  /* Plans */
37
38  +download(X)[source(_)]: true
39     <- -free(X).
40
41  -download(X)[source(_)]: true
42     <- +free(X).
```

Listing 3: agent.asl class

```
1   /**
2    * Agent agent in project Thesis_simulation
3    *
```

```
4   * @author Tina Balke
5   * @version 23-06-2011
6   */
7
8
9
10  /* initial beliefs */
11  movement(O).
12  decisionNumber(O).
13
14  /* initial rules */
15  file_to_small(SIZE) :- SIZE <= 0.
16
17  /* plans */
18
19  /**
20   * calculate random number that is used for determining whether to download a
21   * file or to move
22   */
23  +!calc_random: true
24      <- .random(R);
25         -+decisionNumber(R).
26
27  /**
28   * decision to move if random value of decision number is smaller or equal 0.2
29   */
30  +!move_decision: decisionNumber(R) & R > 0.2
31      <- !check_event.
32
33  +!move_decision: decisionNumber(R) & R <= 0.2
34      <- !move.
35
36  /**
37   * every time the decision is made to download the file itself by the agent
38   * the battery costs for self-download are added (depending on the downloaded
39   * amount (file/chunksize).
40   */
41  +!download_self(NO,SIZE): not file_to_small(SIZE)
42      <- !check_download_channel.
43
44  +!check_download_channel: free(CHANNEL)
45      <- ?assigned_chunk(GM,NO,CHUNKNUMBER,CHUNKSIZE);
46         check_permission_download(GM,CHUNKNUMBER);
47         !download_decision.
48
49  +!check_download_channel: not free(_)
50      <- .send(basestation, askOne, free(_), Reply);
51         !check_download_channel.
52
53  +!download_decision: power(P) & permission(PE) & P == true & PE == true
54      <- ?event(E);
55         ?assigned_chunk(GM,NO,CHUNKNUMBER,CHUNKSIZE);
56         .send(basestation, tell, download);
57         /* updating the download costs */
58         update_download_costs(CHUNKSIZE,E);
59         external_event_download(GM,CHUNKNUMBER,CHANNEL);
60         .send(basestation, untell, download);
61         !move.
62  +!download_decision: power(P) & permission(PE) & P == false & PE == false
63      <- !check_download_channel.
64
65  /**
66   * download self if size is 0 (i.e. file_to_small is true); this method is
67   * used for logging and updating the beliefs, etc. however no additional
68   * download costs accumulate
69   */
70  +!download_self(NO,SIZE): file_to_small(SIZE)
71      <- !move.
72
73  /* further plans */
74  +!update_event: true
75      <- ?event(E);
76         Enew = E+1;
77         -+event(Enew).
78
```

Listing 4: interaction_agent.asl class

```
1   /**
2    * Agent interaction_agent in project Thesis simulation
3    *
4    * @author Tina Balke
5    * @version 23-06-2011
6    */
7
8   /* initial beliefs */
9   event(-1).
10
11  /* initial rules */
12  sufficient_group_size(GSIZE) :- GSIZE > 0.
13  oppCoststooHigh(SIZE) :- oppCosts(R) & R >= SIZE.
14  sufficient_group_size(SIZE) :- GSIZE > 0.
15
16  /* initial goals */
17  +!start: true
18      <- !start.
19
20  /* Plans */
21  +!start: true
22      <- !calc_random;
23         get_number_interactions;
24         /* set agent properties */
25         set_mobility_level;
26         set_location;
27         !move_decision.
28
29  /**
30   * checking whether the agent has performed more then a certain number of
31   * downloads (download events): If this is the case, then it sends a message
32   * to the sim-agent. This agent checks whether all agents have sent him a
33   * message and if so, stops the simulation. If the event threshold is not
34   * met the agent simply continues with download_file.
35   */
36  +!check_event: enoughInteractions
37      <- .send(sim_agent,tell,end).
38
```

```
39  +!check_event: not enoughInteractions
40      <-  downloading.
41
42  /**
43   * in case the percept download_file(NO,SIZE) is made, generate a goal with
44   * the same name
45   */
46  +download_file(NO,SIZE): true
47      <-  !update_event;
48          ?event(E);
49          set_alternative_battery_costs(SIZE,E);
50          get-opportunity_costs;
51          !download_file(NO,SIZE).
52
53  //Mode 1
54  /**
55   * in case the size of the file is so small that a cooperation does not pay off
56   * (opportunity costs are higher then the respective file size) download self
57   */
58  +!download_file(NO,SIZE): oppCoststoHigh(SIZE)
59      <-  !download_self(NO,SIZE).
60
61  /**
62   * in case the opportunity costs are smaller then the size, check for neighbours
63   * (query environment for neighbour percepts and count the total number of
64   * neighbours).
65   */
66  +!download_file(NO,SIZE): not oppCoststoHigh(SIZE)
67      <-  !location(LOCATION);
68          check_neighbours(LOCATION);
69          .count(neighbour(_),NoNeighbours);
70          !cooperate_decision(NO,SIZE,NoNeighbours).
71
72  //Mode 2
73  /**
74   * if no neighbours are in the vicinity, then download self
75   */
76  +!cooperate_decision(NO,SIZE,NoNeighbours): NoNeighbours == 0
77      <-  !download_self(NO,SIZE).
78
79  /**
80   * otherwise, check whether anyone of the neighbours has sent a cooperation
81   * proposal
82   */
83  +!cooperate_decision(NO,SIZE,NoNeighbours): NoNeighbours \== 0
84      <-  .wait(3000);
85          !check_for_coop_proposals(NO,SIZE).
86
87  /**
88   * if this is the case, then check for a signal that the group infos have been
89   * generated
90   */
91  +!check_for_coop_proposals(NO,SIZE): group_joined & infos_generated
92      <-  !check_group_infos(NO,SIZE).
93
94  /**
95   * if the agent hasn't received any group infos so far, it waits till they have
96   * been generated
97   * been generated
98   */
99  +!check_for_coop_proposals(NO,SIZE): group_joined & not infos_generated
100     <-  .wait("+infos_generated");
101         !check_group_infos(NO,SIZE).
102
103 /**
104  * if no proposals have been send search for partners yourself
105  */
106 +!check_for_coop_proposals(NO,SIZE): not group_joined & not coop_request_sent
107     <-  +coop_request_sent;
108         .findall(Name,neighbour(Name),N);
109         .send(N,tell,coop_proposal(NO));
110         ?event(E);
111         /* update energy costs */
112         update_battery_coopRequest(E);
113         .wait(5000);
114         /* deadline is added to the agents belief base, so it can determine
115         * whether it received delayed acceptances */
116         +proposal_deadline;
117         .count(accept(coop_proposal(NO,_))[source(_)],GSIZE);
118         !negotiate_decision(NO,SIZE,GSIZE).
119
120 //reactions the proposals of other players
121 /**
122  * if same goal and not already group joined or request send self,
123  * accept cooperate proposal
124  */
125 @coop2[priority(1)]
126 +coop_proposal(NO)[source(Agent)]: download(NO,SIZE) & not group_joined & not
127     coop_request_sent
128     <-  +group_joined;
129         -my_name(Name);
130         /* send acceptance message to requesting agent */
131         .send(Agent,tell,accept(coop_proposal(NO,Name)));
132         ?event(E);
133         /* update energy costs */
134         update_battery_coopRequest(E).
135
136 /**
137  * if the agent has sent a coop request himself or has joined another group
138  * already, it will decline other requests
139  */
140 @coop3[priority(-1)]
141 +coop_proposal(NO)[source(Agent)]: coop_request_sent | group_joined
142     <-  .abolish(coop_proposal(NO)[source(Agent)]).
143
144 /**
145  * if not same file to download, decline proposal
146  */
147 @coop4[priority(-1)]
148 +coop_proposal(NO)[source(Agent)]: not download(NO,SIZE)
149     <-  .abolish(coop_proposal(NO)[source(Agent)]).
150
151 /**
152  * if finished all interactions, decline proposal with highest priority
153 @coop5[priority(0)]
```

```
154 +coop_proposal(MO)[source(Agent)]: enoughInteractions
155     <-- .abolish(coop_proposal(MO)[source(Agent)]).
156
157 /**
158  * default is to decline any proposals with low priority
159  */
160 @coop8[priority(2)]
161 +coop_proposal(MO)[source(Agent)]: true
162     <-- .abolish(coop_proposal(MO)[source(Agent)]).
163
164 /**
165  * if agents accept cooperation proposals after the group has formed, they are
166  * being told so
167  */
168 +accept(coop_proposal(MO,Name))[source(Name)]: coop_request_sent |
       proposal_deadline
169     <-- .abolish(accept(coop_proposal(MO,Name))[source(Name)]);
170         .send(Name,tell,to_late);
171         .send(Name,tell,infos_generated).
172
173 /**
174  * Mode 4 -- no partner risk is checked for if no enforcement mechanism are used
175  */
176  * if no responses have arrived, then download self
177  */
178 +negotiate_decision(MO,SIZE,GSIZE): not sufficient_group_size(GSIZE)
179     <-- !download_self(MO,SIZE).
180
181 /**
182  * if responses have arrived, then negotiate (get chunks and key)
183  */
184 +negotiate_decision(MO,SIZE,GSIZE): sufficient_group_size(GSIZE)
185     <-- .findall(agent,accept(coop_proposal(MO,Agent)),GROUP);
186         .random(KEY);
187         getGroupNumber(GSIZE);
188         ?groupnumber(GN);
189         +leader(GN);
190         /* calls the chunk assignment method in the environment */
191         assign_chunks(GN,GROUP,KEY,MO);
192         !check_group_infos_leader(MO,SIZE,GROUP).
193
194 /**
195  * checking group infos: if the agent is the leader, then it gives all group
196  * members the information that the group information are available
197  */
198 +!check_group_infos_leader(MO,SIZE,GROUP): true
199     <-- .send(GROUP,tell,infos_generated);
200         !check_group_infos(MO,SIZE).
201
202 /**
203  * reactions of non group leaders to the perception that the group infos are
204  * available
205  */
206 +!check_group_infos(MO,SIZE): assigned_chunk(GN,MO,CHUNKNUMBER,CHUNKSIZE)
207     <-- !assigned_chunk(GN,MO,CHUNKNUMBER,CHUNKSIZE).
208
209 +!check_group_infos(MO,SIZE): to_late | bad_image(MO)
210     <-- !download_self(MO,SIZE).

211 //Mode 6
212 /**
213  * checking whether any other agent has sent some chunks
214  */
215 +!control_reception(GN,CHUNKSIZE): not leader(GN)
216     <-- !check_deadline(GN,CHUNKSIZE).
217
218 /**
219  * methods for non group leader to wait with the cooperation checking till the
220  * deadline event
221  */
222 +!check_deadline(GN,CHUNKSIZE): deadline
223     <-- !update_reception(GN,CHUNKSIZE).
224
225 +!check_deadline(GN,CHUNKSIZE): not deadline
226     <-- .wait("+deadline");
227         !update_reception(GN,CHUNKSIZE).
228
229 /**
230  * checking whether any other agents have sent their chunks. different methods
231  * are used because of the leader role and in case the agents have not sent own
232  * chunks because the sending and reception costs are different (the agent has
233  * include or exclude the own chunk it was supposed to send)
234  */
235 +!control_reception(GN,CHUNKSIZE): leader(GN) & sent
236     <-- .wait(4000);
237         ?collaborators(GN,GROUP);
238         /* setting deadline for sending chunks */
239         .send(GROUP,tell,deadline);
240         ?groupkey(GN,KEY);
241         .count(sharing(GN,_,KEY,_)[source(_)],RECEIVED);
242         /* assumption: each agent sends his fair share if it send sth. at
           all.
243         * received bits. */
244         * i.e. the total amount received is the fair share times the number
             of
245         TotalReceived = CHUNKSIZE * (RECEIVED - 1);
246         ?event(E);
247         /* update energy costs */
248         update_reception_costs(TotalReceived,E);
249         /* get missing bits from basestation */
250         ?total_chunk_number(GN,TCHUNKS);
251         Rest = ((TCHUNKS-1)*CHUNKSIZE)-TotalReceived;
252         !download_self(MO,Rest).
253
254 /**
255  * checking whether any other agent has sent some chunks
256  */
257 +!control_reception(GN,CHUNKSIZE): leader(GN) & not sent
258     <-- .wait(4000);
259         ?collaborators(GN,GROUP);
260         .send(GROUP,tell,deadline);
261         ?groupkey(GN,KEY);
262         .count(sharing(GN,_,KEY,_)[source(_)],RECEIVED);
263         /* assumption: each agent sends his fair share if it send sth. at
           all.
264         * i.e. the total amount received is the fair share times the number
265
```

```
266   * received bits. */
267   TotalReceived = CHUNKSIZE * RECEIVED;
268   ?event(E);
269   /* update energy costs */
270   update_reception_costs(TotalReceived,E);
271   /* get missing bits from basestation */
272   ?total_chunk_number(GM,TCHUNKS);
273   Rest = ((TCHUNKS-1)*CHUNKSIZE)-TotalReceived;
274   !download_self(NO,Rest).
275
276 +!update_reception(GM,CHUNKSIZE): not sent
277   <- ?groupkey(GM,KEY);
278      .count(sharing(GM,_,KEY,_)[source(_)],RECEIVED);
279   /* Assumption: each agent sends his fair share if it send sth. at
      all.
280   * i.e. the total amount received in the fair share times the number
      of
281   * received bits. */
282      TotalReceived = CHUNKSIZE * RECEIVED;
283      ?event(E);
284   /* update energy costs */
285      update_reception_costs(TotalReceived,E);
286   /* get missing bits from basestation */
287      ?total_chunk_number(GM,TCHUNKS);
288      Rest = ((TCHUNKS-1)*CHUNKSIZE)-TotalReceived;
289      .wait(3000);
290      !download_self(NO,Rest).
291
292 +!update_reception(GM,CHUNKSIZE): sent
293   <- ?groupkey(GM,KEY);
294      .count(sharing(GM,_,KEY,_)[source(_)],RECEIVED);
295   /* Assumption: each agent sends his fair share if it send sth. at
      all.
296   * i.e. the total amount received is the fair share times the number
      of
297   * received bits. */
298      TotalReceived = CHUNKSIZE * (RECEIVED - 1);
299   /* update energy costs */
300      update_reception_costs(TotalReceived,E);
301      ?event(E);
302   /* get missing bits from basestation */
303      ?total_chunk_number(GM,TCHUNKS);
304      Rest = ((TCHUNKS-1)*CHUNKSIZE)-TotalReceived;
305      .wait(3000);
306      !download_self(NO,Rest).
307
308 /* general plans */
309 /**
311   * movement: for the method the old location is being used: The old location is
313   * needed to delete it from the belief base and the new location is determined
314   * according to the movement model and added to the belief base of the agent.
315   */
316 +!move: true
317   <- ?location(L);
318      remove_percept(L);
319      !abolish_beliefs;
320      ?movement(M);
321      Mnew = M+1;
322      +movement(Mnew);
323      move_agent(L,Mnew);
324      !calc_random;
325      !move_decision.
326
327 /**
328   * abolishes the beliefs of an agent after the interaction (e.g. peers...):
329   * This method is used to ensure that the agent is not confusing beliefs if it
330   * is trying to download a new file and gets beliefs with of the same form, but
331   * with other content.
332   */
333 +!abolish_beliefs: true
334   <- ?movement(M);
335      ?event(E);
336      ?cheating(C);
337      -abolish(_);
338      +event(E);
339      +movement(M);
340      +cheating(C).
```

Listing 5: honest_agent.asl class

```
1  /**
2   * Agent honest_agent in project Thesis_simulation
3   * This agent always acts honestly, i.e. always tries to fulfill
4   * its commitments.
5   * @author Tina Balke
6   * @version 23-06-2011
7   */
8
9  /* inheritance from other agent classes */
10 {include("agent.asl")}
11 {include("interaction_agent.asl")}
12 /* important comment for Jason Agents cannot override inherited methods, but
13  * but the new methods are added to the inherited methods; this can result in
14  * an agent having several methods called twice, what in turn will result in the
15  * respective plans being called twice!
16  */
17
18 /* initial beliefs */
19 cheating(0).
20
21 /* plans */
22 +!assigned_chunk(GM,NO,CHUNKNUMBER,CHUNKSIZE): true
23   <- !check_download_channel.
24
25 +!check_download_channel: free(_)
26   <- ?assigned_chunk(GM,NO,CHUNKNUMBER,CHUNKSIZE);
27      check_permission_download(GM,CHUNKNUMBER);
28      !download_decision.
29
30 +!check_download_channel: not free(_)
31   <- ?assigned_chunk(GM,NO,CHUNKNUMBER,CHUNKSIZE);
32      .send(basestation, askOne, free(_), Reply);
```

Listing 6: util agent.asl class

```
1   /**
2    * Agent util agent in project Thesis simulation
3    * This agent will always act based on the utility maximizing rational.
4    *
5    * @author Tina Balke
6    * @version 23-06-2011
7    */
8
9   /* inheritance from other agent classes */
10  {include("agent.asl")}
11  {include("interaction_agent.asl")}
12  /* imports the functions of Jason; agents cannot overwrite inherited methods, but
13   * but the new methods are added to the inherited methods; this can result in
14   * an agent having several methods twice, what in turn will result in the
15   * respective plans being called twice!
16   */
17
18  /* initial beliefs */
19

20  cheating(0).
21
22  /* plans */
23
24  +!assigned_chunk(GM,NO,CHUNKNUMBER,CHUNKSIZE): true
25      <-  !check_download_channel.
26
27  +!check_download_channel: free(_)
28      ?assigned_chunk(GM,NO,CHUNKNUMBER,CHUNKSIZE);
29      check_permission.download(GM,CHUNKNUMBER);
30      !download_decision.
31
32  +!check_download_channel: not free(_)
33      .send(basestation, askOne, free(_), Reply);
34      !check_download_channel.
35
36  +!download_decision: power(P) & permission(PE) & P == true & PE == true
37      ?assigned_chunk(GM,NO,CHUNKNUMBER,CHUNKSIZE);
38      ?event(E);
39      .send(basestation, tell, download);
40      /* updating the energy costs */
41      update_download_costs(CHUNKSIZE,E);
42      external_event_download(GM,CHUNKNUMBER);
43      .send(basestation, untell, download);
44      /* determine the costs of sending and defection */
45      ?cheating(C);
46      get_utility_sending(CHUNKSIZE,E,C);
47      ?sending_costs(S);
48      ?defection_costs(D);
49      !sending_decision(CHUNKSIZE,S,D).
50
51  +!download_decision: power(P) & permission(PE) & P == false & PE == false
52      <-  !check_download_channel.
53
54  +!check_download_channel: not free(_)
55      .send(basestation, askOne, free(_), Reply);
56      !check_download_channel.
57
58  //Mode 6
59  /**
60   * if sending is more expensive (battery-wise) then cheating --> cheat
61   */
62  +!sending_decision(GM,CHUNKSIZE,S,D): S > D
63      <-  /* increment cheating counter by 1 */
64      ?cheating(C);
65      Cnew = C + 1;
66      cheat(GM);
67      -+cheating(Cnew);
68      !control_reception(GM,CHUNKSIZE).
69
70  /**
71   * if sending is less expensive (battery-wise) then cheating --> cooperate
72   */
73  +!sending_decision(GM,CHUNKSIZE,S,D): S <= D
74      <-  ?assigned_chunk(GM,NO,CHUNKNUMBER,CHUNKSIZE);
75      ?collaborators(GM,GROUP);
76      ?groupkey(GM,KEY);
77      ?event(E);

34      !check_download_channel.
35
36  +!download_decision: power(P) & permission(PE) & P == true & PE == true
37      ?assigned_chunk(GM,NO,CHUNKNUMBER,CHUNKSIZE);
38      ?event(E);
39      .send(basestation, tell, download);
40      /* updating the energy costs */
41      update_download_costs(CHUNKSIZE,E);
42      external_event_download(GM,CHUNKNUMBER);
43      .send(basestation, untell, download);
44      !sending_decision(GM,CHUNKSIZE).
45
46  +!download_decision: power(P) & permission(PE) & P == false & PE == false
47      <-  !check_download_channel.
48
49  +!check_download_channel: not free(_)
50      .send(basestation, askOne, free(_), Reply);
51      !check_download_channel.
52
53  //Mode 5
54  /**
55   * always cooperate
56   */
57  +!sending_decision(GM,CHUNKSIZE): true
58      ?assigned_chunk(GM,NO,CHUNKNUMBER,CHUNKSIZE);
59      ?collaborators(GM,GROUP);
60      ?groupkey(GM,KEY);
61      ?event(E);
62      .my_name(NAME);
63      /* sending */
64      .send(GROUP,tell,sharing(GM,CHUNKNUMBER,KEY,NAME));
65      +sent;
66      /* update energy costs */
67      update_sending_costs(CHUNKSIZE,E,GM,CHUNKNUMBER);
68      external_event_sending(GM,CHUNKNUMBER);
69      !control_reception(GM,CHUNKSIZE).
```

```
78        .my_name(NAME);
79        /* sending chunks */
80        .send(GROUP,tell,sharing(GM,CHUNKNUMBER,KEY,NAME));
81     -sent;
82        external_event_sending(GM,CHUNKNUMBER);
83        /* update energy costs */
84        update_sending_costs(CHUNKSIZE,E,GM,CHUNKNUMBER);
85        !control_reception(GM,CHUNKSIZE).
```

Listing 7: malicious.agent.asl class

```
1   /**
2    * Agent malicious_agent in project Thesis_simulation
3    * This agents goal is to harm the system, regardless of its own utility
4    *
5    * @author Tina Balke
6    * @version 23-06-2011
7    *
8    */
9
10  /* inheritence from other agent classes */
11  {include("agent.asl")}
12  {include("interaction_agent.asl")}
13  /* important comment for Jason: agents cannot overwrite inherited methods, but
14   * but the new methods are added to the inherited methods; this can result in
15   * an agent having several methods twice, what in turn will result in the
16   * respective plans being called twice!
17   */
18
19  /* initial beliefs */
20  cheating(O).
21  badImageNumber(O).
22
23  /* plans */
24
25  //Mode 5
26  /* never cooperate */
27  +!assigned_chunk(GM,NO,CHUNKNUMBER,CHUNKSIZE): true
28     <- !check_download_channel.
29
30  +!check_download_channel: free(_)
31     ?assigned_chunk(GM,NO,CHUNKNUMBER,CHUNKSIZE);
32     check_permission_download(GM,CHUNKNUMBER);
33     !download_decision.
34
35  +!check_download_channel: not free(_)
36     .send(basestation, askOne, free(_), Reply);
37     !check_download_channel.
38
39  +!download_decision: power(P) & permission(PE) & P == true & PE == true
40     ?event(E);
41     .send(basestation, tell, download);
42     /* updating the energy costs */
43     update_download_costs(CHUNKSIZE,E);
44     external_event_download(GM,CHUNKNUMBER);
45     .send(basestation, untell, download);
46     !control_reception(GM,CHUNKSIZE).
47
48  +!download_decision: power(P) & permission(PE) & P == false & PE == false
49     <- !check_download_channel.
50
51  +!check_download_channel: not free(_)
52     .send(basestation, askOne, free(_), Reply);
53     !check_download_channel.
54
```

Listing 8: image.agent.asl class

```
1   /**
2    * Agent image_agent in project Thesis_simulation
3    * The agents inheriting from this agent use image information.
4    *
5    * @author Tina Balke
6    * @version 12-07-2011
7    *
8    */
9
10  /* initial beliefs and rules */
11  event(-1).
12
13  /* initial rules */
14  enoughInteractions :- event(X) & num_interactions(Y) & X >= Y-1.
15  oppCoststoHigh(SIZE) :- oppCosts(X) & X >= SIZE.
16  sufficient_group_size(GSIZE) :- GSIZE > 0.
17
18  /* initial goals */
19  +start: true <- !start.
20
21
22  /* plans */
23  +!start: true
24     <- !calc_random;
25     get_number_interactions;
26     /*Set Agent properties */
27     set_mobility_level;
28     set_location;
29     !move_decision.
30
31  /**
32   * checking whether the agent has performed more then a certain number of
33   * downloads (download events). if this is the case, then it sends a message
34   * to the sim_agent: This agent checks whether all agents have sent him a
35   * message and if so, stops the simulation. If the event threshold is not
36   * met the agent simply continues with downloading.
37   */
38  +!check_event: enoughInteractions
39     <- .send(sim_agent,tell,end).
40
41  +!check_event: not enoughInteractions
42     downloading.
43
44  /*
45   * in case the percept download(NO,SIZE) is made, generate a goal with
46   * the same name
47   */
48  +download(NO,SIZE): true
```

```
49        !update_event;
50 <-     ?event(E);
51        set_alternative_battery_costs(SIZE,E);
52        get_opportunity_costs;
53        !download_file(NO,SIZE).
54
55 //Node 1
56 /**
57  * In case the size of the file is so small that a cooperation does not pay off
58  * (opportunity costs are higher then the respective file size) download self.
59  */
60 +!download_file(NO,SIZE): oppCoststooHigh(SIZE)
61 <-     !download_self(NO,SIZE).
62
63 /**
64  * In case the opportunity costs are smaller then the size and the perceived
65  * risk not above 0.5 (i.e. 50 percent) check for neighbours (query environment
66  * for neighbour percepts and count the total number of neighbours)
67  */
68 +!download_file(NO,SIZE): not oppCoststooHigh(SIZE)
69 <-     ?location(LOCATION);
70        check_neighbours(LOCATION);
71        .count(neighbour(_),NoNeighbours);
72        !cooperate_decision(NO,SIZE,NoNeighbours).
73
74 //Node 2
75 /**
76  * if no neighbours are in the vicinity, then download self
77  */
78 +!cooperate_decision(NO,SIZE,NoNeighbours): NoNeighbours == 0
79 <-     !download_self(NO,SIZE).
80
81 /**
82  * otherwise, check whether anyone of the neighbours has sent a cooperation
83  * proposal
84  */
85 +!cooperate_decision(NO,SIZE,NoNeighbours): NoNeighbours \== 0
86 <-     .wait(3000);
87        !check_for_coop_proposals(NO,SIZE).
88
89 +!check_for_coop_proposals(NO,SIZE): group_joined(Sender)
90 <-     /* checking image information */
91        check_image(Sender);
92        !check_coop_level(NO,Sender,SIZE).
93
94 /**
95  * if no proposals have been send and no group was joined --> search for
96  * partners yourself
97  */
98 +!check_for_coop_proposals(NO,SIZE): not coop_proposal(NO,Sender) & not
       group_joined(_)
99 <-     +coop_request_sent(NO);
100       .findall(Name,neighbour(Name),N)
101       .send(N,tell,coop_proposal(NO));
102       ?event(E);
103       update_battery_coopRequest(E);
104       .wait(6000);
105       /* deadline belief added to the agents belief base, so it can determine
106       * whether it received delayed acceptances */
107       +proposal_deadline;
108 <-    .findall(Agent,accept(coop_proposal(NO,Agent)),A);
109       !check_for_coop_proposals_answers(NO,SIZE,A).

110 //reactions the proposals of other players
111 /**
112  * If same goal and not already group joined or request send self.
113  * accept cooperate proposal
114  */
115 +coop_proposal(NO,SIZE)[source(Agent)]: download(NO,SIZE) & not group_joined(_) &
       not coop_request_sent(_)
116
117 <-    +group_joined(Agent).
118
119 /**
120  * if the agent has sent a coop request himself or has joined another group
121  * already, it will decline other requests
122  */
123 @coop_imag2[priority(1)]
124 +coop_proposal(NO,SIZE)[source(Agent)]: enoughInteractions
125 <-    .abolish(coop_proposal(NO)[source(Agent)]).
126
127 /**
128  * if not same file to download, decline proposal
129  */
130 @coop_imag3[priority(0)]
131 +coop_proposal(NO,SIZE)[source(Agent)]: not download(NO,SIZE)
132 <-    .abolish(coop_proposal(NO)[source(Agent)]).
133
134 /**
135  * if finished all interactions, decline proposal with highest priority
136  */
137 @coop4[priority(-1)]
138 +coop_proposal(NO,SIZE)[source(Agent)]: enoughInteractions
139 <-    .abolish(coop_proposal(NO)[source(Agent)]).
140
141 /**
142  * default is to decline any proposals with low priority
143  */
144 @coop5[priority(2)]
145 +coop_proposal(NO,SIZE)[source(Agent)]: true
146 <-    .abolish(coop_proposal(NO)[source(Agent)]).
147
148 /**
149  * if agents accept cooperation proposals after the group has formed, they are
150  * being told so
151  */
152 +accept(coop_proposal(NO,Name))[source(Name)]: proposal_deadline | not
       coop_request_sent(NO)
153 <-    .abolish(accept(coop_proposal(NO,Name))[source(Name)]);
154       .send(Name,tell,to_late);
155       .send(Name,tell,infos_generated).
156
157 //Node 4 -- partner risk is determined with image information
158 +!check_for_coop_proposals_answers(NO,SIZE,A): A \== []
159 <-    /* group initiator checks the answers it got */
160       getTrustingPartners(A);
161       /* and counts how many agents that answered are trustworthy */
```

```
162          .count(cooperator(_)[source(percept)],GSIZE).
163    !negotiate_decision(MD,SIZE,GSIZE).
164
165    //Mode 3
166    /* if the number of answers is empty --> download self */
167    +!check_for_coop_proposals_answers(MD,SIZE,A): A == []
168    <-   !download_self(MD,SIZE).
169
170    /**
171     * methods for checking the trustworthiness of the the agent that has sent a
172     * cooperation request
173     */
174    /* if trustworthy --> cooperate */
175    @check_coop_level[priority(1)]
176    +!check_coop_level(MD,Sender,SIZE): coop(Sender)
177    <-   .my_name(NAME);
178         .send(Sender,tell,accept(coop_proposal(MD,NAME)));
179         ?event(E);
180         update_battery_coopRequest(E);
181         check_infos_generated(MD,SIZE).
182
183    /* else decline cooperation proposal and download self */
184    @check_coop_level2[priority(1)]
185    +!check_coop_level(MD,Sender,SIZE): defect(Sender)
186    <-   .send(Sender,tell,bad_proposal(MD));
187         !download_self(MD,SIZE).
188
189    @check_coop_level3[priority(2)]
190    +!check_coop_level(MD,Sender,SIZE): true
191    <-   .abolish(coop_proposal(MD)[source(Sender)]).
192
193    /**
194     * reactions of non group leaders to the perception that the group infos are
195     * available
196     */
197    +!check_infos_generated(MD,SIZE): infos_generated
198    <-   !check_group_infos(MD,SIZE).
199
200    +!check_infos_generated(MD,SIZE): not infos_generated
201    <-   wait(+infos_generated);
202         !check_group_infos(MD,SIZE).
203
204    /**
205     * ceck the number of the agents that have answered and have a high enough
206     * image level (number of agents that meet that criterion = GSIZE=
207     */
208    /* if GSIZE = 0, i.e. no agent meets that criterio, download self */
209    +!negotiate_decision(MD,SIZE,GSIZE): not sufficient_group_size(GSIZE)
210         /* Sending defectors the message that no cooperation due to bad image */
211    <-   .findall(Agent,defector(Agent)[source(percept)],DEFECTORS);
212         /* telling all agents with not sufficient image level of the bad image */
213         .send(DEFECTORS,tell,bad_image(MD));
214         .send(DEFECTORS,tell,infos_generated);
215         /* and download self */
216         !download_self(MD,SIZE).
217

218    /* otherwise, initiate group */
219    +!negotiate_decision(MD,SIZE,GSIZE): sufficient_group_size(GSIZE)
220    <-   .findall(Agent,defector(Agent),DEFECTORS);
221         .send(DEFECTORS,tell,bad_image(MD));
222         .findall(Agent,cooperator(Agent),GROUP);
223         .random(KEY);
224         ?groupNumber(GSIZE);
225         +groupnumber(GN);
226         +leader(GN);
227         /* calls the chunk assignment method in the environment */
228         .send(DEFECTORS,tell,infos_generated);
229         assign_chunks(GN,GROUP,KEY,MD);
230         !check_group_infos_leader(MD,SIZE,GROUP).
231
232    /**
233     * checking group infos: if the agent is the leader, then it gives all group
234     * members the information that the group information are available
235     */
236    +!check_group_infos_leader(MD,SIZE,GROUP): true
237    <-   .send(GROUP,tell,infos_generated);
238         !check_group_infos(MD,SIZE).
239
240    /* if agent was accepted by a group, it waits for the assignment infos */
241    +!check_group_infos(MD,SIZE): assigned_chunk(GN,MD,CHUNKNUMBER,CHUNKSIZE)
242    <-   !assigned_chunk(GN,MD,CHUNKNUMBER,CHUNKSIZE).
243
244    /* if it gets a reject message, it downloads self */
245    +!check_group_infos(MD,SIZE): bad_image(MD) | to_late
246    <-   !download_self(MD,SIZE).
247
248    //Mode 6
249    /**
250     * checking whether any other agent has sent some chunks
251     */
252    +!control_reception(GN,CHUNKSIZE): not leader(GN)
253    <-   !check_deadline(GN,CHUNKSIZE).
254
255    /**
256     * methods for non group leader to wait with the cooperation checking till the
257     * deadline event
258     */
259    +!check_deadline(GN,CHUNKSIZE): deadline
260    <-   !update_reception(GN,CHUNKSIZE).
261
262    +!check_deadline(GN,CHUNKSIZE): not deadline
263    <-   .wait("+deadline");
264         !update_reception(GN,CHUNKSIZE).
265
266    /**
267     * checking whether any other agent has sent some chunks and whether file is
268     * complete as a consequence. different methods
269     * are used because of the leader role and in case the agents have not sent own
270     * chunks because the sending and reception costs are different (the agent has
271     * include or exclude the own chunk it was supposed to send)
272     */
273    +!control_reception(GN,CHUNKSIZE): leader(GN) & sent
274    <-   .wait(4000);
275         ?collaborators(GN,GROUP);
```

```
276         .send(GROUP,tell,deadline);
277         external_event_deadline(GM);
278         ?groupkey(GM,KEY);
279         .count(sharing(GM,_,KEY,_),RECEIVED);
280         /* Assumption: each agent sends his fair share if it send sth. at
281            all..
             * i.e. the total amount received is the fair share times the number
               of
282           * received bits. */
283         TotalReceived = CHUNKSIZE * (RECEIVED - 1);
284         ?event(E);
285         /* update energy costs */
286         update_reception_costs(TotalReceived,E);
287         /* update image information by comparing the groupmembers and the
               agents
288           * that have sent a chunk */
289         .findall(Agent,sharing(GM,_,KEY,Agent),COOPGROUP);
290         updateImageInfo(GROUP,COOPGROUP);
291         /* get missing bits from basestation */
292         ?total_chunk_number(GM,TCHUNKS);
293         Rest = ((TCHUNKS-1)*CHUNKSIZE)-TotalReceived;
294         !download_self(MO,Rest).
295     /**
296      * checking whether any other agent has sent some chunks and whether file is
297      * complete as a consequence
298      */
299     +!control_reception(GM,CHUNKSIZE): leader(GM) & not sent
300         <- .wait(4000);
301         ?collaborators(GM,GROUP);
302         .send(GROUP,tell,deadline);
303         external_event_deadline(GM);
304         ?groupkey(GM,KEY);
305         .count(sharing(GM,_,KEY,_),RECEIVED);
306         /* Assumption: each agent sends his fair share if it send sth. at
307            all..
             * i.e. the total amount received is the fair share times the number
               of
308           * received bits. */
309         TotalReceived = CHUNKSIZE * RECEIVED;
310         ?event(E);
311         /* update energy costs */
312         update_reception_costs(TotalReceived,E);
313         /* update image information by comparing the groupmembers and the
               agents
314           * that have sent a chunk */
315         .findall(Agent,sharing(GM,_,KEY,Agent),COOPGROUP);
316         updateImageInfo(GROUP,COOPGROUP);
317         /* get missing bits from basestation */
318         ?total_chunk_number(GM,TCHUNKS);
319         Rest = ((TCHUNKS-1)*CHUNKSIZE)-TotalReceived;
320         !download_self(MO,Rest).
321     /*
322      * checking whether any other agent has sent some chunks and whether file is
323      * complete as a consequence, beforehand, check whether the deadline has passed
324      */
325
326
327     +!update_reception(GM,CHUNKSIZE): not sent

328         <- ?groupkey(GM,KEY);
329         ?collaborators(GM,GROUP);
330         .count(sharing(GM,_,KEY,_),RECEIVED);
331         /* Assumption: each agent sends his fair share if it send sth. at
332            all..
             * i.e. the total amount received is the fair share times the number
               of
333           * received bits. */
334         TotalReceived = CHUNKSIZE * RECEIVED;
335         ?event(E);
336         /* update energy costs */
337         update_reception_costs(TotalReceived,E);
338         /* update image information by comparing the groupmembers and the
               agents
339           * that have sent a chunk */
340         .findall(Agent,sharing(GM,_,KEY,Agent),COOPGROUP);
341         updateImageInfo(GROUP,COOPGROUP);
342         /* get missing bits from basestation */
343         ?total_chunk_number(GM,TCHUNKS);
344         Rest = ((TCHUNKS-1)*CHUNKSIZE)-TotalReceived;
345         !download_self(MO,Rest).
346
347     +!update_reception(GM,CHUNKSIZE): sent
348         <- ?groupkey(GM,KEY);
349         ?collaborators(GM,GROUP);
350         .count(sharing(GM,_,KEY,_),RECEIVED);
351         /* Assumption: each agent sends his fair share if it send sth. at
352            all..
             * i.e. the total amount received is the fair share times the number
               of
353           * received bits. */
354         TotalReceived = CHUNKSIZE * (RECEIVED - 1);
355         ?event(E);
356         /* update energy costs */
357         update_reception_costs(TotalReceived,E);
358         /* update image information by comparing the groupmembers and the
               agents
359           * that have sent a chunk */
360         .findall(Agent,sharing(GM,_,KEY,Agent),COOPGROUP);
361         updateImageInfo(GROUP,COOPGROUP);
362         /* get missing bits from basestation */
363         ?total_chunk_number(GM,TCHUNKS);
364         Rest = ((TCHUNKS-1)*CHUNKSIZE)-TotalReceived;
365         !download_self(MO,Rest).
366
367     /* further Plans */
368     /**
369      * movement: for the method the old location is being used: The old location is
370      * needed to delete it from the belief base and the new location is determined
371      * according to the movement model and added to the belief base of the agent
372      */
373
374     @move_image[atomic]
375     +!move: true
376         <- ?location(L);
377         remove_percept(L);
378         !abolish_beliefs;
379         ?movement(M);
```

```
380    Mnew = M+1;
381    -!movement(Mnew);
382    move_agent(L,Mnew);
383    !calc_random;
384    !move_decision.
385
386    /*
387     * Abolishes the beliefs of an agent after the interaction (e.g. parters,....);
388     * This is to ensure that the agent is not confusing beliefs if it is trying
389     * to download a new file and gets beliefs with of the same form, but with
390     * other content.
391     */
392    @abolish_beliefs-image[atomic]
393    +!abolish_beliefs: true
394        <- ?movement(M);
395           -movement(M);
396           ?event(E);
397           ?cheating(C);
398           ?badImageNumber(I);
399           .abolish(_);
400           +event(E);
401           +movement(M);
402           +cheating(C);
403           +badImageNumber(I).
404
405    /**
406     * reactions to bad_image and bad_proposal information about the agent itself
407     * by others --> increment the respective counters by one
408     */
409    +bad_image(_): true
410        <- ?badImageNumber(I);
411           Inew = I + 1;
412           -+badImageNumber(Inew).
413
414    +bad_proposal(_): true
415        <- ?badImageNumber(I);
           Inew = I + 1;
           -+badImageNumber(Inew).
```

Listing 9: honest_image_agent.asl class

```
 1    /**
 2     * Agent honest image agent in project Thesis simulation
 3     * This agent always acts honestly, i.e. always tries to fulfill
 4     * its commitments. It uses image information.
 5     *
 6     * @author Tina Balke
 7     * @version 12-07-2011
 8     *
 9     */
10    /* inheritance from other agent classes */
11    {include("agent.asl")}
12    {include("image_agent.asl")}
13    /* important comment for Jason, agents cannot overwrite inherited methods, but
14     * but the new methods are added to the inherited methods; this can result in
15     * an agent having several methods twice, what in turn will result in the
16     * respective plans being called twice!
17     */
18
19
20    /* initial beliefs */
21    cheating(O).
22    badImageNumber(O).
23
24    /* plans */
25
26    /**
27     * if the agent is assigned a chunk, it will download it and the respective
28     * energy consumption costs are added
29     */
30    +!assigned_chunk(GM,NO,CHUNKNUMBER,CHUNKSIZE): true
31        <- !check_download_channel.
32
33    +!check_download_channel: free(_)
34        <- ?assigned_chunk(GM,NO,CHUNKNUMBER,CHUNKSIZE);
35           check_permission_download(GM,CHUNKNUMBER);
36           !download_decision.
37
38    +!check_download_channel: not free(_)
39        <- .send(basestation, askOne, free(_), Reply);
40           !check_download_channel.
41
42    +!download_decision: power(P) & permission(PE) & P == true & PE == true
43        <- ?assigned_chunk(GM,NO,CHUNKNUMBER,CHUNKSIZE);
44           ?event(E);
45           .send(basestation, tell, download);
46           /* updating the energy costs */
47           update_download_costs(CHUNKSIZE,E);
48           external_event_download(GM,CHUNKNUMBER);
49           .send(basestation, untell, download);
50           !sending_decision(GM,CHUNKSIZE).
51
52    +!download_decision: power(P) & permission(PE) & P == false & PE == false
53        <- .send(basestation, askOne, free(_), Reply);
54           !check_download_channel.
55
56    +!check_download_channel: not free(_)
57        <- .send(basestation, askOne, free(_), Reply);
58           !check_download_channel.
59
60    //Node 5
61    /*
62     * always cooperate
63     */
64    +!sending_decision(GM,CHUNKSIZE): true
65        <- ?assigned_chunk(GM,NO,CHUNKNUMBER,CHUNKSIZE);
66           ?collaborators(GM,GROUP);
67           ?group(GM,KEY);
68           ?event(E);
69           .my_name(NAME);
70           /* sending */
71           .send(GROUP, tell, sharing(GM,CHUNKNUMBER,KEY,NAME));
72           +sent;
73           /* update energy costs */
74           update_sending_costs(CHUNKSIZE,E,GM,CHUNKNUMBER);
75           external_event_sending(GM,CHUNKNUMBER);
76           !control_reception(GM,CHUNKSIZE).
```

Listing 10: util.image_agent.asl class

```
1   /**
2    * Agent util_image_agent in project Thesis_simulation
3    * This agent will always act based on the utility maximizing rational and uses
4    * reputation information.
5    *
6    * @author Tina Balke
7    * @version 12-07-2011
8    *
9    */
10  /* inheritance from other agent classes */
11  {include("agent.asl*")}
12  {include("image-agent.asl*")}
13  {include("image-agent.asl*")}
14  /* important comment for Jason, agents cannot overwrite inherited methods, but
15   * but the new methods are added to the inherited methods; this can result in
16   * an agent having several methods twice, what in turn will result in the
17   * respective plans being called twice!
18   */
19
20  /* initial beliefs */
21  cheating(O).
22  badImageNumber(O).
23
24  /* plans */
25  +!assigned_chunk(GN,NO,CHUNKNUMBER,CHUNKSIZE): true
26   <--  !check_download_channel.
27
28  +!check_download_channel: free(_)
29   <--  ?assigned_chunk(GN,NO,CHUNKNUMBER,CHUNKSIZE);
30        check_permission_download(GN,CHUNKNUMBER);
31        !download_decision.
32
33  +!check_download_channel: not free(_)
34   <--  .send(basestation, askOne, free(_), Reply);
35        !check_download_channel.
36
37  +!download_decision: power(P) & permission(PE) & P == true & PE == true
38   <--  ?assigned_chunk(GN,NO,CHUNKNUMBER,CHUNKSIZE);
39        ?event(E);
40        .send(basestation, tell, download);
41        /* updating the energy costs */
42        update_download_costs(CHUNKSIZE,E);
43        external_event_download(GN,CHUNKNUMBER);
44        .send(basestation, untell, download);
45        /* determine the costs of sending and defection */
46        ?badImageNumber(I);
47        ?collaborators(KEY,GROUP);
48        getGroupSize(GROUP);
49        ?groupSize(GS);
50        /* determine the costs of sending and defection based on image */
51        getUtilitysending_image(CHUNKSIZE,R,I,GS);
52        ?sending_costs(S);
53        ?defection_costs(D);
54        !sending_decision(GN,CHUNKSIZE,S,D).
55
56  +!download_decision: power(P) & permission(PE) & P == false & PE == false
57   <--  !check_download_channel.
```

```
58  +!check_download_channel: not free(_)
59   <--  .send(basestation, askOne, free(_), Reply);
60        !check_download_channel.
61
62  //Mode 5
63  /**
64   * if sending is more expensive (battery-wise) then cheating --> cheat
65   */
66  +!sending_decision(GN,CHUNKSIZE,S,D): S > D
67        /* increment cheating counter by 1 */
68   <--  ?cheating(C);
69        Cnew = C + 1;
70        cheat(GN);
71        -+cheating(Cnew);
72        !control_reception(GN,CHUNKSIZE).
73
74  /**
75   * if sending is less expensive (battery-wise) then cheating --> cooperate
76   */
77  +!sending_decision(GN,CHUNKSIZE,S,D): S <= D
78   <--  ?assigned_chunk(GN,NO,CHUNKNUMBER,CHUNKSIZE);
79        ?collaborators(GN,GROUP);
80        ?groupkey(GN,KEY);
81        ?event(E);
82        .my_name(NAME);
83        /* sending chunks */
84        .send(GROUP,tell,sharing(GN,CHUNKNUMBER,KEY,NAME));
85        /* update energy costs */
86        external_event_sending(GN,CHUNKNUMBER);
87        update_sending_costs(CHUNKSIZE,E,GN,CHUNKNUMBER);
88        external_event_sending(GN,CHUNKNUMBER);
89        !control_reception(GN,CHUNKSIZE).
90
91
```

Listing 11: malicious.image_agent.asl class

```
1   /**
2    * Agent honest_image_agent in project Thesis_simulation
3    * This agent always acts honestly, i.e. always tries to fulfill
4    * its commitment. It uses image information.
5    *
6    * @author Tina Balke
7    * @version 12-07-2011
8    *
9    */
10  /* inheritance from other agent classes */
11  {include("agent.asl*")}
12  {include("image-agent.asl*")}
13  /* important comment for Jason, agents cannot overwrite inherited methods, but
14   * but the new methods are added to the inherited methods; this can result in
15   * an agent having several methods twice, what in turn will result in the
16   * respective plans being called twice!
17   */
18
19
20  /* initial beliefs */
21
```

```
22  cheating(0).
23  badImageNumber(0).
24
25  /* plans */
26
27  /**
28   * if the agent is assigned a chunk, it will download it and the respective
29   * energy consumption costs are added
30   */
31  +!assigned_chunk(GM,MO,CHUNKNUMBER,CHUNKSIZE): true
32      <-  !check_download_channel.
33
34  +!check_download_channel: free(_)
35      <-  ?assigned_chunk(GM,MO,CHUNKNUMBER,CHUNKSIZE);
36          check_permission_download(GM,CHUNKNUMBER);
37          !download_decision.
38
39  +!check_download_channel: not free(_)
40      <-  .send(basestation, askOne, free(_), Reply);
41          !check_download_channel.
42
43  +!download_decision: power(P) & permission(PE) & P == true & PE == true
44      <-  ?assigned_chunk(GM,MO,CHUNKNUMBER,CHUNKSIZE);
45          ?event(E);
46          /* updating the energy costs */
47          update_download_costs(CHUNKSIZE,E);
48          external_event_download(GM,CHUNKNUMBER);
49          .send(basestation,untell,download);
50          !control_reception(GM,CHUNKSIZE).
51
52
53  +!download_decision: power(P) & permission(PE) & P == false & PE == false
54      <-  !check_download_channel.
55
56  +!check_download_channel: not free(_)
57      <-  .send(basestation, askOne, free(_), Reply);
58          !check_download_channel.
```

Listing 12: rep_agent.asl class

```
1   /**
2    * Agent rep_agent in project Thesis_simulation.
3    * The agents inheriting from this agent use reputation information.
4    *
5    * @author Tina Balke
6    * @version 12-07-2011
7    */
8
9
10  /* initial beliefs */
11  event(-1).
12  messages(0).
13
14  /* initial rules */
15  enoughInteractions :- event(X) & num_interactions(Y) & X >= Y-1.
16  oppCoststoHigh(SIZE) :- oppCosts(X) & X >= SIZE.
17  sufficient_group_size(GSIZE) :- GSIZE > 0.
18
19  /* initial goals */
20  +start: true
21      <-  !start.
22
23  /* plans */
24  +!start: true
25      <-  !calc_random;
26          get_number_interactions;
27          /* set agent properties */
28          set_mobility_level;
29          set_location;
30          !move_decision.
31
32  /**
33   * checking whether the agent has performed more than a certain number of
34   * to/loads (download events). If this is the case, then it sends a message
35   * to the sim_agent. This agent checks whether all agents have sent him a
36   * message and if so, stops the simulation. If the event threshold is not
37   * met the agent simply continues with downloading.
38   */
39  +!check_event: enoughInteractions
40      <-  .send(sim_agent,tell,end).
41
42  +!check_event: not enoughInteractions
43      <-  downloading.
44
45  /**
46   * in case the percept download(MO,SIZE) is made, generate a goal with
47   * the same name
48   */
49  +download(MO,SIZE): true
50      <-  !update_event;
51          ?event(E);
52          set_alternative_battery_costs(SIZE,E);
53          get_opportunity_costs;
54          !download_file(MO,SIZE).
55
56  //Mode 1
57  /**
58   * in case the size of the file is so small that a cooperation does not pay off
59   * (opportunity costs are bigger than the respective file size) download self.
60   */
61  +!download_file(MO,SIZE): oppCoststoHigh(SIZE)
62      <-  !download_self(MO,SIZE).
63
64  /**
65   * in case the opportunity costs are smaller than the size and the perceived
66   * risk not above 0.5 (i.e. 50 percent) check for neighbours (query environment
67   * for neighbour percepts and count the total number of neighbours).
68   */
69  +!download_file(MO,SIZE): not oppCoststoHigh(SIZE)
70      <-  !location(LOCATION);
71          check_neighbours(LOCATION);
72          .count(neighbour(_),NoNeighbours);
73          !cooperate_decision(MO,SIZE,NoNeighbours).
74
75  //Mode 2
76  /**
```

```
77   * if no neighbours are in the vicinity, then download self
78   */
79  +!cooperate_decision(MD,SIZE,NoNeighbours): NoNeighbours == 0
80  <-- !download_self(MD,SIZE).
81  /**
82   * otherwise, check whether anyone of the neighbours has sent a cooperation
83   * proposal
84   */
85  +!cooperate_decision(MD,SIZE,NoNeighbours): NoNeighbours \== 0
86  <-- .wait(3000);
87      !check_for_coop_proposals(MD,SIZE).
88
89  /**
90   * if joined a group, initiate plans to check reputation of cooperation
91   * requester
92   */
93  +!check_for_coop_proposals(MD,SIZE): group_joined(Target)
94  <-- /* triggering plan to check reputation information */
95      !check_rep(MD,SIZE,Target).
96
97  /* checking reputation information */
98  +!check_rep(MD,SIZE,Target): true
99  <-- /* checking own image information first */
100     check_own_info(Target);
101     !check_coop_level(MD,Target,SIZE).
102
103 /**
104
105  * if no proposals have been send search for partners yourself
106  */
107 +!check_for_coop_proposals(MD,SIZE): not group_joined(_)
108 <-- +coop_request_sent;
109     .findall(Name,neighbour(Name),N);
110     .send(N,tell,coop_proposal(MD));
111     update_battery_coopRequest(E);
112     ?event(E);
113     .wait(4000);
114     +proposal_deadline(MD);
115     .findall(Agent,accept(coop_proposal(MD,Agent)),A);
116     !check_for_coop_proposals_answers(MD,SIZE,A).
117
118 //reactions the proposals of other players
119 /**
120  * if goal to download same file and not engaged in another interaction, then
121  * join
122  */
123 @coop_rep1[priority(1)]
124 +coop_proposal(MD)[source(Agent)]: download(MD,SIZE) & not group_joined(_) &
    not coop_request_sent
125 <-- /* joined flag, set right at the beginning, so agent cannot join another
126     * group while checking reputation information */
127     +group_joined(Agent).
128
129 /** if finished all interactions, decline proposal with highest priority
130  */
131 @coop_rep2[priority(0)]
132 +coop_proposal(MD)[source(Agent)]: enoughInteractions

134 <-- .abolish(coop_proposal(MD)[source(Agent)]).
135 /**
136  * default is to decline any proposals with lowest priority
137  */
138 @coop_rep3[priority(2)]
139 +coop_proposal(MD)[source(Agent)]: true
140 <-- .abolish(coop_proposal(MD)[source(Agent)]).
141
142 /**
143  * if agents accept cooperation proposals after the group has formed, they are
144  * being told so
145  */
146 +accept(coop_proposal(MD,Name))[source(Name)]: proposal_deadline(MD) | not
    coop_request_sent
147 <-- .abolish(accept(coop_proposal(MD,Name))[source(Name)]);
148     .send(Name,tell,to_late);
149     .send(Name,tell,infos_generated).
150
151 //Mode 3
152 +!check_for_coop_proposals_answers(MD,SIZE,A): A \== []
153 <-- /* group initiator checks the answers it got, checking its own
154     * information about the agents having answered */
155     check_own_info_initiator(A);
156     /* all the agents for which it has not information, it needs to ask
157     * information about */
158     !check_coop_level(MD,SIZE).
159
160 /* if the number of answers is empty --> download self */
161 +!check_for_coop_proposals_answers(MD,SIZE,A): A == []
162 <-- !download_self(MD,SIZE).
163
164 /**
165  * requesting information for all agents where no own information exists */
166 +!check_result_empty_image(MD,SIZE): true
167 <-- .findall(Agent,own_information(Agent,empty),Targets);
168     !send_Information_Requests(Targets,MD,SIZE).
169
170 /**
171  * if there are no agents the agent in the group does not have any
172  * image information for, it will directly go into the negotiation decision by
173  * by checking the number of agents in the group it has positive information
174  * about
175  */
176 +!send_Information_Requests(Targets,MD,SIZE): Targets == []
177 <-- .count(own_information(_,"VERY_GOOD")[source(percept)],GSIZE1);
178     .count(own_information(_,"GOOD")[source(percept)],GSIZE2);
179     GSIZE = GSIZE1 + GSIZE2;
180     !negotiate_decision(MD,SIZE,GSIZE).
181
182 /*
183  * otherwise send request for reputation information for the first element of
184  * the list of agents that have agreed to join
185  */
186 +!send_Information_Requests(Targets,MD,SIZE): Targets \== []
187 <-- .nth(0,Targets,Target);
188     .findall(Name,neighbour(Name),N);
189     .send(N,tell,recommendationRequest(Target));
190     ?event(E);
```

```
191          /* update energy consumption */
192          update.battery.coopRequest(E);
193          .wait(2000);
194          /* deadline until the recommendations are accepted */
195          +deadlineRecommendation:
196          /* check how many information have arrived */
197          .findall(Rec_Source,rec_info(Target,_,)[source(Rec_Source)],RC);
198          !check_recomendations_initiator(MO,Target,SIZE,RC).
199
200
201  //Mode 4
202  /** not group leaders check for the reputation information of the agent trying to
203   * initiate a group, if they don't have own image information
204   */
205  +!check_coop_level(MO,Target,SIZE): own_information(Target,empty)
206      <-  .findall(Name,neighbour(Name),N);
207          .send(N,tell,recommendationRequest(Target));
208          ?event(E);
209          /* update energy consumption */
210          update.battery.coopRequest(E);
211          .wait(2000);
212          +deadlineRecommendation;
213          .findall(Rec_Source,rec_info(Target,_,)[source(Rec_Source)],RC);
214          !check_recomendations(MO,Target,SIZE,RC).
215
216
217  /**
218   * if they have own image information and this information is positive, then
219   * they will join the group
220   */
221  +!check_coop_level(MO,Target,SIZE): own_information(Target,"VERY_GOOD") |
222  own_information(Target,"GOOD")
223      <-  !my_name(NAME);
224          /* accept cooperation offer */
225          .send(Target,tell,accept(coop_proposal(MO,NAME)));
226          ?event(E);
227          /* update energy consumption */
228          update.battery.coopRequest(E);
229          !check_infos_generated(MO,SIZE).
230
231  /**
232   * if they have own image information and this information is negative (i.e. not
233   * positive), then they will not join the group and download themselves;
234   * bad_proposal is just a flag that is being used and no actual message,
235   * therefor no energy costs accumulate for it.
236   */
237  +!check_coop_level(MO,Target,SIZE): not own_information(Target,"VERY_GOOD") &
238  not own_information(Target,"GOOD") & not own_information(Target,empty)
239      <-  .send(Target,tell,bad_proposal(MO));
240          !download_self(MO,SIZE).
241
242  /**
243   * waiting of non group leaders to the perception that the group infos are
244   * available
245   */
246  +!check_infos_generated(MO,SIZE): infos_generated
247      <-  !check_group_infos(MO,SIZE).


247  +!check_infos_generated(MO,SIZE): not infos_generated
248      <-  .wait("+infos_generated");
249          !check_group_infos(MO,SIZE).
250
251  /**
252   * checking answer for reputation request sent, if no answers were given, then
253   * the agent (which no information seem to exist about) is being given a chance
254   * and the agent accepts the cooperation proposal
255   */
256  +!check_recomendations(MO,Target,SIZE,RC): RC == []
257      <-  !my_name(NAME);
258          /* accept proposal */
259          .send(Target,tell,accept(coop_proposal(MO,NAME)));
260          ?event(E);
261          /* update energy consumption */
262          update.battery.coopRequest(E);
263          !check_infos_generated(MO,SIZE).
264
265  /**
266   * checking answer for reputation request sent, if answers are being given, then
267   * the agent checks the first recommendation for the target agent only
268   */
269  +!check_recomendations(MO,Target,SIZE,RC): RC \== []
270      <-  .nth(0,RC,RECOMMENDER);
271          !check_recommendation(MO,Target,SIZE,RECOMMENDER).
272
273  /**
274   * checking answer for reputation request sent, if the group initiator does not
275   * any answers, then it will give the agents (which no information seem to exist
276   * about) a chance by adding them to the group and deciding about the negotiating
277   * based on the size of the potential group
278   */
279  +!check_recomendations_initiator(MO,Target,SIZE,RC): RC == []
280      <-  .count(own_information(_,"VERY_GOOD")[source(percept)],GSIZE1);
281          .count(own_information(_,"GOOD")[source(percept)],GSIZE2);
282          .count(own_information(_,empty)[source(percept)],GSIZE3);
283          GSIZE = GSIZE1 + GSIZE2 + GSIZE3;
284          !negotiate_decision(MO,SIZE,GSIZE).
285
286  /**
287   * checking answer for reputation request sent, if answers are being given, then
288   * the group initiator checks the first recommendation for the target agent only
289   */
290  +!check_recomendations_initiator(MO,Target,SIZE,RC): RC \== []
291      <-  .nth(0,RC,RECOMMENDER);
292          !check_recommendation_initiator(MO,Target,SIZE,RECOMMENDER).
293
294  /* checking reliability of recommender */
295  +!check_recommendation(MO,Target,SIZE,RECOMMENDER): rec_info(Target,Value)[
296  source(RECOMMENDER)]
297      <-  check_recommender(Target,RECOMMENDER,Value);
298          !coop_decision(MO,Target,SIZE,RECOMMENDER).
299
300  /* checking reliability of recommender (group initiator) */
301  +!check_recommendation_initiator(MO,Target,SIZE,RECOMMENDER): rec_info(Target,
302  Value)[source(RECOMMENDER)]
```

```
303
304 /**
305  * if recommender reliability is low, don't trust the recommendation and count
306  * number of answer from targets the agent has positive or unknow image
307  * information for in order to check whether the group size is sufficient for
308  * a cooperation
309  */
310 +!coop_decision_initiator(MO,Target,SIZE,RECOMMENDER): recommender_evaluation(
    RECOMMENDER,bad)
311   <- .count(own_information(_,"VERY_GOOD")[source(percept)],GSIZE1);
312      .count(own_information(_,"GOOD")[source(percept)],GSIZE2);
313      .count(own_information(_,empty)[source(percept)],GSIZE3);
314      GSIZE = GSIZE1 + GSIZE2 + GSIZE3;
315      !negotiate_decision(MO,SIZE,GSIZE).
316 /**
317  * if recommender reliability is good, or no information is available, trust
318  * recommendation and go on with the cooperation decision depending on what
319  * recommender says about target
320  */
321 +!coop_decision_initiator(MO,Target,SIZE,RECOMMENDER): recommender_evaluation(
    RECOMMENDER,good) | recommender_evaluation(RECOMMENDER,empty)
322   <- !coop_decision_with_recommendation_initiator(MO,Target,SIZE,
    RECOMMENDER).
323
324 /**
325  * if recommender reliability was high enough and the recommender gives a
326  * negative feedback about the target, then don't cooperate with the target
327  */
328 +!coop_decision_with_recommendation_initiator(MO,Target,SIZE,RECOMMENDER):
    rec_info(Target,"BAD")[source(RECOMMENDER)] | rec_info(Target,"VERY_BAD")[
    source(RECOMMENDER)]
329   <- .send(Target,tell,bad_image(MO));
330      .abolish(own_information(Target,empty));
331      .count(own_information(_,"VERY_GOOD")[source(percept)],GSIZE1);
332      .count(own_information(_,"GOOD")[source(percept)],GSIZE2);
333      .count(own_information(_,empty)[source(percept)],GSIZE3);
334      GSIZE = GSIZE1 + GSIZE2 + GSIZE3;
335      !negotiate_decision(MO,SIZE,GSIZE).
336
337
338 /**
339  * if recommender reliability was high enough and the recommender gives a
340  * positive feedback about the target, then cooperate with the target
341  */
342 +!coop_decision_with_recommendation_initiator(MO,Target,SIZE,RECOMMENDER): not
    rec_info(Target,"BAD")[source(RECOMMENDER)] & not rec_info(Target,"VERY_BAD"
    )[source(RECOMMENDER)]
343   <- .count(own_information(_,"VERY_GOOD")[source(percept)],GSIZE1);
344      .count(own_information(_,"GOOD")[source(percept)],GSIZE2);
345      .count(own_information(_,empty)[source(percept)],GSIZE3);
346      GSIZE = GSIZE1 + GSIZE2 + GSIZE3;
347      !negotiate_decision(MO,SIZE,GSIZE).
348
349 /**
350  * if recommender reliability is bad -- and agent is not group initiator --
351  * then don't trust the recommendation and give the agent having send the
352  * cooperation request a chance, by accepting the request
353  */
354 +!coop_decision(MO,Target,SIZE,RECOMMENDER): recommender_evaluation(RECOMMENDER,
    bad)
355   <- .my_name(NAME);
356      /* accepting the request */
357      .send(Target,tell,accept(coop_proposal(MO,NAME)));
358      ?event(E);
359      /* update energy consumption */
360      update_battery_coopRequest(E);
361      !check_infos_generated(MO,SIZE).
362 /**
363  * if recommender reliability is not bad, then check the recommendation and use
364  * it in the decision process
365  */
366 +!coop_decision(MO,Target,SIZE,RECOMMENDER): recommender_evaluation(RECOMMENDER,
    good) | recommender_evaluation(RECOMMENDER,empty)
367   <- !coop_decision_with_recommendation(MO,Target,SIZE,RECOMMENDER).
368 /**
369  * if recommender reliability was high enough and the recommender gives a
370  * negative feedback about the target, then don't cooperate, but download self
371  */
372 +!coop_decision_with_recommendation(MO,Target,SIZE,RECOMMENDER): rec_info(Target
    ,"BAD")[source(RECOMMENDER)] | rec_info(Target,"VERY_BAD")[source(
    RECOMMENDER)]
373   <- .send(Target,tell,bad_image(MO));
374      !download_self(MO,SIZE).
375
376
377
378 /**
379  * if recommender reliability was high enough and the recommender gives a
380  * positive feedback about the target, then cooperate with the target by
381  * accepting the cooperation proposal
382 */
383 +!coop_decision_with_recommendation(MO,Target,SIZE,RECOMMENDER): not rec_info(
    Target,"BAD")[source(RECOMMENDER)] & not rec_info(Target,"VERY_BAD")[source(
    RECOMMENDER)]
384   <- .my_name(NAME);
385      /* accepting the cooperation proposal */
386      .send(Target,tell,accept(coop_proposal(MO,NAME)));
387      ?event(E);
388      /* update energy consumption */
389      update_battery_coopRequest(E);
390      !check_infos_generated(MO,SIZE).
391
392 //general handling of reputation information
393 /**
394  * reputation recommendation arrives before the deadline, then add it to the
395  * belief base
396  */
397 +rec_info(Target,Value)[source(Rec_Source)]: not deadlineRecommendation
398   <- +rec_info(Target,Value)[source(Rec_Source)];
399      updateRecommendation(Target,Rec_Source,Value).
400 /**
401  * reputation recommendation arrives after the deadline, abolish the information
402  */
403 +rec_info(Target,Value)[source(Rec_Source)]: deadlineRecommendation
404   <- .abolish(rec_info(Target,Value)[source(Rec_Source)]).
405
```

```
406
407
408   //Mode 4 (agents will interact with agents that have passed the reputation check
409   )
      /**
410    * check the number of the agents that have answered and have a high enough
411    * image/reputation level (number of agents that meet that criterion = GSIZE)
412    */
413
414   /* if GSIZE = 0, i.e. no agent meets that criteria --> download self */
415   +!negotiate_decision(MO,SIZE,GSIZE): not sufficient_group_size(GSIZE)
416       <-  /* sending defectors the message that no cooperation due to bad
              image */
417       .findall(Agent,own_information(Agent,"BAD"),DEFECTORS1);
418       .findall(Agent,own_information(Agent,"VERY_BAD"),DEFECTORS2);
419       .send(DEFECTORS1,tell,bad_image(MO));
420       .send(DEFECTORS2,tell,bad_image(MO));
421       /* send flag, that these agents can check the group infos due to bad
422       .findall(Agent,accept(coop_proposal(MO,Agent)),ANSWERED);
423       .send(ANSWERED,tell,reputation_generated);
424       !download_self(MO,SIZE).
425
426
427   /* otherwise, initiate group */
428   +!negotiate_decision(MO,SIZE,GSIZE): sufficient_group_size(GSIZE)
429       <-  /* finding all potential cooperators */
430       .findall(Agent,own_information(Agent,"VERY_GOOD"),GSIZE1);
431       .findall(Agent,own_information(Agent,"GOOD"),GSIZE2);
432       .findall(Agent,own_information(Agent,"empty"),GSIZE3);
433       /* finding all agent that have a bad image */
434       .findall(Agent,own_information(Agent,"BAD"),DEFECTORS1);
435       .findall(Agent,own_information(Agent,"VERY_BAD"),DEFECTORS2);
436       /* sending agent with bad image message that no cooperation possible
437       .send(DEFECTORS1,tell,bad_image(MO));
438       .send(DEFECTORS2,tell,bad_image(MO));
439       .concat(GSIZE1,GSIZE2,GSIZE3,GROUP);
440       .findall(Agent,accept(coop_proposal(MO,Agent)),ANSWERED);
441       /* generate group infos */
442       .random(KEY);
443       getGroupNumber(GSIZE);
444       ?groupnumber(GN);
445       +leader(GN);
446       /* assign chunks */
447       assign_chunks(GN,GROUP,KEY,MO);
448       /* send flag to all agents that they can check the group infos */
449       .send(ANSWERED,tell,reputation_generated);
450       check_group_infos(MO,SIZE).
451
452   /* if agent was accepted by a group, it waits for the assignment infos */
453   +!check_group_infos(MO,SIZE): assigned_chunk(GN,MO,CHUNKNUMBER,CHUNKSIZE)
454       <-  !assigned_chunk(GN,MO,CHUNKNUMBER,CHUNKSIZE).
455
456   /* if it gets a reject message, it downloads self */
457   +!check_group_infos(MO,SIZE): bad_image(MO) | to_late
458       <-  !download_self(MO,SIZE).
459
460   /**

461    * methods for non group leader to wait with the cooperation checking till the
462    * deadline event
463    */
464   +!check_deadline(GM,CHUNKSIZE): deadline
465       <-  !update_reception(GM,CHUNKSIZE).
466
467   +!check_deadline(GM,CHUNKSIZE): not deadline
468       <-  .wait("+deadline");
469       !update_reception(GM,CHUNKSIZE).
470
471   //Mode 6
472   /* updating reputation information */
473   /*
474    * checking whether any other agent has sent some chunks and whether file is
475    * complete as a consequence
476    */
477   +!control_reception(GM,CHUNKSIZE): leader(GM) & sent
478       <-  .wait(3000);
479       ?collaborators(GM,GROUP);
480       .send(GROUP,tell,deadline);
481       external_event_deadline(GM);
482       ?groupkey(GM,KEY);
483       .count(sharing(GM,_,KEY,_),RECEIVED);
484       /* assumption: each agent sends his fair share if it send sth. at
              all.
485        * i.e. the total amount received is the fair share times the number
              of
486        * received bits. */
487       TotalReceived = CHUNKSIZE * (RECEIVED - 1);
488       !event(E);
489       /* update battery costs */
490       update_reception_costs(TotalReceived,E);
491       /* update reputation information by comparing the groupmembers and
              the
492       /* agents that have sent a chunk */
493       .findall(Agent,sharing(GM,_,KEY,Agent),COOPGROUP);
494       updateReputationInfos(GROUP,COOPGROUP);
495       /* get missing bits from basestation */
496       ?total_chunk_number(GM,TCHUNKS);
497       Rest = ((TCHUNKS-1)*CHUNKSIZE)-TotalReceived;
498       !download_self(MO,Rest).
499
500   /**
501    * checking whether any other agent has sent some chunks and whether file is
502    * complete as a consequence
503    */
504   +!control_reception(GM,CHUNKSIZE): leader(GM) & not sent
505       <-  .wait(3000);
506       ?collaborators(GM,GROUP);
507       .send(GROUP,tell,deadline);
508       external_event_deadline(GM);
509       ?groupkey(GM,KEY);
510       .count(sharing(GM,_,KEY,_),RECEIVED);
511       /* assumption: each agent sends his fair share if it send sth. at
              all.
512        * i.e. the total amount received is the fair share times the number
513        * received bits. */
```

```
514      TotalReceived = CHUNKSIZE * RECEIVED;
515      ?event(E);
516      /* update battery costs */
517      update_reception_costs(TotalReceived,E);
518      /* update reputation information by comparing the groupmembers and
           the
519      * agents that have sent a chunk */
520      .findall(Agent,sharing(GM_..,KEY,Agent),COOPGROUP);
521      updateReputationInfo(GROUP,COOPGROUP);
522      /* get missing bits from basestation */
523      ?total_chunk_number(GM,TCHUNKS);
524      Rest = ((TCHUNKS-1)*CHUNKSIZE)-TotalReceived;
525      !download_self(HO,Rest).
526
527   /** checking whether any other agent has sent some chunks and whether file is
528    * complete as a consequence, beforehand, check whether the deadline has passed
529    */
530
531   +!control_reception(GM,CHUNKSIZE): not leader(GM)
532      <- !check_deadline(GM,CHUNKSIZE).
533
534   +!update_reception(GM,CHUNKSIZE): not sent
535      ?groupkey(GM,KEY);
536      ?collaborators(GM,GROUP);
537      .count(sharing(GM..,KEY,_),RECEIVED);
538      /* Assumption: each agent sends his fair share if it send sth. at
           all.
         * i.e. the total amount received is the number fair share times the number
539      * received bits.*/
540      TotalReceived = CHUNKSIZE * RECEIVED;
541      ?event(E);
542      /* update battery costs */
543      update_reception_costs(TotalReceived,E);
544      /* update reputation information by comparing the groupmembers and
           the
545      * agents that have sent a chunk */
546      .findall(Agent,sharing(GM_..,KEY,Agent),COOPGROUP);
547      updateReputationInfo(GROUP,COOPGROUP);
548      /* get missing bits from basestation */
549      ?total_chunk_number(GM,TCHUNKS);
550      Rest = ((TCHUNKS-1)*CHUNKSIZE)-TotalReceived;
551      !download_self(HO,Rest).
552
553   +!update_reception(GM,CHUNKSIZE): sent
554      ?groupkey(GM,KEY);
555      ?collaborators(GM,GROUP);
556      .count(sharing(GM..,KEY,_),RECEIVED);
557      /* assumption: each agent sends his fair share if it send sth. at
           all.
         * i.e. the total amount received is the number fair share times the number
           of
558      * received bits. */
559      TotalReceived = CHUNKSIZE * (RECEIVED - 1);
560      ?event(E);
561      /* update battery costs */
562      update_reception_costs(TotalReceived,E);
563      /* update reputation information by comparing the groupmembers and
           the
564      * agents that have sent a chunk */
565      .findall(Agent,sharing(GM_..,KEY,Agent),COOPGROUP);
566      updateReputationInfo(GROUP,COOPGROUP);
567      /* get missing bits from basestation */
568      ?total_chunk_number(GM,TCHUNKS);
569      Rest = ((TCHUNKS-1)*CHUNKSIZE)-TotalReceived;
570      !download_self(HO,Rest).
571   /* further Plans */
572
573   /**
574    * movement: for the method the old location is being used. The old location is
575    * needed to delete it from the BB and the new location is determined
576    * according to the movement model and added to the BB of the agent.
577    */
578   @move_rep[atomic]
579   +!move: true
580      <- ?location(L);
581      remove_percept(L);
582      !abolish_beliefs;
583      ?movement(M);
584      Mnew = M-1;
585      -+movement(Mnew);
586      move_agent(L,Mnew);
587      !calc_random;
588      !move_decision.
589
590   /**
591    * abolishes the beliefs of an agent after the interaction (e.g. partners....)
592    * This is to ensure that the agent is not confusing beliefs if it is trying
593    * to download a new file and gets beliefs with of the same form, but with
594    * other content.
595    */
596   @abolish_beliefs_reputation[atomic]
597   +!abolish_beliefs: true
598      <- ?messages(ME);
599      ?movement(M);
600      ?event(E);
601      ?cheating(C);
602      ?badImageNumber(I);
603      -abolish(_);
604      +event(E);
605      +messages(ME);
606      +movement(M);
607      +cheating(C);
608      +badImageNumber(I).
609
610   /* reactions to bad_image information about the agent itself by others */
611   +bad_image(_): true
612      <- ?badImageNumber(I);
613      Inew = I + 1;
614      -+badImageNumber(Inew).
615
616   +bad_proposal(_): true
617      <- ?badImageNumber(I);
618      Inew = I + 1;
619      -+badImageNumber(Inew).
```

Listing 13: honest_rep_agent.asl class

```
1   /**
2    * Agent honest_rep_agent in project Thesis-simulation
3    * This agent acts honestly, i.e. always tries to fulfill
4    * its commitments. It uses reputation information.
5    *
6    * @author Tina Balke
7    * @version 12-07-2011
8    *
9    */
10
11  /* inheritance from other agent classes */
12  {include("agent.asl")}
13  {include("rep_agent.asl")}
14  /* important comment for Jason, agents cannot overwrite inherited methods, but
15   * but the new methods are added to the inherited methods; this can result in
16   * an agent having several methods twice, what in turn will result in the
17   * respective plans being called twice!
18   */
19
20  /* initial beliefs */
21  cheating(0).
22  badImageNumber(0).
23
24  /* plans */
25  +!assigned_chunk(GM,NO,CHUNKNUMBER,CHUNKSIZE): true
26     <-  !check_download_channel.
27
28  +!check_download_channel: free(_)
29     <-  ?assigned_chunk(GM,NO,CHUNKNUMBER,CHUNKSIZE);
30         check_permission_download(GM,CHUNKNUMBER);
31         !download_decision.
32
33  +!check_download_channel: not free(_)
34     <-  .send(basestation, askOne, free(_), Reply);
35         !check_download_channel.
36
37  +!download_decision: power(P) & permission(PE) & P == true & PE == true
38         ?assigned_chunk(GM,NO,CHUNKNUMBER,CHUNKSIZE);
39         ?event(E);
40         .send(basestation, tell, download);
41         /* updating the energy costs */
42         update_download_costs(CHUNKNUMBER,E);
43         external_event_download(GM,CHUNKNUMBER);
44         .send(basestation, untell, download);
45         !sending_decision(GM,CHUNKSIZE).
46
47  +!download_decision: power(P) & permission(PE) & P == false & PE == false
48         <-  !check_download_channel.
49
50  +!check_download_channel: not free(_)
51         .send(basestation, askOne, free(_), Reply);
52         !check_download_channel.
53
54  //Mode 5
55  /*
56   * always cooperate
57   */
58  +!sending_decision(GM,CHUNKSIZE): true
59     <-  ?assigned_chunk(GM,NO,CHUNKNUMBER,CHUNKSIZE);
60         ?collaborators(GM,GROUP);
61         ?groupkey(GM,KEY);
62         ?event(E);
63         .my_name(NAME);
64         /* sending */
65         .send(GROUP,tell,sharing(GM,CHUNKNUMBER,KEY,NAME));
66         +sent;
67         /* update energy costs */
68         update_sending_costs(CHUNKSIZE,E,GM,CHUNKNUMBER);
69         external_event_sending(GM,CHUNKNUMBER);
70         !control_reception(GM,CHUNKSIZE).
71
72  //reaction to recommendation requests of other agents
73  /** if the agent has answered more requests then events have taken place it will
74   * not answer (to save battery). it will also not answer if it is currently
75   * answering a message (i.e. it is blocked) or if it is being asked about itself
76   */
77
78  @rec_answer_true[priority(1)]
79  +!recommendationRequest(Target)[source(Agent)]: messages(M) & event(E) & .my_name
        (M) & M \== Target & not blocked & M<=E
80     <-  +blocked;
81         -messages(M);
82         +messages(Mnew);
83         check_own_info(Target);
84         !info_sending_decision(Target,Agent).
85
86  @rec_answer_abolish[priority(2)]
87  +!recommendationRequest(Target)[source(Agent)]: true
88     <-  .abolish(recommendationRequest(Target)[source(Agent)]).
89
90  /**
91   * otherwise (if it has information) it will find its own information about the
92   * target and report honestly
93   */
94  +!info_sending_decision(Target,Agent): own_information(Target,VALUE) & VALUE \==
        empty
95     <-  .send(Agent,tell,rec_info(Target,VALUE));
96         ?event(E);
97         /* costs added for sending reputation information */
98         update_battery_coopRequest(E);
99         .abolish(recommendationRequest(Target)[source(Agent)]).
100
101 /**
102  * if the agent doesn't have information about the target it will not answer
103  */
104 +!info_sending_decision(Target,Agent): own_information(Target,VALUE) & VALUE ==
        empty
105    <-  .abolish(recommendationRequest(Target)[source(Agent)]).
106
107 +!info_sending_decision(Target,Agent): not own_information(Target,VALUE)
108    <-  .abolish(recommendationRequest(Target)[source(Agent)]).
110
```

Listing 14: util.rep.agent.asl class

```
1   /**
2    * Agent util.image_agent in project Thesis_simulation
3    * This agent will always act based on the utility maximizing rational and uses
4    * image information.
5    *
6    * @author Tina Balke
7    * @version 12-07-2011
8    *
9    */
10
11  /* inheritance from other agent classes */
12  {include("agent.asl")}
13  {include("rep-agent.asl")}
14  /* important comment for Jason, agents cannot overwrite inherited methods, but
15   * but the new methods are added to the inherited methods; this can result in
16   * an agent having several methods called twice, what in turn will result in the
17   * respective plans being called twice!
18   */
19
20  /* initial beliefs */
21  cheating(0).
22  badImageNumber(0).
23
24  /* plans */
25  +!check_download_channel: true
26      <-  !check_download_channel.
27  +!assigned_chunk(GN,NO,CHUNKNUMBER,CHUNKSIZE): true
28      <-  ?assigned_chunk(GN,NO,CHUNKNUMBER,CHUNKSIZE);
29          check_permission.download(GN,CHUNKNUMBER);
30          !download_decision.
31  +!check_download_channel: not free(_)
32      <-  .send(basestation, askOne, free(_), Reply);
33          !check_download_channel.
34  +!download_decision: power(P) & permission(PE) & P == true & PE == true
35      <-  ?event(E);
36          .send(basestation, tell, download)
37          /* updating the energy costs */
38          update_download_costs(CHUNKSIZE,E);
39          external_event.download(GN,CHUNKNUMBER);
40          /* determine the costs of sending and defection */
41          ?badImageNumber(I);
42          ?collaborators(KEY,GROUP);
43          getGroupSize(GROUP);
44          ?groupSize(GS);
45          /* checking for the energy costs associated with cooperation and
46           * defection */
47          get_utility_sending_image(CHUNKSIZE,E,I,GS);
48          ?sending_costs(S);
49          ?defection_costs(D);
50          ?sending_decision(GN,CHUNKSIZE,S,D).
51  +!download_decision: power(P) & permission(PE) & P == false & PE == false
```

```
58      <-  !check_download_channel.
59  +!check_download_channel: not free(_)
60      <-  .send(basestation, askOne, free(_), Reply);
61          !check_download_channel.
62
63  //Mode 5
64  /**
65   * if sending is more expensive (battery-wise) then cheating --> cheat
66   */
67  +!sending_decision(GN,CHUNKSIZE,S,D): S > D
68      <-  /* increment cheating counter by 1 */
69          ?cheating(C);
70          Cnew = C + 1;
71          Chat(GN);
72          -+cheating(Cnew);
73          !control_reception(GN,CHUNKSIZE).
74
75  /**
76   * if sending is less expensive (battery-wise) then cheating --> cooperate
77   */
78  +!sending_decision(GN,CHUNKSIZE,S,D): S <= D
79      <-  ?assigned_chunk(GN,NO,CHUNKNUMBER,CHUNKSIZE);
80          ?collaborators(GN,GROUP);
81          ?groupKey(GN,KEY);
82          ?event(E);
83          .my_name(NAME);
84          /* sending chunks */
85          .send(GROUP,tell,sharing(GN,CHUNKNUMBER,KEY,NAME));
86          +sent;
87          external_event_sending(GN,CHUNKNUMBER);
88          /* update energy costs */
89          update_sending_costs(CHUNKSIZE,E,GN,CHUNKNUMBER);
90          !control_reception(GN,CHUNKSIZE).
91
92  //reaction to recommendation requests of other agents
93  /* do not answer in order to save energy */
94  +recommendationRequest(Sender)[source(Agent)]: true
95      <-  .abolish(recommendationRequest(Sender)[source(Agent)]).
```

Listing 15: malicious.rep.agent.asl class

```
1   /**
2    * Agent malicious.rep.agent in project Thesis_simulation
3    * This agent goal is to harm the system, regardless of its own utility
4    *
5    * @author Tina Balke
6    * @version 23-06-2011
7    *
8    */
9
10  /* inheritance from other agent classes */
11  {include("agent.asl")}
12  {include("rep-agent.asl")}
13  /* important comment for Jason, agents cannot overwrite inherited methods, but
14   * but the new methods are added to the inherited methods; this can result in
15   * an agent having several methods called twice, what in turn will result in the
```

```
16      * respective plans being called twice!
17      */
18
19      /* initial beliefs */
20      cheating(0).
21      badImageNumber(0).
22
23      /* plans */
24
25      //Mode 5
26      /** never cooperate and never download right at the beginning */
27      +!assigned_chunk(GM,NO,CHUNKNUMBER,CHUNKSIZE): true
28          <-    !check_download_channel.
29
30      +!check_download_channel: free(_)
31          <-    ?assigned_chunk(GM,NO,CHUNKNUMBER,CHUNKSIZE);
32                check_permission.download(GM,CHUNKNUMBER);
33                !download_decision.
34
35      +!check_download_channel: not free(_)
36          <-    .send(basestation, askOne, free(_), Reply);
37                !check_download_channel.
38
39      +!download_decision: power(P) & permission(PE) & P == true & PE == true
40          <-    ?assigned_chunk(GM,NO,CHUNKNUMBER,CHUNKSIZE);
41                ?event(E);
42                /* updating the energy costs */
43                .send(basestation, tell, download);
44                update.download_costs(CHUNKSIZE,E);
45                external_event.download(GM,CHUNKNUMBER);
46                .send(basestation, untell, download);
47                !control_reception(GM,CHUNKSIZE).
48
49      +!download_decision: power(P) & permission(PE) & P == false & PE == false
50          <-    !check_download_channel.
51
52      +!check_download_channel: not free(_)
53          <-    .send(basestation, askOne, free(_), Reply);
54                !check_download_channel.
55
56      // reaction to reputation information requests
57      /** if the agent has answered more request then events have taken place it will
58       * not answer (to save battery), it will also not answer if it is currently
59       * answering a message (i.e. it is blocked) or if it is being asked about itself
60       */
61      @rec_answer_false[priority(1)]
62      +recommendationRequest(Target)[source(Agent)]: messages(M) & event(E) & .my_name
63          (M) & M \== Target & not blocked & M<=E
64          <-    +blocked;
65                .send(Agent,tell,rec_info(Target,"VERY_BAD"));
66                update_battery_coopRequest(E);
67                .abolish(recommendationRequest(Target)[source(Agent)]);
68                .abolish(blocked).
69
70      @rec_answer_abolish[priority(2)]
71      +recommendationRequest(Target)[source(Agent)]: true
72          <-    .abolish(recommendationRequest(Target)[source(Agent)]).
```

Appendix B

Implementing the Basic Wireless Mobile Grid Scenario

Looking at the actual implementation for representing the mobile phones users and their individual behavioural specifications as agents in the WMG scenario, we used five different agent classes: `agent.asl`, `interaction_agent.asl`, `honest_agent.asl`, `util_agent.asl`, and `malicious_agent.asl` all shown in Figure B.1[1]. The class `agent.asl` contains general agent behaviour. `interaction.asl` specifies the interaction information that is common to all agents (regardless of their behavioural type), whereas the final three classes define the behavioural-specific reasoning. We refer to the latter four classes as general interaction agents, because they contain the information for the agents that engage in WMG sharing interaction. In the simulation experiments, for each agent of one particular behavioural type, an instance of the respective agent class is created. This instance is then parameterised with agent specific parameters (e.g. the initial location of the agent).

Figure B.1 shows two more agent classes that we use in the simulation experiments: `base_station.asl` and `sim_agent.asl`. The class `base_station.asl` contains the reasoning for the base stations (i.e. their internal check on whether they are busy or not). The class `sim_agent.asl` is used to generate an agent that directs the simulation by initialising all other agents as well as finishing the simulation after all interaction agents have performed their actions. The `sim_agent.asl` class is only instantiated once in each simulation experiment.

Figure B.2 contains a code snippet from the `interaction_agent.asl` class that shows the initial plans and beliefs of the agent as well as its decision considerations for decision node 1 in Figure 7.1, i.e. the checking of the opportunity costs of

---

[1] The extension `.asl` stands for *AgentSpeak language*, which is used is in Jason to to define agent behaviour.

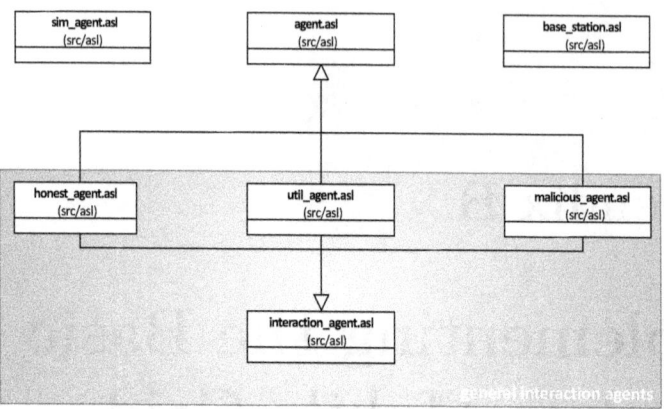

Figure B.1: Class Diagram of Implemented Jason Agent Classes

sharing/downloading a file[2].

The three central elements of the AgentSpeak code are: (i) goals the agent is supposed to achieve which are specified by an exclamation mark (e.g. `!start` in line 16). (ii) beliefs the agent has about its environment which are expressed as literals (e.g. `event(-1)` in line 10)[3], and (iii) plans (i.e. the steps an agent has to take to achieve its current intention) that have the following form `+/- !intention : precondition ← plan steps` (e.g. `+!download_file(NO,SIZE): oppCosts(X) & X >= SIZE → !download_self(NO,SIZE)` in lines 57–58). A plan can have several plan steps. These steps are separated by the ";" operator.

In the code itself all characters (as well as any sequence of them) starting with a lower-case letter are atoms[4], whereas all the ones starting with upper case letters are interpreted as logical variables. These variables are initially free and are grounded in the process of the simulation. The "_" operator in AgentSpeak is the anonymous variable. This means that it can unify with any value and is therefore often used if the respective value represented is irrelevant for the rest of the plan. The *not* operator, as in ASP, is understood in terms of negation-as-failure, whereas classical negation is denoted by the "∼" operator. Finally, the +/- operators in front of events and plans refer to the addition/deletion of the respective beliefs/goals and specify when the agents will execute a certain plan or act on a certain belief. `+!download_file(NO,SIZE): precondition ← plan steps` for example specifies that an agent will execute the plan steps only if beforehand

---

[2]In order not to distract from the presentation of the description of implementation of the agent behaviour, we only present extracts of the agent code, and do not go into detail about the technicalities of the complete code. The interested reader is referred to Appendix A where all the agent code (including explanatory comments) is presented.

[3]It should be noted, that beliefs in Jason are based on the *modalities of truth* as understood in the modal logic literature. As distinct from a statement of absolute truth, a belief simply expresses the fact that the agent currently believes the statement to be true.

[4]We understand the terms literal and atom as explained on page 93.

```
 9    /* Initial beliefs and rules */
10    event(-1).
11
12    /* Initial goals */
13
14    /* Plans */
15    +start: true
16             <-           !start.
17
18    +!start: true
19             <-              !calc_random;
20                          get_number_interactions;
21                          /*Set Agent properties */
22                          set_mobility_level;
23                          set_location;
24                          !move_decision.
25
26    /**
27     * Checking whether the agent has performed more then a certain number of
28     * downloads (download events). If this is the case, then it sends a message
29     * to the sim_agent. This agent checks whether all agents have sent him a
30     * message and if so, stops the simulation. If the event threshold is not
31     * met the agent simply continues with download_file.
32     */
33    +!check_event: enoughInteractions
34             <-           .send(sim_agent,tell,end).
35
36    +!check_event: not enoughInteractions
37             <-           downloading.
38
39    enoughInteractions :- event(X) & num_interactions(Y) & X >= Y-1.
40
41    /**
42     * In case the percept download_file(NO,SIZE) is made, generate a goal with
43     * the same name.
44     */
45    +download(NO,SIZE): true
46             <-              !update_event;
47                          ?event(E);
48                          set_alternative_battery_costs(SIZE,E);
49                          get_opportunity_costs;
50                          !download_file(NO,SIZE).
51
52    //Node 1
53    /**
54     * In case the size of the file is so small that a cooperation does not pay off
55     * (opportunity costs are higher then the respective file size) download self.
56     */
57    +!download_file(NO,SIZE): oppCosts(X) & X >= SIZE
58             <-           !download_self(NO,SIZE).
59
60    /**
61     * In case the opportunity costs are smaller then the size
62     * check for neighbours (query environment
63     * for neighbour percepts and count the total number of neighbours).
64     */
65    +!download_file(NO,SIZE): oppCosts(X) & X < SIZE
66             <-           ?location(LOCATION);
67                          check_neighbours(LOCATION);
68                          .count(neighbour(_),NoNeighbours);
69                          !cooperate_decision(NO,SIZE,NoNeighbours).
```

Figure B.2: Code Fragments of the interaction_agent.asl class

the intention `download_file(NO,SIZE)` was added to its belief base. Interpreting the belief/goal addition/deletion as events taking place, this corresponds well with our event-driven notion of time described in Section 4.3.2 and is one of the reasons that we opted for the Jason Simulation Platform.

Looking at the code fragment shown in Figure B.2 in more detail, what it expresses is the following: initially the agent has the belief `event(-1)` which it uses to check how many interactions it has already performed[5], by comparing the value with the number of actions it is meant to perform (line 39) whenever the `enoughInteractions` criterion is checked as a precondition to executing a plan (e.g. in line 33). At the beginning of a simulation experiment the agents are sense the belief `start`. Whenever they do so, they add the intention `!start`to their belief base (lines 15–16). This addition results in the execution of the related plan (lines 18–24) and the step-wise execution of the plan steps. These steps include the basic configuration of the agents (e.g. an initial location). The actual interaction considerations described in Section 7.1.3.2 start whenever the agent acquires the belief that it wants to obtain a file (line 45). If this happens, it will increment the event *belief* by one (line 46) and use its environment[6] to determine which opportunity costs might be incurred if it decided to cooperate to obtain the particular file (line 49). These costs are added as the belief `oppCosts(X)` to the agent's belief base, with X denoting the value of the opportunity costs. After finishing this plan step the intention to download the file is triggered (line 50) and the plans for that intention are called, and their preconditions are checked (lines 57–69). In the case where the opportunity costs of the download are larger than the size of the file itself (line 57) the agent will chose to download the file by executing the respective plan (line 58). In the opposite case (i.e. `oppCosts(X) & X < SIZE`, line 65), the agent will check its current location and look for neighbours in its vicinity, (lines 66–67)[7] and count them to then check whether enough neighbours are in its vicinity to consider cooperation (line 68). As a result of the agents' bounded rationality, they can only detect (and communicate with) agents which are within their mobile phone range (we refer to these agents as neighbours). Any agent that is not in the searching agent's neighbourhood therefore is not found and cannot be considered for any cooperation.

Similar to the pattern for Node 1, the remaining nodes of the agent decision process are encoded in Jason. The only point where the code differs is when agents consider whether to cooperate or to defect in Node 5. This is why for this node, each behavioural group file has different plans. These can be seen in Figures B.4–B.5.

Looking at the differences in code, after being assigned a chunk all three

---

[5]We limited the number of interaction per agent to the value of 50 in the simulation experiments. This value was chosen for statistical reasons. Thus Field (2009) suggests that a minimum of 50 samples are required in order to be able to apply tests focusing on mean comparison.

[6]Plan steps in which the agents perceive their environment to obtain some information do not have any preceding symbol such as !, ? or .

[7]If called, any plan that fulfills its preconditions is executed. For the opportunity costs we therefore chose the preconditions to have no intersection at all, resulting in only either of the two `!download_file(NO,SIZE)` being executed.

```
18   //Node 5
19   /* Never cooperate and never download right at the beginning */
20   +!assigned_chunk(GN,NO,CHUNKNUMBER,CHUNKSIZE): true
21           <-          ...
22                       ?event(E);
23                       update_download_costs(CHUNKSIZE,E);
24                       external_event_download(GN,CHUNKNUMBER);
25                       ...
26                       !control_reception(GN,CHUNKSIZE).
```

Figure B.3: Cooperation Decision Code Fragments of the malicious_agent.asl class

```
17   +!assigned_chunk(GN,NO,CHUNKNUMBER,CHUNKSIZE): true
18           <-          ...
19                       ?event(E);
20                       update_download_costs(CHUNKSIZE,E);
21                       external_event_download(GN,CHUNKNUMBER);
22                       /* determine the costs of sending and defection */
23                       ?cheating(C);
24                       get_utility_sending(CHUNKSIZE,E,C);
25                       ?sending_costs(S);
26                       ?defection_costs(D);
27                       !sending_decision(GN,CHUNKSIZE,S,D).
28
29   //Node 5
30   /*
31    * If sending is more expensive (battery-wise) then cheating --> cheat
32    */
33   +!sending_decision(GN,CHUNKSIZE,S,D): S > D
34           <-          ?cheating(C);
35                       Cnew = C + 1;
36                       cheat(GN);
37                       -+cheating(Cnew);
38                       !control_reception(GN,CHUNKSIZE).
39
40   /*
41    * If sending is less expensive (battery-wise) then cheating --> cooperate
42    */
43   +!sending_decision(GN,CHUNKSIZE,S,D): S <= D
44           <-          ?assigned_chunk(GN,NO,CHUNKNUMBER,CHUNKSIZE);
45                       ?collaborators(GN,GROUP);
46                       ?groupkey(GN,KEY);
47                       ?event(E);
48                       .my_name(NAME);
49                       .send(GROUP,tell,sharing(GN,CHUNKNUMBER,KEY,NAME));
50                       +sent;
51                       external_event_sending(GN,CHUNKNUMBER);
52                       /* update battery costs */
53                       update_sending_costs(CHUNKSIZE,E,GN,CHUNKNUMBER);
54                       !control_reception(GN,CHUNKSIZE).
```

Figure B.4: Cooperation Decision Code Fragments of the util_agent.asl class

```
15   //Node 5
16   /**
17    * Always cooperate
18    */
19   +!sending_decision(GN,CHUNKSIZE):true
20          <-           ?assigned_chunk(GN,NO,CHUNKNUMBER,CHUNKSIZE);
21                       ?collaborators(GN,GROUP);
22                       ?groupkey(GN,KEY);
23                       ?event(E);
24                       .my_name(NAME);
25                       .send(GROUP,tell,sharing(GN,CHUNKNUMBER,KEY,NAME));
26                       +sent;
27                       /* update battery costs */
28                       update_sending_costs(CHUNKSIZE,E,GN,CHUNKNUMBER);
29                       external_event_sending(GN,CHUNKNUMBER);
30                       !control_reception(GN,CHUNKSIZE).
```

Figure B.5: Cooperation Decision Code Fragments of the honest_agent.asl class

agent types will download the assigned chunks. In doing so they use some battery that is added to their download costs. Afterwards, the malicious agents will not send their chunks, but directly check which chunks they have received from the other agents in their cooperation group (line 26 in Figure B.3). In contrast to the malicious agents, the honest and utility agents decide about sending their downloaded chunks to the other agents in the cooperation group. The honest agents will always decide to send the chunks and encrypt[8] them with a group key, to broadcast them together with the ID of the chunk as well as their own ID (line 25 in Figure B.5). The utility agents will consider the utility gained from sending first. Therefore, they calculate the costs associated with sending as well as the costs that are associated with defecting for them (e.g. in form of sanctions or the loss of future interaction possibilities) and weigh one against the other. Hence, they have two plans for their sending decision intention: one for the case where the costs of sending are higher then the expected defection costs (line 33 in Figure B.4) and one for the case where the sending costs are lower then the defection costs (line 43 in Figure B.4). In the former case, the utility agents will decide not to send their share and directly continue by observing their reception of chunks from other partners (line 38 in Figure B.4), whereas in the latter case they will send their share to the other group members (line 38 B.4). For sending, the agents are charged sending costs depending on the size of the chunk they have sent (line 53 in Figure B.4 and line 28 in Figure B.5 respectively).

As mentioned earlier, in the experiments conducted for the basic WMG scenario we did not employ any enforcement mechanisms. This will result in defection costs close or equal to zero, leading the utility agents to cheat instead to cooperate. We show that the inclusion of enforcement mechanisms increases the possibility of having to pay defection costs for the agents, so that for the utility agents the decision between defecting and cooperating can sometimes be

---

[8]The encryption key is agreed upon by the group members beforehand. It is used to ensure that only members of the cooperation group can use the broadcasted chunks of the group members and just obtaining chunks by receiving broadcast messages is useless.

tipped towards cooperation.

Appendix C

The Run-Time Inst*AL* Specifications

This appendix contains the Inst*AL* specifications of the run-time model. Figures C.2–C.7 show the code presented in Section 5.4 again to illustrate the differences between the design and the run-time model in detail. For this purpose, in contrast to the design-time specifications, we have printed all the rules that remain when transitioning from the design to the run-time model in bold print and thereby indirectly point out which rules can be removed from the design-time model (i.e. all rules not in bold print). To enable an easier reading of the final run-time specifications, Figures C.8 and C.9 present the complete run-time model without showing the design-time rules.

```
1   % This is dynamically created in the run-time case
2   Handset: alice bob
3   Chunk: x1 x2 x3 x4
4   Channel: c1 c2
5   Time: 1 2 3 4 5 6 7 8 9 10 11 12 13 14 15 16 17 18 19 20 21
```

Figure C.1: The Domain File of the Run Time WMG Model

```
 1    institution grid;
 2
 3    type Handset;
 4    type Chunk;
 5    type Channel;
 6    type Time;
 7
 8    %% exogenous events %%
 9    exogenous event download(Handset,Chunk,Channel);
10    exogenous event send(Handset,Chunk);
11    exogenous event deadline
12    exogenous event clock;
13
14    %% creation event %%
15    create event creategrid;
16
17    %% normative events %%
18    inst event intDownload(Handset,Chunk,Channel);
19    inst event intSend(Handset,Chunk);
20    inst event intReceive(Handset,Chunk);
21    inst event transition;
22    inst event intDeadline;
23
24    %% violation event %%
25    violation event misuse(Handset);
26
27    %% fluents %%
28    fluent downloadChunk(Handset,Chunk);
29    fluent hasChunk(Handset,Chunk);
30    fluent areceive(Handset,Time); % receiving from handset
31    fluent asend(Handset,Time); % sending by handset
32    fluent creceive(Handset,Time); % receiving from basestation
33    fluent transmit(Channel,Time); % sending by basestation
34
35    %% fluents for time-related aspects %%
36    fluent previous(Time,Time);
37    fluent countdown(Time);
38
39    %% non-inertial fluents %%
40    fluent busyHSending(Handset);
41    % indicates that the handset is sending to a peer
42    fluent busyHReceiving(Handset);
43    % indicates that the handset is receiving from a peer
44    fluent busyBReceiving(Handset);
45    % indicates that the handset is receiving from the base
46    fluent busyChannel(Channel);
47    % indicates that the channel is busy
```

Figure C.2: Declaration of Types and Events in the Run Time WMG Model

```
49    %-----------------------------------------------------------------
50    % noninertial rules
51    %-----------------------------------------------------------------
52
53    busyHSending(Handset) when asend(Handset,Time);
54    busyHReceiving(Handset) when areceive(Handset,Time);
55    busyBReceiving(Handset) when creceive(Handset,Time);
56    busyChannel(Channel) when transmit(Channel,Time);
```

Figure C.3: Specification of the Noninertial Rules in the Run Time WMG Model

```
58  %-------------------------------------------------------------------
59  % countdown rules
60  %-------------------------------------------------------------------
61
62  transition initiates countdown(T2) if countdown(T1), previous(T1,T2);
63  transition generates intDeadline if countdown(1);
64  misuse(A) terminates pow(intReceive(A,X));
65  misuse(A) terminates perm(intReceive(A,X));
```

Figure C.4: Generation and Consequence Relations for Deadline-Countdown in the Run Time WMG Model

```
67  %-------------------------------------------------------------------
68  % rules for downloading
69  %-------------------------------------------------------------------
70
71  % handset A requests a block from the base station on channel C
72  download(A,X,C) generates intDownload(A,X,C) if not hasChunk(A,X);
73           not busyChannel(C), not busyBReceiving(A);
74
75  intDownload(A,X,C) initiates hasChunk(A,X);
76  intDownload(A,X,C) initiates perm(send(A,X));
77  intDownload(A,X,C) initiates creceive(A,4), transmit(C,4);
78  % handset and channel are busy for 4 time-units when a chunk
79  % is downloaded from the base station
80
81  intDownload(A,X,C) terminates pow(intDownload(A,X1,C1));
82  intDownload(A,X,C) terminates pow(intDownload(B,X1,C));
83  intDownload(A,X,C) terminates downloadChunk(A,X);
84  intDownload(A,X,C) terminates perm(download(A,X,C1));
85
86  download(A,X,C) generates transition;
87  clock generates transition;
88
89  transition initiates transmit(C,T2) if transmit(C,T1), previous(T1,T2);
90  transition initiates creceive(A,T2) if creceive(A,T1), previous(T1,T2);
91  transition initiates pow(intDownload(A,X,C)) if creceive(A,1);
92
93  transition terminates creceive(A,Time);
94  transition terminates transmit(C,Time);
```

Figure C.5: Generation and Consequence Relations for Downloading in the Run Time WMG Model

```
96   %---------------------------------------------------------------------
97   % rules for sharing
98   %---------------------------------------------------------------------
99
100  send(A,X) generates intSend(A,X) if hasChunk(A,X), not busyHSending(A),
101          not busyHReceiving(A);
102
103  send(A,X) generates intReceive(B,X) if not hasChunk(B,X),
104          not busyHSending(B), not busyHReceiving(B), hasChunk(A,X),
105          not busyHSending(A), not busyHReceiving(A);
106
107  intSend(A,X) initiates asend(A,2);
108  intSend(A,X) terminates pow(intSend(A,X));
109  intSend(A,X) terminates perm(intSend(A,X));
110
111  intReceive(A,X) initiates hasChunk(A,X);
112  intReceive(A,X) initiates areceive(A,2);
113  intReceive(A,X) terminates pow(intReceive(A,X));
114  intReceive(A,X) terminates perm(intReceive(A,X));
115
116  send(A,X) generates transition;
117  clock generates transition;
118
119  transition initiates asend(A,T2) if asend(A,T1), previous(T1,T2);
120  transition initiates areceive(A,T2) if areceive(A,T1), previous(T1,T2);
121  transition initiates pow(intReceive(A,X)) if areceive(A,1);
122  transition initiates pow(intSend(A,X)) if asend(A,1);
123
124  transition terminates asend(A,Time);
125  transition terminates areceive(A,Time);
```

Figure C.6: Generation and Consequence Relations for Sharing in the Run Time
WMG Model

```
127  %----------------------------------------------------------------------
128  % countdown
129  %----------------------------------------------------------------------
130
131  initially countdown(20), pow(transition), perm(transition),
132          perm(clock), pow(intDeadline), perm(intDeadline); perm(deadline);
133
134
135  %----------------------------------------------------------------------
136  % downloading - dynamically created at run-time
137  %----------------------------------------------------------------------
138
139  initially pow(transition), perm(transition), perm(clock),
140          pow(intDownload(A,B,C)), perm(intDownload(A,B,C)),
141          perm(download(alice,x1,C)), perm(download(alice,x3,C)),
142          perm(download(bob,x2,C)), perm(download(bob,x4,C)),
143          downloadChunk(alice,x1), downloadChunk(alice,x3),
144          downloadChunk(bob,x2), downloadChunk(bob,x4);
145
146  %----------------------------------------------------------------------
147  % sharing - dynamically created at run-time
148  %----------------------------------------------------------------------
149
150  initially pow(transition), perm(transition),
151          perm(clock), pow(intReceive(Handset,Chunk)),
152          perm(intReceive(Handset,Chunk)),
153          pow(intSend(Handset,Chunk)),
154          perm(intSend(Handset,Chunk)),
155          obl(send(alice,x1), intDeadline, misuse(alice)),
156          obl(send(alice,x3), intDeadline, misuse(alice)),
157          obl(send(bob,x2), intDeadline, misuse(bob)),
158          obl(send(bob,x4), intDeadline, misuse(bob));
159
160  %----------------------------------------------------------------------
161  % time
162  %----------------------------------------------------------------------
163
164  initially previous(20,19);
165  initially previous(19,18);
166  initially previous(18,17);
167  initially previous(17,16);
168  initially previous(16,15);
169  initially previous(15,14);
170  initially previous(14,13);
171  initially previous(13,12);
172  initially previous(12,11);
173  initially previous(11,10);
174  initially previous(10,9);
175  initially previous(9,8);
176  initially previous(8,7);
177  initially previous(7,6);
178  initially previous(6,5);
179  initially previous(5,4);
180  initially previous(4,3);
181  initially previous(3,2);
182  initially previous(2,1);
```

Figure C.7: Initial State of the Run Time WMG Model, Post-Negotiation Phase

```
 1  institution grid;
 2
 3  type Handset;
 4  type Chunk;
 5  type Channel;
 6
 7  %% exogenous events %%
 8  exogenous event download(Handset,Chunk,Channel);
 9  exogenous event send(Handset,Chunk);
10  exogenous event deadline
11
12  %% creation event %%
13  create event creategrid;
14
15  %% normative events %%
16  inst event intDownload(Handset,Chunk,Channel);
17  inst event intSend(Handset,Chunk);
18  inst event intReceive(Handset,Chunk);
19
20  %% violation event %%
21  violation event misuse(Handset);
22
23  %% fluents %%
24  fluent downloadChunk(Handset,Chunk);
25  fluent hasChunk(Handset,Chunk);
26
27  %----------------------------------------------------------------
28  rules for downloading
29  %----------------------------------------------------------------
30
31  handset A requests a block from the base station on channel C
32  download(A,X,C) generates intDownload(A,X,C) if not hasChunk(A,X);
33
34  intDownload(A,X,C) initiates hasChunk(A,X);
35  intDownload(A,X,C) initiates perm(send(A,X));
36
37  intDownload(A,X,C) terminates downloadChunk(A,X);
38  intDownload(A,X,C) terminates perm(download(A,X,C1));
39
40  %----------------------------------------------------------------
41  % rules for sharing
42  %----------------------------------------------------------------
43
44  send(A,X) generates intSend(A,X) if hasChunk(A,X);
45
46  send(A,X) generates intReceive(B,X) if not hasChunk(B,X), hasChunk(A,X);
47
48  intReceive(A,X) initiates hasChunk(A,X);
49  intReceive(A,X) terminates perm(intReceive(A,X));
50
```

Figure C.8: The Run Time WMG Model – Part 1

```
51    %-------------------------------------------------------------------
52    % countdown
53    %-------------------------------------------------------------------
54
55    initially perm(deadline);
56
57    %-------------------------------------------------------------------
58    % downloading - dynamically created at run-time
59    %-------------------------------------------------------------------
60
61    initially
62          pow(intDownload(A,B,C)), perm(intDownload(A,B,C)),
63          perm(download(alice,x1,C)), perm(download(alice,x3,C)),
64          perm(download(bob,x2,C)), perm(download(bob,x4,C)),
65          downloadChunk(alice,x1), downloadChunk(alice,x3),
66          downloadChunk(bob,x2), downloadChunk(bob,x4);
67
68    %-------------------------------------------------------------------
69    % sharing - dynamically created at run-time
70    %-------------------------------------------------------------------
71
72    initially pow(intReceive(Handset,Chunk)),
73          perm(intReceive(Handset,Chunk)),
74          pow(intSend(Handset,Chunk)),
75          perm(intSend(Handset,Chunk)),
76          obl(send(alice,x1), intDeadline, misuse(alice)),
77          obl(send(alice,x3), intDeadline, misuse(alice)),
78          obl(send(bob,x2), intDeadline, misuse(bob)),
79          obl(send(bob,x4), intDeadline, misuse(bob));
```

Figure C.9: The Run Time WMG Model – Part 2

Appendix D

Comparing Single- and Multi-Threaded Jason Simulations

In order to run simulations with high numbers of agents, the Jason simulation environment allows for the running the simulation in a multi-threaded mode, which reduces the response time of individual agents as well as allows more agents to take part in the simulation with in total. When using a single-threaded simulation it was only feasible to work with 200 agents at most, whereas running the same simulation in a multi-threaded fashion allowed for four times as many agents (i.e. 800). Since the number of agents participating in a system can have a significant impact and low agent numbers are inappropriate for representing large scale telecommunication systems, in thi dissertation we opted for the multi-threaded simulation runs. To test that in Jason multi-threaded simulation runs do not yield different simulation results then the respective single-threaded execution, we run both both execution modes of the simulation (i.e. single- and multi-threaded) with exactly the same setup for 50, 100 and 200 agents (i.e. up to the maximum number of the single-threaded mode) and compared the results with the two-sample Kolmogorov-Smirnov test.

The two-sample Kolmogorov-Smirnov test is a non-parametric tesst that used to determine whether two underlying probability distributions (resulting from finite samples) significantly differ from each other. The resulting goodness-of-fit statistic defines the largest absolute difference between two cumulative distribution functions as a measure of disagreement. The classical two-sided Kolmogorov-Smirnov test thereby checks whether the null hypothesis that the distribution of values in the samples are the same.

For the simulation this means that the single-threaded results are taken as control sample, against which the observed multi-threaded simulation results are tested. The significance level at which the test was conducted was 0.05

To specify single- or multi-threaded runs we alter the parameter `infrastructure`

: in the Jason project file, by setting it to `Centralised(pool,X)`. For single-threaded runs we set $X = 1$ and for multi-threaded runs $X = 4$. This configuration is only possible when Jason is run on a single machine.

The results of the Kolmogorov-Smirnov showed no significant evidence against the null hypothesis (with all p-values > 0.6), which allows us to assume that multi-threaded experiments will not have significantly different results from single-threaded ones.

CV Tina Balke

10/2011 – present	**University of Surrey, UK:** Research Fellow / Centre for Research in Social Simulation
04/2008 – 09/2011	**University of Bayreuth, Germany:** Research Assistant & PhD Candidate / Chair of Information System Management
07/2009 – 9/2011	**University of Bath:** Visiting Student / Department of Computer Science
04/2007 – 03/2008	**BFM, Germany:** Research Assistant / Bayreuth Research Institute for Small and Medium-Sized Enterprises
10/2002 – 03/2007	**University of Bayreuth:** Graduate Studies / Business Administration
09/1994 – 06/2002	**Johann-Gottfried-Herder-Gymnasium:** Secondary School / Abitur
09/1999 – 07/2000	**Westwood St. Thomas' School:** Secondary School / A-Level